THE FRONTIERS COLLECTION

THE FRONTIERS COLLECTION

Series Editors:

A.C. Elitzur L. Mersini-Houghton M. Schlosshauer M.P. Silverman R. Vaas H.D. Zeh

The books in this collection are devoted to challenging and open problems at the forefront of modern science, including related philosophical debates. In contrast to typical research monographs, however, they strive to present their topics in a manner accessible also to scientifically literate non-specialists wishing to gain insight into the deeper implications and fascinating questions involved. Taken as a whole, the series reflects the need for a fundamental and interdisciplinary approach to modern science. Furthermore, it is intended to encourage active scientists in all areas to ponder over important and perhaps controversial issues beyond their own speciality. Extending from quantum physics and relativity to entropy, consciousness and complex systems – the Frontiers Collection will inspire readers to push back the frontiers of their own knowledge.

Other Recent Titles

Weak Links
Stabilizers of Complex Systems from Proteins to Social Networks
By P. Csermely

The Biological Evolution of Religious Mind and Behaviour
Edited by E. Voland and W. Schiefenhövel
Particle Metaphysics

A Critical Account of Subatomic Reality
By B. Falkenburg

The Physical Basis of the Direction of Time
By H.D. Zeh

Mindful Universe
Quantum Mechanics and the Participating Observer
By H. Stapp

Decoherence and the Quantum-To-Classical Transition
By M. Schlosshauer

The Nonlinear Universe
Chaos, Emergence, Life
By A. Scott

Symmetry Rules
How Science and Nature are Founded on Symmetry
By J. Rosen

Quantum Superposition
Counterintuitive Consequences of Coherence, Entanglement, and Interference
By M.P. Silverman

Series home page – springer.com

Maximilian Schlosshauer

Editor

ELEGANCE AND ENIGMA

The Quantum Interviews

 Springer

Editor
Maximilian Schlosshauer
Institute for Quantum Optics and Quantum Information
Boltzmanngasse 3
A-1090 Wien
Austria
maximilian.schlosshauer@univie.ac.at

Series Editors:
Avshalom C. Elitzur
Bar-Ilan University, Unit of Interdisciplinary Studies, 52900 Ramat-Gan, Israel
email: avshalom.elitzur@weizmann.ac.il

Laura Mersini-Houghton
Dept. Physics, University of North Carolina, Chapel Hill, NC 27599-3255, USA
email: mersini@physics.unc.edu

Maximilian Schlosshauer
Institute for Quantum Optics and Quantum Information, Boltzmanngasse 3,
A-1090 Wien, Austria
email: maximilian.schlosshauer@univie.ac.at

Mark P. Silverman
Trinity College, Dept. Physics, Hartford CT 06106, USA
email: mark.silverman@trincoll.edu

Rüdiger Vaas
University of Giessen, Center for Philosophy and Foundations of Science, 35394 Giessen,
Germany
email: ruediger.vaas@t-online.de

H. Dieter Zeh
Gaiberger Straße 38, 69151 Waldhilsbach, Germany
email: zeh@uni-heidelberg.de

ISSN 1612-3018
ISBN 978-3-642-26981-3   ISBN 978-3-642-20880-5 (eBook)
DOI 10.1007/978-3-642-20880-5
Springer Heidelberg Dordrecht London New York

Book design and typesetting: Maximilian Schlosshauer

Cover design: KünkelLopka GmbH

Printed on acid-free paper

Springer is part of Springer Science+Business Media (www.springer.com)

For Kari and Eli

— M. S.

The truth about the world, he said, is that anything is possible. Had you not seen it all from birth and thereby bled it of its strangeness it would appear to you for what it is, a hat trick in a medicine show, a fevered dream, a trance bepopulate with chimeras having neither analogue nor precedent, an itinerant carnival, a migratory tentshow whose ultimate destination after many a pitch in many a muddled field is unspeakable and calamitous beyond reckoning.

The universe is no narrow thing and the order within it is not constrained by any latitude in its conception to repeat what exists in one part in any other part. Even in this world more things exist without our knowledge than with it and the order in creation which you see is that which you have put there, like a string in a maze, so that you shall not lose your way. For existence has its own order and that no man's mind can compass, that mind itself being but a fact among others.

—CORMAC McCARTHY, *Blood Meridian*

PREFACE

Why approach the foundations of quantum mechanics, this vast and fascinating subject area, through interviews? Why not a proper textbook instead? After all, many of the classic titles, such as Max Jammer's *The Philosophy of Quantum Mechanics* and Bernard d'Espagnat's *Conceptual Foundations of Quantum Mechanics*, have aged visibly. A complete, up-to-date account of the field—one that would also pay appropriate attention to recent developments like quantum information, experiments, and reconstructions of quantum theory—is arguably lacking. I'm well aware of this situation. So it is not as if the idea of writing a textbook hadn't ever occurred to me, or as if I simply shied away from the effort, however enormous I suspect the investment would have to be.

Opting for the interview format instead is, as I see it, neither a cheap cop-out nor merely a temporary substitute. Rather, it is a uniquely effective way of laying out the field of quantum foundations as it stands today. It won't be news to you that this field is no cut-and-dried solid-state physics: just attend any conference devoted to quantum foundations, and you'll know that the debates at such events have the zeal of a political convention. How could a single author do the field full justice without coloring her story? I do think it could be done, but you'd have to be a card-carrying member of the Party of Utterly Neutral Quantum Scholars ("Punqs") not to be accused of supporting, however subtly, the line of a particular foundational program or mindset.

The interview approach has diversity built in from the outset. It allows you to perceive the subject through the eyes of the field's leading practitioners. You won't need to go through stacks of research papers to get a representative cross-section of views, or trust any one author to faithfully reproduce all the shades of gray. Last but not least, interviews lend themselves to an informal and personal style. After all, we read books for enjoyment. They shouldn't be a slog.

Of course, if not handled judiciously, the interview approach can also go astray. One obvious make-or-break issue is the choice of questions. There's always a danger of bias, of putting spin on the questions. Here are some of the goals I set out.

I wanted the questions to cover a wide range of topics, so that this book would provide a comprehensive reflection on the field. There are, of course, the standard themes: interpretations of quantum theory, the measurement problem, quantum states, probabilities, issues of nonlocality and completeness, and the like. But included as well are questions on newer areas, such as quantum information and reconstructions, and about interdisciplinary aspects, such as the role of philosophy and the implications of our quest for a unified theory.

Another goal was to phrase the questions in reasonably broad terms, because I didn't want the respondents to get caught up in technical details, nor did I want to unduly restrict the range of possible answers. At the same time, I tried to keep each question focused on a well-defined topic, so that answers can be compared side by side and don't turn into blanket statements and clichéd generalizations.

I also wanted to leave room for personal stories among all the heavy going. How did people originally become interested in quantum foundations? What would it take for them to embrace a rivaling view? What role do they attribute to individual temperament when it comes to the choice of foundational agenda? Those kind of questions.

Picking participants for the interviews can be treacherous territory as well. On a practical level, we may tend to choose people we already know well, thereby running the risk of inadvertently excluding the up-and-coming talent or the recluse. On a psychological level, we may gravitate toward people who share our own worldview. I tried my best to assemble a cast that would do justice to the diversity of the field. The interviewees for this book come in all foundational stripes: agnostics, informationalists, Bohrians, Everettians, Bohmians, Bayesians, collapsists, ensemblists, reconstructionists—you name it. They come from physics, philosophy, and mathematics departments, and they range in age from the budding young academic to the distinguished emeritus professor who might have shaken hands with Einstein and Bohr. A serious lack of diversity, however, occurs in the gender department, as all participants are men. Another reflection of how regrettably male-dominated the world of physics (and the philosophy of physics) is! I lament this situation as much as you do, and if you'd like to suggest suitable female participants for a future edition of this book, please let me know.

A few words on how the book is organized. Biographical sketches introduce the participants at the beginning of the book. I put the same seventeen questions to each of the seventeen participants (the identical numbers are pure coincidence). All interviews were conducted in writing. Answers were limited to about one page in length, *on average*, and nothing has been omitted here.

A minimal background in quantum mechanics should be all you need for this book. There's a glossary at the end of the book (page 295) that explains some of the technical terms repeatedly appearing in the interviews. Have a look there if you're new to the field.

As for the grouping of the interview answers, there are two obvious alternatives: by respondent or by question. Organization by respondent emphasizes autobiographical coherence, allowing respondents to build a continuous argument. Organization by question stresses thematic coherence, allowing you to easily compare the different positions on a particular issue. I decided that it was this possibility of direct comparison that mattered most. So I chose organization by question. This format also means that you won't have to make your way through seventeen separate interviews that each tick off the same list of questions—something that could quickly become tiresome. And the certain amount of autobiographical discontinuity inherent in the grouping-by-question approach can also turn into an asset, because it compels the respondents to treat each question as an independent entity, thus making answers more self-contained.

Each chapter is devoted to one particular interview question. It kicks off with a few opening remarks to whet your appetite. These teasers are not meant to amount to any kind of in-depth review. Obviously, a question like "What single question about the foundations of quantum mechanics would you put to an omniscient being?" wouldn't anyway lend itself to much of a technical survey. In other instances, when a particularly juicy question comes along—say, concerning the Bell inequalities or the meaning of quantum states—I provide a highly compressed introduction to the subject. To go any further would be to infringe on the interviewees' territory.

This is not the first interview book on quantum mechanics. In the early 1990s, Julian Brown, a radio producer with the BBC Science Unit, teamed up with Paul Davies to do a series of interviews with physicists interested in the foundations of quantum mechanics. Davies presented these interviews in the form of a program for BBC Radio 3, featuring conversations with Alain Aspect, John Bell, John Wheeler, Rudolf Peierls, David Deutsch, John Taylor, David Bohm, and Basil Hiley. The program found enthusiastic listeners, including at least one of our interviewees (see Lucien Hardy's story, page 29). Buoyed by this success, Brown and Davies decided to publish the transcripts of the interviews in book form. *The Ghost in the Atom: A Discussion of the Mysteries of Quantum Physics* came out in 1993.

It's a delightful little book, and I recommend checking it out when you have the chance. Two decades on, it feels a little dated, though a good number of the issues it discusses are as fresh as ever. In many ways, *The Ghost in the Atom* is rather different from the book you're holding in your hands. It is organized by respondent, and the questions change from interview to interview and focus on the respondents' individual foundational research programs. Curiously, it so happens that none of the people interviewed in *The Ghost in the Atom* appear in this book. So the two books are perhaps best regarded as complementary.

My thanks so, first and foremost, to the participants themselves. This book would not exist without their generous offer to lend their time and voice to the project, and it would be worthless without their insight and wisdom. When I first sent out the interview invitations, something miraculous happened: not a single person declined. These consistently positive initial responses were enormously encouraging and got the project off to a good start. And as the interviews came trickling in, one by one over the course of several months, I was amazed by the depth and diversity of the responses, and humbled by the effort and thought that had gone into them. Special thanks go to Chris Fuchs and David Mermin, who, besides their own answers, contributed a number of helpful comments and suggestions.

An essential part in making this book a reality was played by Angela Lahee, editor at Springer. Angela is the kind of editor the doomsayers tell us no longer exists. For several years now, she's been a trusted friend and confidante who is always willing to share her expertise and lend a sympathetic ear. Right from the moment when I first put the idea of this book to her, Angela threw her wholehearted support behind it. Her feedback accompanied the making of the book from start to finish. In particular, she provided thoughtful comments on the interview questions and on a draft of some of the chapter introductions.

At the end of the day, what really enables us to do what we do is the nourishment we get by being around the people dearest to our hearts. I'm most grateful to my wife, Kari, and to my son, Eli, who was born last year, for all the love and happiness we share every day. And thank you, Kari, for all your untiring patience and encouragement while your man is working on yet another weighty tome.

March 2011   MAXIMILIAN SCHLOSSHAUER

CONTENTS

THE PARTICIPANTS

Guido Bacciagaluppi is a senior lecturer in philosophy at the University of Aberdeen, U.K. Born in Milan, Italy, in 1965, he started out studying physics at ETH in Zurich, Switzerland, but switched over to mathematics and earned his diploma degree. He then did a master's degree in history and philosophy of science at Cambridge (under Michael Redhead). He went on to get his Ph.D. in philosophy, also at Cambridge. His thesis, supervised by Jeremy Butterfield, examined modal interpretations of quantum mechanics. After stints as British Academy Postdoctoral Fellow at Balliol College, Oxford, and junior lecturer in philosophy of physics at Oxford, in 2000 Guido became assistant professor in philosophy at the University of California at Berkeley. He then worked as an Alexander von Humboldt Fellow at the University of Freiburg, Germany, and at IGPP, Freiburg. He was also a researcher at IHPST, Paris. Seeing that there was yet another continent to explore, in 2006 he voyaged to Sydney, Australia, to take up a position as senior research fellow at the Centre for Time. He returned to Europe in 2009 to settle in Aberdeen, where he now lives in a small part of an Aberdeenshire castle with his wife, Jennifer Bailey, his daughter, Katie, and two dogs, Yabby and Stella.

He teaches various undergraduate and postgraduate courses at Aberdeen, mainly in philosophy of science. He is the coauthor, with Antony Valentini, of *Quantum Theory at the Crossroads: Reconsidering the 1927 Solvay Conference* (Cambridge, 2009). He is an associate member at IHPST, and a trustee and secretary at the Archive for Mathematical Sciences and Philosophy. Guido's principal spare-time interest is music. At the tender age of two years, he exclaimed in his playpen, "Mahler finished, put on Schubert!" Participants of a conference in Oxford will also forever remember Guido as the third slave in *The Magic Flute* that his wife was conducting. He had also prepared the English translation of the opera ("Papageno, frisch hinauf: ende deinen Lebenslauf!" turned into "Papageno, up the tree: put an end to your CV!"). Sometime he will need to finish writing that cello sonata.

ČASLAV BRUKNER is an associate professor in the Faculty of Physics at the University of Vienna and a visiting professor at the University of Belgrade. Born in Novi Sad, Serbia, in 1967, he earned a master's degree in physics from the University of Vienna in 1995 and a doctorate of technical sciences from the Vienna University of Technology in 1999. His Ph.D. thesis, titled "Information in Individual Quantum Systems," was advised by Anton Zeilinger. In 2003 he completed his *Habilitation* at the University of Vienna. In 2004 he was a Marie Curie Fellow at Imperial College, London, and from 2005 to 2007 he was a senior researcher at the Institute for Quantum Optics and Quantum Information, Austrian Academy of Sciences, Vienna. From 2005 to 2008 he was also chair professor at Tsinghua University in Beijing, China.

His research articles have been awarded "top pick" honors by the Institute of Physics, *Europhysics News*, *Physical Review Letters*, and *Physics Today*. He is a member of the editorial board at *New Journal of Physics*, member of the John Templeton Foundation's Eurasian board of advisers, and member of the Foundational Questions Institute's consortium. His current teaching schedule includes courses on theoretical classical mechanics, quantum information, and quantum foundations.

Časlav lives in Vienna with his wife, Zorica Mitrovic-Brukner, and their twins, Isidora and Sergej, who were born in 1999. He likes the silence at a small lake on the island of Mljet in Croatia, the voice of Nick Cave on the track "Zero Is Also a Number," waking up with short black Turkish coffee, the book *The Use of Man* by Alexander Tisma, a pint of London Pride at The Queens Arms pub in London's South Kensington district, and the movie *My Night at Maud's* by Eric Rohmer.

JEFFREY BUB is a Distinguished Professor with the Department of Philosophy and the Institute for Physical Science and Technology at the University of Maryland, College Park. Born in Cape Town, South Africa, in 1942, he started out studying chemical and electrical engineering at the University of Cape Town, but then switched to pure science and majored in math and physics. He got his graduate education from Birkbeck College, University of London, where in 1966 he earned a Ph.D. in mathematical physics. His dissertation, advised by David Bohm, was titled "The Problem of Measurement in Quantum Mechanics."

Ironically, Jeff's first job was as a postdoc in the chemistry department at the University of Minnesota, where he was working with Alden Mead, a physical chemist. The chemistry department was just across the Mall from Ford Hall, where Herbert Feigl ran the Center for Philosophy of Science with Grover Maxwell, and Jeff was able to get a two-year postdoc with Feigl at the Center. His first teaching position was at Yale, initially a joint physics–philosophy appointment, then just in philosophy. From 1971 to 1986 Jeff held a position at the University of Western Ontario. He has been at the University of Maryland since 1986, with stints as visiting professor at various places (Princeton in 1989, Yale in 1993, University of California–Irvine in 1994, University of California–San Diego in 1999). Jeff is the author of *The Interpretation of Quantum Mechanics* (Reidel, 1974) and *Interpreting the Quantum World* (Cambridge, 1997; revised paperback edition, 1999), which won the Lakatos Award in 1998. In 2005 he was awarded the Kirwan Faculty Research and Scholarship Prize. *Foundations of Physics* recently published a Festschrift in Jeff's honor, and his work has been celebrated in the collection *Physical Theory and Its Interpretation: Essays in Honor of Jeffrey Bub* (Springer, 2006, edited by W. Demopoulos and I. Pitowsky).

He lives in Washington, DC, in a narrow, three-story Victorian with his wife, Robin. In the summer, they live in their house in Quinson, a small village in the Alpes d'Haute Provence, where they enjoy hiking the hills.

ARTHUR FINE is a professor of philosophy and an adjunct professor of history and physics at the University of Washington, Seattle. Born in Lowell, Massachusetts, in 1937, he earned a bachelor's degree in mathematics from the University of Chicago, where he also studied physics, systems theory, and philosophy. Undecided about whether to pursue math or philosophy, he did both, starting with a master's degree in mathematics from Illinois Institute of Technology. He then earned a Ph.D. in philosophy from the University of Chicago. His dissertation was on the quantum theory of measurement. In physics, he worked with Gregor Wenzel from the Fermi Institute, and in philosophy with Henry Mehlberg, who was his supervisor. He was a postdoc in history and philosophy of science at Cambridge University.

He has taught at the University of Illinois, Cornell, Northwestern, and now at the University of Washington. He served as president of the Philosophy of Science Association and the American Philosophical Association. He has also been a Guggenheim Fellow, a fellow at the Center for Advanced Study in the Behavioral Sciences, and the recipient of a senior fellowship from the National Endowment for the Humanities. He teaches undergraduates and graduate students in logic and philosophy of science and supervises graduate students in physics and in history of science. He enjoys doing public education about science and has given lectures at the Adler Planetarium in Chicago and other outlets. He is the author of *The Shaky Game: Einstein, Realism, and the Quantum Theory* (Chicago, 1986; Japanese translation, 1992; second edition, 1996) and coeditor, with James T. Cushing and Shelly Goldstein, of *Bohmian Mechanics and Quantum Theory: An Appraisal* (Kluwer, 1996).

He lives with his wife, Micky Forbes, in Port Townsend, a small town on a peninsula across Puget Sound from Seattle. Ocean on three sides, and mountains to the east and west. They have a cat, Qiao, a beautiful tabby. His son Dana, a mathematician, and daughter Sharon, a physician, live on the East Coast. Arthur enjoys hiking, and he and Micky spend time in France (Paris or the Vaucluse) whenever possible.

CHRISTOPHER FUCHS is a researcher at the Perimeter Institute for Theoretical Physics in Waterloo, Canada. He is also an adjunct professor of physics and applied mathematics at the University of Waterloo, an adjunct professor of physics and astronomy at the University of New Mexico, and an affiliate of the Institute for Quantum Computing at the University of Waterloo. While writing his interview, he was secondarily affiliated with the Stellenbosch Institute for Advanced Study, South Africa.

Born in Cuero, Texas, in 1964, Chris studied physics and mathematics at the University of Texas at Austin. There he met John Wheeler, who became his research supervisor. In 1996 he got his Ph.D. in physics from the University of New Mexico, Albuquerque; his dissertation, "Distinguishability and Accessible Information in Quantum Theory," was advised by Carlton Caves. He was the Lee A. DuBridge Prize Postdoctoral Fellow at the California Institute of Technology (1996–9) and a Director-Funded Postdoctoral Fellow at Los Alamos (1999–2000). From 2000 to 2007 he was a research staff member at Bell Labs, Murray Hill, New Jersey.

Two collections of Chris's selected email correspondence on quantum mechanics have been published as books: *Notes on a Paulian Idea: Foundational, Historical, Anecdotal & Forward-Looking Thoughts on the Quantum* (Växjö, 2003) and *Coming of Age with Quantum Information: Notes on a Paulian Idea* (Cambridge, 2010). In 2010 Chris received the International Quantum Communication Award. He is associate editor of *Quantum Information and Computation* (Rinton Press) and member of the editorial board of the Springer series Fundamental Theories of Physics and The Western Ontario Series in Philosophy of Science. Besides (co)organizing numerous conferences, he has given over 170 invited lectures in all corners of the world, heeding his stepfather's advice that "travel is the best form of education" and earning Executive Platinum status with American Airlines. He lives in Waterloo with his wife, Kristen ("Kiki"), and their two daughters, Emma and Katie.

GIANCARLO GHIRARDI is a professor emeritus of physics at the University of Trieste, Italy. Born in Milan, Italy, in 1935, he studied physics at the University of Milan and went on to earn his doctoral degree in physics in 1959. He worked as a research associate at the Ispra Center and the National Institute of Nuclear Physics before joining the University of Trieste in 1963 as an assistant professor in theoretical physics and becoming full professor in 1976. For seventeen years, he was the director of the Department of Theoretical Physics at Trieste. He has taught a plethora of courses, mostly on quantum mechanics, and has supervised over fifty doctoral theses.

GianCarlo is the coauthor, with L. Fonda, of *Symmetry Principles in Quantum Physics* (Marcel Dekker, 1970) and the author of *Sneaking a Look at God's Cards* (Princeton, 2003), for which he received the Primo Rovis Prize. He has organized several conferences and edited their proceedings. He has been a member of the editorial boards of *European Journal of Physics*, *Studies in History and Philosophy of Modern Science*, *Journal of Physics A*, *Nuovo Cimento B*, and *African Physical Review*. His seventieth birthday, in 2005, was celebrated with the meetings "Are There Quantum Jumps?" (Trieste) and "On the Present Status of Quantum Mechanics" (Mali Lošinj, Croatia), as well as with the special issue "The Quantum Universe" published by *Journal of Physics A*. He is a Membre Titulaire de L'Académie Internationale de Philosophie des Sciences, Bruxelles, and a member of the New York Academy of Sciences. He is also a consultant for the Abdus Salam International Centre for Theoretical Physics, Trieste, and a member of its academic board.

GianCarlo is passionate about popularizing science and regularly delivers talks at schools and social events. He cherishes music and the visual arts, owns a vast collection of banknotes depicting scientists, and loves hiking in the mountains, particularly in the Dolomites. He is working on completing an encyclopedic book about symmetry at all levels, from nature to the arts and the sciences. He lives in Trieste with his wife and is proud father of three daughters.

SHELLY GOLDSTEIN is a professor of mathematics, physics, and philosophy at Rutgers University. Born in Augusta, Georgia, in 1947, he received a bachelor's degree in physics from Yeshiva College and a Ph.D., also in physics, from Yeshiva University. His doctoral thesis was concerned with the application of ergodic theory to infinite-particle systems, with the aim of understanding the process of convergence to thermodynamic equilibrium of macroscopic systems. Shelly now believes that both infinite systems and ergodic theory have far less relevance to the foundations of statistical mechanics, and to convergence to thermodynamic equilibrium in particular, than he thought then. He did a postdoc at the Institute for Advanced Study in Princeton, and he then became assistant professor at Cornell before joining the faculty at Rutgers thirty-five years ago.

He teaches courses on the foundations of quantum mechanics, Bohmian mechanics, quantum reality, the foundations of statistical mechanics, and probability theory. He is the coeditor, with James T. Cushing and Arthur Fine, of *Bohmian Mechanics and Quantum Theory: An Appraisal* (Kluwer, 1996) and the coauthor, with Detlef Dürr and Nino Zanghì, of the forthcoming Springer volume *Quantum Physics Without Quantum Philosophy*. He lives in an old house in Highland Park, New Jersey, with Io and Agnes, his cats. He loves going to the movies and opera. Watching baseball and basketball also rank high on his list of favorite spare-time activities.

DANIEL GREENBERGER is the Mark W. Zemansky Professor of Physics at the City College of New York. Born in the Bronx, New York, in 1933, he knew he wanted to be a physicist before he knew what physics was. It probably had something to do with the atomic bomb, which ended the war when he was eleven. Before that, he wanted to be a chemist. He knew what that was because he had a nice chemistry set.

He went to the Bronx High School of Science, a very good school for interesting nerds. The class included Shelly Glashow, Steve Weinberg, and some other well-known physicists. Subsequently, he went to MIT, and then to the University of Illinois for grad school. He started working for Francis Low, who moved to MIT, so he tagged along with him and wrote a thesis on nucleon–pion scattering. He went to Ohio State for a year and won an NSF postdoc at Berkeley. After that, he joined the City College of New York, where he has been for the past forty-seven years.

He held a Fulbright fellowship to Vienna and won a Humboldt award to Munich (Garching). On his sixty-fifth birthday, *Foundations of Physics* devoted a Festschrift to him, and on his seventy-fifth birthday, the University of Vienna devoted a conference to him and Helmut Rauch. He is a fellow of the American Physical Society and an honorary member of the Austrian Academy of Sciences. He is on the editorial board of two physics journals, and managing editor of another. He has organized and chaired several quantum-foundations meetings and (co)edited their proceedings, including "New Techniques and Ideas in Quantum Measurement Theory" (1986), "Fundamental Problems in Quantum Theory" (1994), and "Recent Theoretical and Experimental Advances in Quantum Physics" (1998). He is coeditor of the *Compendium of Quantum Physics* (Springer, 2009). He has had fellowships or sabbaticals at MIT, Berkeley, Oxford, Ulm, Vienna, the Max Planck Institute for Quantum Optics in Garching, the Atominstitut in Vienna, and NSU in Singapore.

He lives in Queens with his wife, Suzanne, and her twenty-five turtles. (Suzanne is president of the NY Turtle and Tortoise Society.)

LUCIEN HARDY is a faculty member at the Perimeter Institute for Theoretical Physics in Waterloo, Canada, and an adjunct professor in the Department of Physics at the University of Waterloo. Born in Birmingham, U.K., in 1966, he earned a bachelor of science in physics from Imperial College and a Ph.D., also in physics, from the University of Durham. His dissertation, titled "Nonlocality, Lorentz Invariance, and Wave–Particle Duality in Quantum Theory," was concerned with quantum foundations and supervised by Euan Squires.

He was a lecturer in mathematical physics at Maynooth College, The National University of Ireland (1992–3); a Royal Society–funded postdoc in the group of Anton Zeilinger, Innsbruck, Austria (1993–4); a lecturer in the mathematical-sciences department at the University of Durham (1994–6); a postdoc with the group of Francesco De Martini, La Sapienza University, Rome (1996–7); and a Royal Society University Research Fellow, University of Oxford (1997–2002). He joined the Perimeter Institute in 2002. Lucien is married and has a daughter.

ANTHONY LEGGETT is John D. and Catherine T. MacArthur Professor of Physics at the University of Illinois at Urbana–Champaign. Born in London in 1938, he attended Balliol College, Oxford, where he majored in Literae Humaniores (classical languages and literature, philosophy, and Greco-Roman history). He then went to Merton College, Oxford, where he took a second undergraduate degree in physics. He completed a Ph.D. degree in theoretical physics, supervised by Dirk ter Haar. After postdoctoral research at Urbana, in Kyoto, and elsewhere, he joined the University of Sussex, U.K., in 1967, being promoted to reader in 1971 and to professor in 1978. In 1983 he took up his current position at Urbana–Champaign.

He is a member of the Royal Society, the American Philosophical Society, and the American Academy of Arts and Sciences. He is also a fellow of the Institute of Physics and of the American Physical Society. In 2003 he was awarded the Nobel Prize in physics, together with A. A. Abrikosov and V. L. Ginzburg, for "pioneering contributions to the theory of superconductors and superfluids." He was knighted by Queen Elizabeth II for "services to physics." Among his honors are the Wolf Prize in physics (2003); Eugene Feenberg Memorial Medal (1999); John Bardeen Prize (1994); Paul Dirac Medal and Prize (1992); Simon Memorial Prize (1981); Fritz London Memorial Award (1981); and the Maxwell Medal and Prize (1975). He has been a visiting professor at Cornell, Illinois, Minnesota, and Waterloo, as well as at ENS in Paris. He has been, just to name a few appointments, a Royal Society of Japan Fellow; Bethe Lecturer, Cornell; Morris Loeb Lecturer, Harvard; Astor Lecturer, Oxford; and Mueller Lecturer, Penn State. Since 2006 he is also the Mike and Ophelia Lazaridis Distinguished Visiting Professor at the University of Waterloo, Canada. He has been on the editorial board of *PNAS*, *New Journal of Physics*, and *Science*. He is the author of *The Problems of Physics* (Oxford, 1987, reissued 2006) and *Quantum Liquids: Bose Condensation and Cooper Pairing in Condensed Matter Systems* (Oxford, 2006).

TIM MAUDLIN is a professor of philosophy at Rutgers University. Born in Washington, DC, in 1958, he earned a bachelor's degree in physics and philosophy from Yale and a Ph.D. in history and philosophy of science from the University of Pittsburgh. His doctoral thesis, "Reasonable Essentialism and Natural Kinds," was advised by Clark Glymour. His whole academic career has been at Rutgers, with the exception of one semester as a visiting associate professor at Harvard. He has been a Guggenheim Fellow, and he is a member of the Académie Internationale de Philosophie des Sciences. He is the author of *Quantum Non-Locality and Relativity* (Blackwell, with a third edition forthcoming in 2011), *Truth and Paradox: Solving the Riddles* (Oxford, 2004), and *The Metaphysics Within Physics* (Oxford, 2007).

Tim and his wife and fellow philosopher, Vishyna, live in Princeton, together with master counterexampler and future philosopher Maxwell (age twelve). Their daughter, Clio, is a budding architect at Barnard. And their Bedlington terrier, Ki, a furry bundle of enthusiasm. Tim's spare-time passions are Croatia and cruciverbalism.

DAVID MERMIN is Horace White Professor of Physics Emeritus at Cornell University. Born in New Haven, Connecticut, in 1935, he studied mathematics and then physics at Harvard, where he received his Ph.D. in physics in 1961. He then worked as an NSF postdoctoral fellow in Rudolf Peierls's department at the University of Birmingham and as a postdoctoral associate with Walter Kohn at the University of California–San Diego before joining the faculty at Cornell University in 1964.

Since 1988 he has written about thirty "Reference Frame" columns for *Physics Today* on aspects of physics and the profession of physics. He is also the author of five books: *Space and Time in Special Relativity* (McGraw–Hill, 1968), *Solid State Physics* (Holt, Rinehart & Winston, 1976, with N. W. Ashcroft), *Boojums All the Way Through: Communicating Science in a Prosaic Age* (Cambridge, 1990), *It's About Time: Understanding Einstein's Relativity* (Princeton, 2005), and *Quantum Computer Science: An Introduction* (Cambridge, 2007). David has been a Guggenheim Fellow. He was awarded the Julius Edgar Lilienfeld Prize of the American Physical Society, the Klopsteg Memorial Award of the American Association of Physics Teachers, and the Russell Distinguished Teaching Award from Cornell University. He is also a fellow of the American Physical Society and a member of the American Academy of Arts and Sciences and of the National Academy of Sciences.

He lives on nearly eight acres of woods, meadows, and gardens with his wife, Dorothy. No current pets but plenty of wildlife. Their daughter is a documentary filmmaker working in London (visit merminfilm.com), and their son is a lawyer in Portland, Maine. When not tending the woods and garden, David works on the piano. He has two Steinway grands at home, with very different personalities. While writing his thoughts for this book, he has been learning the Beethoven C Minor Variations and the third Chopin Ballade. He hates large physics conferences, being given a schedule of people to see when he visits a physics department, and the level of political discourse in the United States. He loves to write.

OLIVIA MIZZI, TORONTO

LEE SMOLIN is a senior faculty member at the Perimeter Institute for Theoretical Physics in Waterloo, Canada, and an adjunct professor of physics at the University of Waterloo. Born in New York City in 1955, he graduated from Hampshire College with a bachelor of arts in physics and philosophy. In 1979 he earned a Ph.D. in theoretical physics from Harvard with a dissertation on quantum gravity. He did postdocs at the Institute for Advanced Study in Princeton, the Institute for Theoretical Physics at the University of California–Santa Barbara, and the Enrico Fermi Institute at the University of Chicago. He then held faculty positions at Yale, Syracuse, and Penn State, where he was a founding member of the Center for Gravitational Physics and Geometry. From 1999 to 2001 he was a visiting professor at Imperial College. In 2001 he relocated to Canada to help set up the Perimeter Institute.

Lee is the author of three books: *The Life of the Cosmos* (Oxford/Weidenfeld & Nicolson, 1997), *Three Roads to Quantum Gravity* (Basic Books/Weidenfeld & Nicolson, 2001), and *The Trouble with Physics* (Houghton Mifflin/Penguin, 2006). In 2009 *The Trouble with Physics* was listed by *Newsweek* among the fifty "Books for Our Time." Lee has given a large number of interviews and public lectures, and he has written numerous popular-science essays, which have appeared in *The New York Review of Books*, *Forbes*, *Physics Today*, *Nature*, and *The Times Literary Supplement*, among others. In 2009, he was honored with the Klopsteg Memorial Award for "extraordinary accomplishments in communicating the excitement of physics to the general public." He is also recipient of the 2007 Majorana Prize. He has been named one of the "100 most influential public intellectuals" by *Prospect* and *Foreign Policy*. He has been profiled in *New Scientist*, *Scientific American*, *Nature*, *The New York Times Magazine*, the *Guardian*, and *The Washington Post*. In 1998 Dutch television aired a documentary about Lee's work.

Lee is married to Dina; their son, Kai Misha William, was born in 2006. Lee enjoys dingy sailing in Toronto's Outer Harbor. He is also a jazz-guitar aficionado.

ANTONY VALENTINI is a professor of physics in the Department of Physics and Astronomy at Clemson University, South Carolina. Born in London in 1965, he received his undergraduate education in mathematics and physics at Cambridge University. He earned his Ph.D. at the International School for Advanced Studies in Trieste, Italy. His dissertation, supervised by Dennis Sciama, was titled "On the Pilot-Wave Theory of Classical, Quantum and Subquantum Physics." For many years he worked independently and taught privately. He did postdocs at the University of Rome "La Sapienza" and at Imperial College. He was a visiting professor at the Perimeter Institute and has recently been supported by the Foundational Questions Institute. He was a research associate with the Theoretical Physics Group at Imperial College, London, before joining Clemson University at the start of 2011. In 1982 he was awarded first prize in the City of London Mathematical Problem Solving Competition. He is currently teaching a graduate course on de Broglie–Bohm theory. He is the coauthor, with Guido Bacciagaluppi, of *Quantum Theory at the Crossroads: Reconsidering the 1927 Solvay Conference* (Cambridge, 2009).

DAVID WALLACE is a lecturer in philosophy at Oxford University and a tutorial fellow in philosophy of science at Balliol College, Oxford. Born in San Rafael, California, in 1976, he earned his undergraduate degree in physics from Merton College, Oxford. In 2002 he received, also from Oxford, his Ph.D. in physics, working with Artur Ekert on the foundations of quantum theory. He then crossed over into the territory of Oxford's philosophy department, earning a master's degree in philosophy in 2004 and a Ph.D. in philosophy in 2010. His doctoral research was advised by Simon Saunders and focused on the Everett interpretation. David did the last five years of this research concurrently with a full-time teaching position, and he actually only bothered finishing the Ph.D. because it basically coincided with his forthcoming book, *The Emergent Multiverse: Quantum Theory According to the Everett Interpretation* (Oxford, 2011). He is also the coauthor, with Simon Saunders, Jon Barrett, and Adrian Kent, of *Many Worlds? Everett, Quantum Theory, and Reality* (Oxford, 2010). He teaches courses on the philosophy of physics, philosophy of science, philosophy of mathematics, philosophy of mind, and on logic. He lives in central Oxford with his wife.

JACQUELINE GODANY

ANTON ZEILINGER is a professor of experimental physics at the University of Vienna. He is also the scientific director at the Institute for Quantum Optics and Quantum Information, Austrian Academy of Sciences, Vienna. He was born in Ried, Austria, in 1945 and studied physics and mathematics at the University of Vienna, where he received, in 1971, his doctorate in physics. His Ph.D. thesis, "Neutron Depolarization in Dysprosium Single Crystals," was supervised by Helmut Rauch. From 1972 to 1981 he worked as a senior research assistant at the Atominstitut in Vienna, and in 1979 he completed his *Habilitation* at the University of Vienna. Before taking up his current position in 1999, he was a visiting associate professor of physics at MIT (1981–3), associate professor at the Technical University of Vienna (1983–90), professor of physics at the Technical University of Munich (1988–9, *Lehrstuhlvertretung*), and professor of experimental physics at the University of Innsbruck (1990–9).

He was a visiting researcher at, among others, Humboldt University, Berlin; Merton College, Oxford; Collège de France, Paris; and Hampshire College, Amherst, Massachusetts. He has also held numerous distinguished lectureships and is a member of several scientific academies. His many awards include the Wolf Prize in physics (2010); Isaac Newton Medal (2008); King Faisal Prize (2005); Descartes Prize (2005); Lorenz Oken Medal (2004); Klopsteg Award (2004); Order Pour le Mérite for Sciences and Arts (2001); and a Senior Humboldt Fellow Prize (2000). He has coedited nine books, most recently *Quantum [Un]speakables, From Bell to Quantum Information* (Springer, 2002), *Quantum Computation and Quantum Information Theory* (World Scientific, 2001), and *Quantum Information: An Introduction to Basic Theoretical Concepts and Experiments* (Springer, 2001). He has also written two popular-science books (in German, with many foreign-language translations): *Einsteins Schleier* (C. H. Beck, 2003) and *Einsteins Spuk* (Bertelsmann, 2005; english version: *Dance of the Photons: From Einstein to Quantum Teleportation*, Farrar, Straus & Giroux, 2010).

WOJCIECH ZUREK is a Laboratory Fellow at Los Alamos and an external professor at the Santa Fe Institute, New Mexico. Born in Bielsko-Biała, Poland, in 1951, he received a master of science degree from the Technical University of Kraków in 1974 and a doctorate from the University of Austin, Texas, in 1979. He was a Tolman Fellow at Caltech for two years before coming to Los Alamos as an Oppenheimer Fellow. He led the Theoretical Astrophysics Group at Los Alamos from 1991 until he was elected a Laboratory Fellow in the Theory Division in 1996.

He founded the network "Complexity, Entropy, and Physics of Information" at the Santa Fe Institute. He was a visiting professor at the University of California at Santa Barbara, where he co-organized the programs "Quantum Coherence and Decoherence" and "Quantum Computing and Chaos" at UCSB's Institute for Theoretical Physics. In 2005 he was awarded the Alexander von Humboldt Prize, and in 2009 the Marian Smoluchowski Medal. In the 2004–5 academic year he was a Phi Beta Kappa Visiting Lecturer, and in 2010 he was appointed the Albert Einstein Prize Visiting Professor at the University of Ulm, Germany. He has coedited, with John Wheeler, the volume *Quantum Theory and Measurement* (Princeton, 1983), and he is the editor of *Complexity, Entropy, and the Physics of Information* (Addison Wesley, 1990).

PROLOGUE

An encounter with quantum mechanics is not unlike an encounter with a wolf in sheep's clothing. Disguised in sleek axiomatic appearance, at first quantum mechanics looks harmless enough. But beware: a moment later, it may sneak up from behind and whack you over the head with some thoroughly mind-boggling questions. Indeed, it's hard to imagine how we could ever cook up another physical theory that's as simultaneously innocuous and cunning. A theory whose formalism can be written down on a napkin whilst attempts to interpret it fill entire libraries. A theory that has seen astonishing experimental confirmation yet leaves us increasingly perplexed the more we think about it. How can we know so well how to apply this theory but disagree so vehemently about what it is telling us?

It is often said, and quite rightly so I think, that quantum mechanics is the most important intellectual achievement of the twentieth century. It tugs at the very roots of our convictions and instincts about what the world around us should reasonably be like, look like, and feel like. Of course, the recognition that extrapolating our subjective experiences and intuitions to sweeping statements about universal principles and the nature of reality may well lead us astray is as old as natural philosophy. And quantum mechanics is not the first physical theory to post a warning sign. Just think of Einstein's theory of relativity. The heliocentric worldview is also worth mentioning, since it made us realize that there can be a disconnect between what we feel—namely, that the earth is at rest—and what is actually happening. Yet quantum mechanics seems to be in a league of its own, taking the game of abstraction and counterintuitiveness to a whole new level.

Grounded in a stunningly elegant formalism that has stood the test of time like a lighthouse towering in the ocean surf, quantum theory is not only unabashedly silent on questions of ontology, but it also gives rise to puzzles that have inspired and been appropriated by audiences ranging from scientists and philosophers to New Age transcendentalists and Marxists. Schrödinger's cat has become a pop-culture icon on a par with the image of Einstein sticking out his tongue. To think about the foundations of quantum mechanics means to deliberate not only about how we can resolve—or dissolve—those puzzles, but also about how, in the process, we can make this cherished theory conceptually stronger from the ground up.

Interest in the field of quantum foundations has waxed and waned over time. The formative years of quantum mechanics—the 1920s, but also the maturing phase of the early 1930s—shone in a unity of purpose: to develop, apply, understand, and interpret quantum theory was, for the most part, one and the same thing. There was little separation yet into the hands-on practitioner on one side of the fence and the philosophically-minded foundationalist on the other. Instead, everybody chipped in ideas for how to come to terms with the new theory's meaning and peculiar implications.

Rather quickly—in fact, probably as early as 1927, in the wake of the fifth Solvay congress—the first cracks appeared in this happy union. The sentiment that interpretive debates had been mostly settled started to take root, and many took the view that Bohr, Heisenberg, von Neumann, and company had already provided satisfactory answers to the confounding questions quantum mechanics had generated. It was time to move on. On top of that, after World War II a pragmatist mindset spilled over from the United States into academic institutions everywhere. Spending one's precious time contemplating foundational questions became frowned upon and distinctly unpopular.

At the same time, however, the 1950s also marked the beginning of a slow resurgence of foundational efforts. For instance, Bohm revisited de Broglie's pilot-wave theory—motivating, in turn, Bell's seminal work on nonlocality in the 1960s—and Everett developed his relative-state interpretation. The 1950s and 1960s also saw Bohr devotees, such as Heisenberg, Rosenfeld, and Weizsäcker, trying to elucidate their master's pronouncements and supply physical and formal foundations for some of Bohr's more hand-waving arguments. New interpretations and versions of quantum theory continued to spring up throughout the 1970s and 1980s. Van Fraassen, as part of his philosophical program of constructive empiricism, introduced the concept of modal interpretations. DeWitt and Graham's 1973 book *The Many-Worlds Interpretation of Quantum Mechanics* popularized and advanced Everett's ideas. Collapse theories appeared that weren't just interpretations but made new predictions. And the consistent-histories approach launched yet another attempt at curing quantum theory of what some had perceived as an unhealthy fixation on the observer. (More on interpretations in Question 3, *My Favorite Interpretation*.)

The technical developments of the 1980s and 1990s also added their share. Decoherence theory, for example, made us understand why Schrödinger cats are such elusive creatures, and it plugged some leaks in interpretations such as Everett's. And depending on who you ask, the recent quantum-information buzz has been either a true eye-opener for foundations or mainly an obfuscation of the real problems (see Question 9, *Quantum Information*).

And so it came to pass that the last two or three decades have witnessed a stunning renaissance of quantum foundations. Today we find ourselves in the fortunate situation where occupation with quantum-foundational matters is no longer inconceivably far from reaching mainstream status. As an indication perhaps, already a good decade ago *Physics Today*—a mainstream physics magazine if there ever was one—ran what seemed like a torrent of articles and opinion pieces on the interpretation (and apparent inconsistency) of quantum mechanics. The job situation of your average foundationalist has also somewhat improved. To be sure, foundational pursuits will probably still elicit reactions ranging from polite disinterest to outright disapproval at many physics departments. But like the teenager who runs away from his bigoted parents in Omaha, Nebraska, to become an openly gay musician in Brooklyn, today a foundationally-minded scholar can start a new academic life at places such as the Perimeter Institute, where he can indulge his interests and still bring home a paycheck at the end of the month. (No less than three of our interviewees call Perimeter their home.)

To this day, the traditional controversies continue to provide most of the intellectual fuel that keeps foundationalists' engines humming: the meaning of quantum states (Question 4), the issue of indeterminism (Question 5), the interpretation of quantum probabilities (Question 6), the measurement problem (Question 7), and the questions posed by the EPR–Bell scenario (Question 8). At the same time, the kind of analytical tweezers we have now at our disposal for revealing and dissecting even the most delicate layers of these problems have enabled us to address foundational issues with unprecedented rigor and detail. Feeling elated by the views from Mount Quantum Information, many foundationalists are also busy putting a new spin on the time-worn game of interpretation: reconstructions of quantum theory, combined with efforts aimed at understanding what precisely delineates quantum mechanics from other probabilistic theories, meet John Wheeler's question "Why the quantum?" head-on (see Question 10, *Reconstructions*). Finally, with experimental techniques reaching stunning levels of refinement, our inquiries are no longer confined to the ivory tower of gedankenexperiments, theoretical examination, and philosophical contemplation (see Question 11, *The Experiment of My Dreams*). This ability to put foundational questions to the jury of the laboratory bench has also political benefits. It makes pragmatic physicists reappraise the relevance and legitimacy of such questions from the point of view of science. And it boosts the image of foundational research in the eyes of funding agencies keen on getting no-nonsense results in return for their generosity.

The field of quantum foundations is back with a vengeance. It's here to stay, and its future looks brighter than ever. May this claim find support in this book, which—so I hope—will also contribute to propelling the field forward.

FIRST ENCOUNTERS

———

What first stimulated your interest in the
foundations of quantum mechanics?

IN THE WAKE of the *Deepwater Horizon* disaster in the Gulf of Mexico, *The New York Times* ran a story about a deckhand named Randy Jones living and working on Bayou Petit Caillou, Louisiana. The bulk of the news piece predictably centered on how Jones had been put out of work by the oil spill. But it also detailed how not only Jones himself but every single generation of his family had made a living fishing on Bayou Petit Caillou—since the 1700s, in fact, which is when Jones's Cajun ancestors had first arrived in the area.

I have yet to hear a similar story about the profession of the quantum foundationalist: about someone who got himself into the field because it was, so to speak, in line with the family business. The foundations of quantum mechanics won't be among the accredited university degrees you learn about at your local college's open house. In the corridors of academia, quantum foundations represents a true niche, a hybrid that's wedged between the classic disciplines of physics and philosophy. Yet it receives little practical support from either of these two disciplines. Hard-nosed scientists tend to give the subject the cold shoulder: nothing but feeble philosophical babble, they proclaim. And many philosophers are intimidated: too much tough physics and mathematics, they fear.

So it takes not only a particular kind of intellect and personality to comfortably navigate the seas of quantum foundations, but also lucky happenstance to embark on the voyage in the first place. You have to discover the subject for yourself, and more often than not, such discovery is a matter of having been in the right place at the right time. Lucky to have picked up, by chance, that one book that got you hooked. Lucky to have caught that one talk that sent sparks flying. Lucky to have met that one luminary who became your inspirational father figure. And if you were really sharp-witted: to have sensed, perhaps, that a Pandora's box of unspoken concerns might lurk beneath your professor's breezy textbook presentation of quantum mechanics.

So it is only fitting to begin this book with stories of first encounters. These recollections are a beautiful testament to the varied and deeply personal paths that have brought together physicists, philosophers, and mathematicians in shared passion for the contents of that Pandora's box. And this passion frequently turns into a lifelong love affair: while there are plenty of former lawyers and business people and bus

drivers and fishermen out there in this world, I have never met anybody whose ardor
for quantum foundations has gone cold. And for many in the field, years gone by
equate not to more answers but to more questions!

GUIDO BACCIAGALUPPI · I suppose I've always had some interest in fundamental issues about the world, and that is why I decided to study physics as an undergraduate. The more specific—and eventually much more serious—interest in the
foundations of quantum mechanics came quite by accident.

I had gone to Zürich, to ETH, for my studies, and it was compulsory for any
student there to take at least one course in the humanities in each semester. Sometime during my first week, I was told about some excellent philosophy lectures by a
visiting professor from Boston. This was Abner Shimony. It turned out that apart
from his lecture course, he was also giving a seminar on the foundations of quantum mechanics. The seminar was co-organized by Hans Primas and featured guest
appearances by John Bell, Nicolas Gisin, Anton Zeilinger, and others. At the time,
most of what was discussed was beyond my technical comprehension (I remember
an "aha" experience when we finally discussed the trace of a matrix in linear algebra). But I was fascinated, and I faithfully attended all sessions. I kept in touch with
Abner even after that first semester.

A few years later, after I had switched subjects and completed my degree in mathematics—and was wondering what to do with it—I eventually decided I should combine my interests and my training and devote myself to the foundations of quantum
mechanics. Again I was extremely lucky and heard about the group led by Michael
Redhead in the Cambridge HPS department, where I eventually wrote my Ph.D.
with Jeremy Butterfield.

The Cambridge group has been one of the main hotbeds of the field in Britain,
and I still feel it is unsurpassed in terms of the richness and productivity of the
research culture it supported. My experience in the Cambridge foundations group
is what made me a researcher, and I could not have wished any better. Still, it is to
Abner I owe my first introduction to the field, as very many others in fact do.

ČASLAV BRUKNER · Already as a young boy, I was more interested in the question "Why?" than the question "How?" Taking apart a watch and putting it back
together never got me as excited as thinking about what makes the watch tick in the
first place. The reason for this wasn't just that often there were a few pieces left after
I had put the watch back together. It was more a form of deep emotional excitement
I experienced when thinking about basic questions.

There were two events, however, which definitely determined the choice of my
future profession. The first one occurred on my thirteenth birthday. As a present, my
brother had given me Vladimir Paar's book *Što se Zbiva u Atomskoj Jezgri?* (English:
What's Going On in the Atomic Nucleus?) The book had all the makings of a good

science book for children: grace of language, storytelling skill, and that extra spark of inspiration that can challenge and enrich the reader's mind. I still remember how entertaining I found the illustrations of electrons, protons, neutrons, and quarks: small colored balls jumping from one flight of stairs to another while changing their energy states. But soon after reading the book, I started to struggle with the meaning of the book's illustrations, according to which the components of ordinary things behave as, and are as real as, the things they make.

The second event determined my future research field. I was in my third year of physics studies at the University of Belgrade, attending a course on quantum mechanics taught by Fedor Herbut, a superb teacher. Both in his classes and his textbook, Herbut succeeded in maintaining rigor while giving the students an intuitive understanding of the subject. In one section of his textbook, he briefly commented on Freedman and Clauser's 1972 experiment, gave the experimental curve of the observed quantum correlations, and remarked that it contradicted predictions of local hidden-variables models. Needless to say, I did not get any real understanding of what this was all about. I probably thought that nobody in his right mind could possibly disagree that properties of objects exist independently of whether they are being observed, and that they cannot be influenced by distant actions. So I started to play around with some models of what I thought would be quantum correlations. I took an ensemble of pairs of (separable) antiparallel quantum spins, evenly distributed over the entire space, and I tried to recover the experimental curve. Alas, I failed! I took another model, and another, and another—and none of them worked. I was deeply shocked. It is fair to say that I have never recovered from this shock.

JEFFREY BUB · As an undergraduate at the University of Cape Town in the early 1960s, I wrote a review of the EPR paper, Bohr's reply, and Bohm's hidden-variables theory. EPR argued that the correlations of entangled quantum states demonstrate the incompleteness of quantum mechanics as a description of physical reality. What struck me at the time was the very idea that there could be a serious dispute in the foundations of physics about an absolutely basic question like the existence of hidden variables underlying the quantum statistics. I don't think my understanding of the problem went very deep, but the experience motivated me to become Bohm's student. Later, as a graduate student, I discovered an intriguing paper by Henry Margenau on the measurement problem in quantum mechanics, with a reference to the book by London and Bauer, *La Théorie de l'Observation en Mécanique Quantique*. The book was not then available in English—now there is a translation in Wheeler and Zurek's *Quantum Theory and Measurement*—but I was able to read it with my limited French at the time. The discussion by London and Bauer was clear and compelling, and I knew I'd found my dissertation topic. I eventually wrote a dissertation with Bohm on hidden variables and the measurement problem, from which a couple of papers were published in *Reviews of Modern Physics* in 1966. Both papers were coauthored with Bohm; the first was a proposed solution to the measurement problem by a "dynamical collapse" hidden-variables theory, closer in spirit to the later Ghirardi–Rimini–Weber theory than to Bohm's 1952 hidden-variables theory; the second was a refutation of the "no go" hidden-variables theorem of Jauch and Piron.

Arthur Fine · My interest in the foundations of quantum theory was overde-
termined. First, as an undergraduate student of mathematics, one of my advisers was
Irving Segal, then engaged in his seminal work on C^*-algebras. Although that topic
was certainly beyond my skills, nevertheless I got a whiff of something exciting and
provocative that I wanted to understand. Later, as a graduate student, I wanted to
study functional analysis, which wasn't being offered as a course. I approached the
mathematician Karl Menger about doing a reading course, and he agreed, but only
if we worked through von Neumann's treatise on the mathematical foundations of
quantum mechanics, which had recently been translated into English. We made our
way through the mathematical background to Hilbert space, the spectral theorem,
and then just kept going—to quantum statistics, the famous no-go theorem, and
all. Menger, who had been a member of the Vienna Circle, had a serious interest in
foundational issues and also a charming but provocative style. For example, he always referred to the quantum theorists as "metaphysicians," which in his lexicon was
not a term of praise. He challenged me over several of the well-known "paradoxes"
and set me on a path that I was fortunate to be able to pursue in my dissertation. The
thesis was on quantum measurement and my mentors there were Henry Mehlberg, a
product of the famous Polish school in the methodology of science, and also Gregor
Wentzel, then the director of the Fermi Institute at the University of Chicago. At
first, Wentzel was very reluctant to entertain the idea that there was any "problem"
over quantum measurement. (He put it to me that quantum theory represented his
youth and he did not like to hear it criticized.) Once I caught his attention over
the issue of collapse, however, he became very interested and helpful, especially in
showing me how a talented physicist thinks about his craft in connection with foundational questions.

Christopher Fuchs · How do you answer a question like this without thinking of Dr. Evil in the *Austin Powers* movies: "My childhood was typical. Summers
in Rangoon. Luge lessons. In the spring we'd make meat helmets ... pretty standard
really." Maybe you've already had some answers like this! It takes a strange type to
get involved in quantum foundations.

   In my case, it all had to do with science fiction and growing up in a small town in
Texas. If you've ever seen the movie *The Last Picture Show*, you'll know the kind of
place I mean. We had three television stations we could pick up from San Antonio,
and the main things they'd show on Friday and Saturday late nights in the early
seventies were science-fiction and horror movies. I gained a kind of taste for surreality
from it—a weird world was a good world to me. Still more important were the after-
school showings of *Star Trek* that I would race home to see; they started when I was
in third grade. I wanted to live the life of Captain Kirk; I wanted to fly to the stars
and have great adventures exploring strange new worlds.

   So in junior high I thought, "Let's just see how to make that happen." I started
to read everything I could on physics. Boy was I disappointed when I learned the
speed of light was in fact a speed limit. At least I was compensated by learning of
black holes and wormholes and tachyons. In the seventh grade I borrowed a copy of

John Archibald Wheeler's book *Geometrodynamics* by interlibrary loan (it came all the way from Texas Tech in Lubbock). I read the words, skipped the equations, and didn't understand most of it, but I tried. Mostly I dreamed.

There was a thought I had then that stands out overpoweringly now. I would tell my friends, "If the laws of physics won't let me go to the stars, then they must be wrong!" Looking back on it, I think that made me quite receptive to two things that would happen in college and set the tone for my whole career. The first was that I read Heinz Pagels's book *The Cosmic Code* on the mysteries of quantum theory my very first week there. The second was that within the year I would meet the real John Wheeler; he wasn't just a myth in a book from Lubbock. I had gone to college undecided on a major, but those two events capped the decision—it had to be physics.

In those days, John walked around not saying "It from Bit"—that came with a later turn of mind—but instead "Law without Law!" As he put it, "Nature conserves nothing; there is no constant of physics that is not transcended ... mutability is a law of nature. *The only law is that there is no law.*" And it all came together for me. If I were to study quantum theory, I might just find a way to make my wish happen. John saw the quantum as a chink in the armor of law. It would take years and years for me to come to grips with the idea, but I found the thought so exciting, so alluring—almost as if it were made for me—that I couldn't keep my eyes off it. Thus, I did what I needed to do; I endured a physics degree, with classical this and solid-state that, so that one day I might make a contribution to quantum foundations. To be sure, it was an endurance contest: I really didn't like physics the way most physics students do, and I suppose more than one hiring committee noticed that.

So, blame it on Heinz Pagels's prose, John Wheeler's inspiration, and the San Antonio, Texas, television stations. Quantum theory has taken my heart since the beginning because of its spicy mélange of law ... without law. In a world where the laws of nature are as mutable as the laws of legislatures, most anything might happen. Imagine that! If it doesn't make your heart flutter, then you're probably looking for a different interpretation of quantum theory than I am. In any case, that's how I first got interested in quantum foundations.

GianCarlo Ghirardi · I had been fascinated by philosophy, logic, and science since my high-school times. The first event that played a concrete role in my interest in the foundations of quantum mechanics goes back to 1959, the year in which I got my degree in physics. I attended a seminar by Giovanni Maria Prosperi on the so-called Daneri–Loinger–Prosperi proposal for overcoming the measurement problem. That seminar was a sort of revelation for me. In fact, even though I had taken an excellent course in quantum mechanics, I had not been so smart as John Bell and Bernard d'Espagnat, who—as they later recalled—had immediately spotted the unsatisfactory status of the theory when they were first exposed to it. In my case, it was a reflection on what I had heard from Prosperi that led me to realize two things.

First, that if one requires that all natural processes obey the linear laws of the theory, it turns out to be impossible to account for the way the theory postulates measurement processes to take place, in particular for wave-packet reduction.

And second, that the solution proposed by Daneri, Loinger, and Prosperi did not represent a satisfactory way out of the difficulties of the formalism. It belonged to the line of thought that some years later Bell appropriately characterized as a FAPP position ("For All Practical Purposes"): the very basic rules of the theory are logically inconsistent and give rise to unacceptable situations, but because of difficulties of practical nature, one can use the formalism as it stands if one is exclusively interested in making practical predictions. In the specific case under consideration, Daneri, Loinger, and Prosperi assumed that there is a level (which is not made precise in their mathematical treatment) at which it becomes *in principle* impossible to measure noncommuting observables on a specific quantum system: the observables quantities for a macro-object form an Abelian set. This amounts to accepting that quantum mechanics does not have universal validity, but without identifying the theory's precise field of application—which is the typical limitation of most of the proposed ways out of the problem.

SHELLY GOLDSTEIN · When, as a physics student, I first learned about quantum mechanics, before I had any real knowledge of its technical content, I was intrigued by the notion that with the quantum revolution classical determinism had been overturned. I was fascinated by the idea that with quantum mechanics, determinism had been replaced by some sort of irreducible, unavoidable randomness in the laws of nature. But one thing about this worried me: that Einstein had not accepted this conclusion. I was really upset about Einstein's resistance to the great quantum revolution. Maybe I was also upset with Einstein as well.

But the first responses to Einstein's reservations that I learned about, for example of Bohr and Heisenberg, were hard for me to follow. They certainly did not seem entirely persuasive.

It was therefore with considerable relief that I learned that the great mathematician John von Neumann was supposed to have proven that Einstein was wrong, that the sort of restoration of determinism that Einstein desired was incompatible with the experimental predictions of quantum mechanics. Von Neumann's proof, it seemed, replaced the vague, incomprehensible arguments of Bohr and Heisenberg by clear rigorous arguments grounded in sharp mathematics. (It was only much later that I came to understand that what von Neumann had accomplished with his argument was in fact much, much less.)

Having learned about this work of von Neumann, I felt much more secure in my belief that contra Einstein, the quantum revolution was genuine. But I still needed to understand exactly what quantum mechanics says. Of course, I had every reason to expect that I would learn this—and more—from the courses that I was then taking and that I would be taking.

This, however, did not happen. As I learned more about the technical details of quantum mechanics, it seemed that I understood less and less about its genuine physical content. The more I understood quantum mechanics as mathematics, the less I understood it as physics. I found the whole business of learning quantum mechanics a peculiar one indeed!

DANIEL GREENBERGER · When I first was exposed to quantum mechanics as an undergraduate, I was thoroughly fascinated by it and read all the books on it I could find, like Heisenberg's little book on the uncertainty principle, and Reichenbach, and Margenau. At MIT, where I was, I asked my undergraduate adviser whether I could get into the field of foundational problems. That was the 1950s, long before Bell's papers. Then the only people worried about it were the older giants like Schrödinger, Bohr, and Einstein. Everybody thought the issues were based purely on opinions and preferences, and nobody thought that experiments were possible. My adviser told me there was an old Chinese proverb, "When elephants fight, it's the grass that gets trampled," and advised me to let the giants fight it out, and to stand clear so I wouldn't get stepped on. I took his advice and went into high-energy physics. (By the way, I never found a Chinese physicist who had heard of the proverb. But I recently met an African one who had heard a version of it, which makes sense, since it uses elephants for its metaphor.)

It was many years later, when I was thinking about the equivalence principle and the possibility of testing it with neutrons, that I met Mike Horne, Anton Zeilinger, and Cliff Shull at a conference, and that was a life-changing experience. It got me into neutron interferometers and quantum foundations, and later into quantum optics.

LUCIEN HARDY · When I was seventeen, my father recorded a BBC Radio documentary called *The Ghost in the Atom* for me (the transcript of this has since been turned into a book). In this program, Paul Davies interviewed various people about the foundations of quantum theory. I listened to the recording over and over. It made very little sense to me back then. Nonetheless, it sounded tremendously exciting. I had already decided I wanted to pursue a career in physics. But except for a few vague items in the school syllabus, I had not previously been exposed to quantum theory. From this radio program, it was clear that here was something very different. The usual way of thinking about reality did not apply. Something could be in two places at the same time. Furthermore, the theory was strangely nonlocal. The experts did not agree on how to interpret the theory. John Bell was interviewed talking about his theorem proving that quantum theory is nonlocal. Alain Aspect talked about his experiment to test Bell's inequalities. David Deutsch talked about the many-worlds interpretation, David Bohm and Basil Hiley discussed the pilot-wave interpretation, Rudolf Peierls and John Wheeler advanced versions of the Copenhagen interpretation, and John Taylor promoted the ensemble interpretation. At the time, these names were unknown to me. I have since met some of them. I never met John Bell (though I saw him lecture twice). He died in 1990, just before I published my own version of his theorem. This was also before the quantum-teleportation paper, which marked the beginning of the quantum-information age, in which his work found applications he probably had never dreamed of.

In any case, I owe Paul Davies (and Julian Brown, the BBC producer) a debt of gratitude for making that program. Perhaps future researchers into quantum foundations will be similarly grateful to Maximilian Schlosshauer for the present volume.

ANTHONY LEGGETT · During my postdoctoral years, and for the first couple of years as tenured faculty member at the University of Sussex, I had little interest in foundational issues. I think I probably took the view attributed by the late John Bell to the "typical" physicist with respect to the quantum measurement problem: that it had long ago been solved, and that I would easily understand the solution if I could ever spare twenty minutes for the exercise.

What converted me was a minilecture series on the measurement problem by a Sussex colleague, Brian Easlea, who had trained as a nuclear theorist but later switched his interests to the history and sociology of science. He had been much influenced by Bell's views, and I remember that he cited in particular the paper with Nauenberg on the "moral aspect" of quantum mechanics. Brian was very enthusiastic and persuasive. Although my initial reaction was that what was needed to solve the problem was a more philosophically sophisticated formulation, I gradually concluded that one could not dispose of it so easily. In fact, by the early summer of 1972, I had pretty much made up my mind to quit doing conventional low-temperature physics and devote myself full-time to the foundations of quantum mechanics. In the event, the spectacular Cornell experiments on what we now know as helium-3—experiments that suggested (at least briefly, to me) that quantum mechanics itself might be breaking down—sidetracked this program for most of the next decade.

But I kept an interest in foundational issues. In fact, while on secondment in Africa in the fall of 1976, I wrote a paper on nonlocal hidden-variables theories, which I eventually revised and published in 2003, and which then, to my surprise, motivated some experiments in this area. I finally managed to get back to my foundational enthusiasms around 1979, at which point I was able to connect them with my more "conventional" interests in the condensed-matter area. This has been one of my main lines of research ever since.

TIM MAUDLIN · My interest in physics generally comes from the same source as my interest in philosophy: to understand what exists at the most fundamental level. With respect to the physical world, the most general specifications of this foundational question are: "What are space and time?" and "What is matter?" Any serious investigation of the first question leads one to study relativity, and any serious investigation of the second to quantum theory.

My own initial exposure to quantum theory was in a standard physics introductory course, which, of course, did not recognize or discuss any foundational questions. I suppose I just accepted the received wisdom: we had learned that nature is indeterministic and somehow "fuzzy." I certainly did not extract any clear account of the nature of matter from the physics, nor any precise sense of even how to go about trying to make things comprehensible. The clearest memory I have concerning quantum theory from college did not come from a physics course, but rather from reading Bernard d'Espagnat's account of Bell's theorem and Aspect's experiments in the *Scientific American*. Bell's argument was crisp and clear, and the experimental results appeared decisive. My roommates at the time still recall seeing me pacing the room, article in hand, going around and around in circles. The violations of Bell's inequality were my lodestar, a sharp-edged result that could guide us to a deeper understanding of space and time.

The interests of my professors in graduate school ran more to space-time theory than quantum theory. The most important lesson I derived from the study of relativity was how much effort it takes to get entirely clear about what a theory asserts, even when there are no obvious foundational problems. But relativity becomes clearer and clearer the more one works with it, so one also gets a sense for what it feels like to really understand a physical theory. I did not, however, leave graduate school with a sharp understanding of how to think about quantum theory. And the tension produced by Bell's result was even more acute.

The decisive turning point came in conversations with Shelly Goldstein about the foundations of quantum theory. Like most everyone else, I had absorbed some ill-defined disdain for Bohm's theory somewhere along the way, although I had never even seen the theory explicitly presented. The theory was supposed to be ad hoc and unnatural (I do not recall being told it was impossible or refuted by experiment). Shelly patiently demonstrated that I did not know what I was talking about (an important service!), but more significantly provided a shining exemplar of a physically comprehensible theory. Whether one likes it or not, Bohmian mechanics is *clear*. As Bell insists, it provides a standard of what an acceptably precise physics that accounts for quantum phenomena can be. Foundations of quantum theory ought to be largely the quest to make every proposal for understanding the physical world equally clear.

DAVID MERMIN · I've always been more fascinated by physics as a conceptual structure than by physics as a set of rules for calculating the behavior of the natural world—what Suman Seth calls the "physics of principles," as opposed to the "physics of problems." My text with Neil Ashcroft on solid-state physics is a success because Neil is as focused on the physics of problems as I am on the physics of principles. Somehow we managed to produce a book that combines both views.

The conceptual structure of quantum mechanics is stranger and lovelier than any perspective on the world that I know of, so I've been fascinated and worried about it from the beginning of my career in physics. Indeed, I became interested in my early teens in the late 1940s, long before I knew enough mathematics to learn the quantum formalism, through the popular writings of George Gamow, Arthur Eddington, and James Jeans. In college, I put these interests on hold, majoring in mathematics and taking only a few courses in (classical) physics.

But I returned to physics in graduate school, where my revived curiosity about quantum foundations was actively discouraged by my teachers. To my disappointment, the Harvard physicists all believed that a preliminary training in the physics of problems was a prerequisite to any understanding of the physics of principles. So, for a quarter of a century, I was deflected full-time into statistical physics, low-temperature physics, and solid-state physics, using (and teaching) quantum mechanics as a beautiful and effective body of rules for manipulating symbols on a page to get answers to questions about experiments in the laboratory. "Shut up and calculate!"

Early in graduate school, Gordon Baym, a fellow student, told me at the Hayes–Bickford cafeteria about Bohm's spin-½ version of Einstein, Podolsky, and Rosen.

EPR was never mentioned in any official academic setting. I immediately concluded that the quantum-mechanical description of physical reality was incomplete, and I made a note to think about completing it when I got tired of the serious pursuits my teachers had set me on. (After my oral qualifying exam, Roy Glauber advised me to stop spending so much time with Gordon Baym. The senior members of my committee, Wendell Furry and Julian Schwinger, seemed to agree.)

More than two decades later, in 1979, fifteen years after John Bell's now-famous paper appeared, I learned about Bell's theorem through the pages of the *Scientific American*. I believe Tony Leggett had tried to tell me about it a few years earlier, but I was too busy with real physics to pay attention. I realized to my astonishment that the more complete theory that EPR had convinced me would someday be found to underly quantum mechanics, resolving all its mysteries, either did not exist or, if it did, would be at least as mysterious. In the three decades since then, I've devoted a significant fraction of my intellectual efforts to pondering such puzzles, mainly trying to boil them down to their simplest possible forms.

LEE SMOLIN · I got interested in physics during a period when I had dropped out of my last year of high school. I was interested in architecture, particularly in developing Buckminster Fuller's geodesic domes. I was somewhat advanced in mathematics, so I studied differential geometry to try to develop a method for triangulating arbitrary curved surfaces. Every book on this subject I consulted had a chapter on general relativity, and this led me to take out a collection of essays about Einstein. This included an essay by Einstein himself, titled "Autobiographical Notes." In it, he described his motivation for becoming a physicist. Reading this essay one evening, I had a vivid impression that I too could become a physicist. He also related his unhappiness with the statistical nature of quantum mechanics and his belief that quantum mechanics is a provisional theory to be replaced by something different that gives a description of individual processes. So I absorbed his unhappiness with quantum mechanics. I then went back to the library and took out books by Bohr, Heisenberg, and de Broglie, which introduced me to their views on the foundations of quantum mechanics. But ever since that time I felt captured by Einstein's view that quantum mechanics is incomplete and needs replacement by a deeper theory.

After that, I was extremely fortunate to go to Hampshire College, where I had been accepted, despite being a high-school dropout, on the strength of my interest in architecture. The physics professor there was Herbert Bernstein, who had developed a new university-physics curriculum that started with the quantum mechanics of finite-state systems. The textbooks for this first-year college course included Dirac's volume on quantum mechanics and Gordon Baym's graduate textbook, with Feynman's Volume III as a supplement.

Herb emphasized the foundational issues in his first-year quantum mechanics course. The EPR paper and Bohr's reply, as well as John Bell's paper on his theorem, were taught toward the end of the course. I recall vividly studying the proof of Bell's theorem during my first spring of college, and staring into the corner of my room and taking in the fact that my eyes were entangled with atoms in the wall. I also wrote a poem to a girl I had met, on the theme that each time we touched our atoms became entangled.

I was seventeen at that time, and already then very focused on the foundational issues in quantum mechanics.

ANTONY VALENTINI · It so happens that I studied general relativity before I studied quantum theory. This seems to have affected my reaction to the latter. It surprised me that quantum theory uses the same elementary degrees of freedom as classical physics—position, momentum, angular momentum, and so on. Nor are any radically new degrees of freedom introduced (with the possible exception of spin). Classical variables are promoted to operators. The values are restricted to eigenvalues. But the set of elementary variables is the same as in classical physics. General relativity, in contrast, abandons Newtonian gravitational force and introduces the new concept of curved space-time. The basic ontology changes. I had expected quantum theory to be in a sense much more novel than it is. But the basic variables are unchanged, only subject to an operator calculus that is itself constructed by analogy with classical physics. In this sense, the theory talks about nonclassical systems as if they were still classical.

My experience with general relativity had led me to expect that quantum theory would be based on, as I would now put it, a different ontology from that of classical physics. I was disappointed that there was no new, distinctively quantum entity. There is the wave function, of course, but it seemed just to give probabilities for values to be taken by the usual set of classical variables. I was also impressed by the conceptual leap that Faraday had made in developing the concept of the electromagnetic field. This was a novel and non-Newtonian ontology, a fundamentally new kind of thing. What comparable new concept did quantum theory introduce? None that I could see.

Then I remember the first time I encountered an explicit statement of the collapse postulate. The book I read said something about how when a system is observed it is "thrown into" an eigenstate. How could merely looking at something make it jump around like that?

As an undergraduate, I read Ballentine's famous 1970 review of the statistical interpretation, and I was impressed to see so many muddles cleared up. It pointed to something deeper, however, along the lines of hidden variables, and by itself could not be satisfactory. I also read a bit about de Broglie–Bohm theory, but unfortunately the papers I read contained obvious mistakes and this led me to think the theory was wrong or at best incomplete.

Later on, during some years I spent outside academia after graduating, I became deeply impressed by quantum nonlocality, and in particular by the puzzle of why we can't use it for instantaneous signaling. In the 1980s some people were still proposing ways to use EPR correlations for signaling, and of course it would always be shown by someone else that the proposals were wrong. The no-cloning theorem, for example, arose in response to such a proposal. I couldn't shake off the feeling that there was something nonlocal going on behind the scenes, and that quantum-uncertainty noise was preventing us from seeing it. By means of Bell's theorem, we could deduce that the nonlocality was there, but the uncertainty principle stopped us from using it for signaling. This was my strong impression. Shimony referred to this sort of thing as a "peaceful coexistence" between relativity and quantum theory. To me, it seemed like a dark and uneasy conspiracy that cried out for an explanation.

I started to form the vague idea that the hidden-variable level must be in a state of some sort of statistical equilibrium, in which the nonlocal effects average to zero. In ordinary statistical mechanics, thermal equilibrium yields finely tuned relations such as detailed balancing, the fluctuation–dissipation theorem, and so on. I was also fascinated by an analogy with Maxwell's demon, who is unable to sort fast and slow gas molecules only because he is in thermal equilibrium with the gas and is therefore subject to the same thermal noise as the molecules themselves. I started to think that something similar must be going on at the hidden-variable level—that we are unable to control the details of hidden variables because we are ourselves stuck in an equilibrium state. I had vague ideas about a hypothetical "subquantum demon," who could predict outcomes of spin measurements more accurately than quantum theory allows, and who could thereby use EPR correlations for nonlocal signaling. But I didn't have a real theory. This must have been around 1988. It was only when I studied de Broglie–Bohm theory properly, in early 1990, that I saw a concrete way to realize the idea.

DAVID WALLACE · I can't really separate that from learning quantum mechanics itself. When I originally came across it as a first-year undergraduate, it just felt like a confusing mess. I think the way most students (in the U.K., at least) start off with quantum theory is a bunch of unrelated stuff about wave equations, uncertainty principles, wave–particle duality, and the like. At that point, the whole subject looked so messy and confused, I struggled to get very interested in it.

But then, a bit later, I learned it properly and saw how elegant it really was from a mathematical and conceptual point of view—and that just made the quantum measurement problem stand out. If the theory was that elegant, it had to be possible to make sense of it somehow. And to a substantial extent, I found I wasn't really able to work on other problems within quantum mechanics as long as I didn't understand how the measurement problem was to be resolved.

Fifteen years later, I now feel pretty confident that the Everett interpretation satisfactorily resolves the measurement problem—and ironically, that's somewhat reduced my interest in the foundations of quantum mechanics. But there are plenty of foundational, conceptual, and philosophical questions left in quantum theory, most of which are easier rather than harder to make progress on from the point of view of having a definite interpretation.

ANTON ZEILINGER · When I first learned quantum mechanics, I was immediately struck by its mathematical beauty, particularly the beauty of the Dirac notation and the Hilbert-space formalism. It was just mind-boggling that such a clear mathematical language could work so nicely in nature. On the other hand, hearing about the different interpretations through discussions of the two-slit experiment, I wanted to know more and understand more. Luckily enough, I happened to work with Helmut Rauch at the time, who had built the first working neutron interferometer. He invited me to participate in an experiment to observe the sign change of a neutron-spin wave function upon a full rotation. Like many of my later fundamental experiments, this sign change of a two-state system—today called a qubit—upon a full cycle between two states has become important in quantum optics and quantum information science.

WOJCIECH ZUREK · This was more a process than love at first sight, but there were several significant moments. Perhaps the first serious one was when I was in high school. I had talked my father into buying Feynman's *Lectures*. They had just appeared in a Polish translation, and I noticed the volumes on the shelf of a bookstore when we were on a hunting trip in northern Poland. The days were largely free; elk hunting happens at dawn or dusk. So I read large fragments of the *Lectures*. I thought that Feynman's discussion of the double-slit experiment was fascinating.

I also took an excellent course on quantum mechanics when I was a student at Kraków's Technical University in the section of nuclear physics. There were just seven students taking our class, and so the teacher (Professor Guła) decided to go for it. It was a very advanced and personalized course. We did not have a textbook, but later, when I was in the U.S., I recognized a similarity between Guła's approach and Dirac's *Quantum Theory*. I also listened to lectures at Jagiellonian University, whose physics department was across the street from the Technical University. There, the approach was more classical, like Schiff's text, and detailed calculations obscured the big picture, but I remember lectures on nonseparability and the EPR paradox, and feeling really happy about the mystery.

In any case, quantum mechanics was not what drew me to physics in the first place. Rather, it was special relativity. There were several semipopular books I read in high school that were very good. The whole idea of being able to deduce so much from so little—constant speed of light for everybody—made an enormous impression on me. After that, I was hooked on physics.

What initially attracted me to science was the way the theory of evolution organized animals present and past. I remember "getting it"—the basic idea of evolution—when I was still a child. My parents were both medical doctors. My father was also a hunter and, as one would say today, an environmentalist. We spent lots of time in the forests or near the water, and I was always fascinated with animals. In fact, I learned to read early because I wanted to read *Life of Animals* by Brehm, an illustrated book that discusses "all" animals with lots of pictures. I pored over those pictures and attempted to read the captions. This is how I (eventually) learned to read.

As for quantum foundations, until I met John Wheeler in Austin, Texas, I had assumed that all of the deep questions were understood—or, in any case, not an appropriate subject for a student. Wheeler changed that. He taught a "Quantum

Measurements" course that, over the two years it was offered, turned into a seminar. We read Bohr and Einstein, but we also discussed connections between quantum theory and information, and we played with ideas. Wigner, Peres, Everett, Unruh, Deutsch, DeWitt, and Sudarshan were there in Austin for long visits or simply on location. I shared an office with Bill Wootters. We took the course together, sparked off each other, and discussed quantum foundations for hours on end. Our term paper on the Einstein–Bohr double-slit debate became our first quantum paper.

Still in Texas, where I stayed as Wheeler's postdoc, I became gradually convinced that questions about quantum mechanics, the role of the observer, and the nature of information in physics are important and largely open. So it was John who encouraged my interest in quantum-physics research, and later—when I went to Caltech to do a postdoc in Kip Thorne's group, where nondemolition measurements were being devised to help with gravity-wave detection—Wheeler insisted I meet regularly with Feynman. At that point, I was spending a lot of time on quantum measurements. I also attended Feynman's course on physics and information, taught jointly with John Hopfield.

Feynman was skeptical but interested in quantum foundations. He insisted on making things very concrete. I would say, for instance, "Consider a quantum system," and he would chime in, "Like a bucket of liquid helium," so that I had to explain whether this example—or, perhaps, a spin-½ system—was something that fit the context. He would also get impatient when mathematics obscured physics. Moreover, he generally assumed he knew better whatever you were trying to explain to him, and tried to explain it to you instead.

For instance, I wanted to try on Feynman the proof of "no cloning." That was one of the few times when I knew I had the right answer before he did, and I wanted to tell it to him. But he insisted that I do not: he wanted to figure it out for himself. He appreciated the severity of the problem—the danger of superluminal communication—but tried other ways of dissolving it. I could not rule out the possibility that there might be other ways of preventing violations of causality ("no cloning" was still a fresh subject matter in my mind and, in fact, did not yet have that name—John Wheeler suggested it later, as a substitute for the more boring title we had devised with Bill Wootters). So I just watched him trying various approaches and told him when I saw a dead end coming—I had earlier visited some of these dead ends myself. When I eventually gave him the proof, he seemed convinced but also frustrated that he did not get it on his own. This was, by the way, very different from how Wheeler would have reacted. With him, one sometimes had the feeling that he was letting you explain something he already knew just to make you feel good.

Being able to explain something to Feynman did wonders for my confidence, especially since he genuinely did not know the answer beforehand, and since it was a simple explanation—exactly the sort he liked. The other confidence boost happened when, while at Caltech, I received invitations to several meetings that came as a result of my "Pointer Basis" paper of 1981. One of these meetings was especially good. It was organized by Ed Fredkin on his island in the Caribbean. Charlie Bennett, Greg Chaitin, Rolf Landauer, Ken Wilson, and Stephen Wolfram were among the twenty or so participants. The level of discussion, the atmosphere of intellectual fellowship and excitement, the questions I got during my talk: this was all a pleasant surprise. Until then, I had been under the impression that quantum foundations was a kiss of death to a physicist's career. During my student days, this was the message I had gotten from essentially everyone, with the notable exception of Wheeler. So getting invited to meetings on the basis of my foundations research was solid evidence that times were a-changin'. From then on, quantum theory was always near the top of my research agenda, although I usually also had some other, more down-to-earth interests, such as astrophysics.

When I first got to Los Alamos, astrophysics, and especially cosmology, took up much of my time. I still enjoy thinking about it, and talking about it with Stirling Colgate was exciting. But eventually, quantum matters started taking up more and more of my time. The arrival of Juan Pablo Paz—who came as my postdoc and became a friend and collaborator—was, in a way, the turning point.

BIG ISSUES

*What are the most pressing problems in the
foundations of quantum mechanics today?*

WHILE SCIENCE CHURNS OUT a relentless series of quantum leaps within a matter of years (if not months), philosophy is accustomed to a much more leisurely ride. As a philosopher friend of mine recently remarked, "Major advances in philosophy happen in units of centuries, and even that might be an optimistic assessment." And indeed, by their very nature, many of the questions that perplexed Kant or even Plato continue to engage the contemporary philosopher. Clearly, the pace of progress is a matter of perspective.

The foundations of quantum mechanics occupy a comfortable middle ground between these two extremes. The field is relatively young and dynamic. And because its object of interest is a physical theory, the field is rooted quite firmly in science, despite the host of metaphysical questions quantum mechanics seems to generate. At the same time, the issues that the founders of the theory already agonized over have not visibly aged in the passing decades. Schrödinger's cat is alive and well fed and not inclined to having its fate decided anytime soon. The ripples of EPR are still felt everywhere. Bohr's interpretation of quantum mechanics keeps flexing its muscles, inspiring a new generation of epistemic and informational viewpoints while sending other people scrambling for an antidote.

But to say that the time-honored themes of quantum theory's first generation are on everyone's lips today as ever is not to suggest that the field of quantum foundations has turned stagnant, or that it has become akin to a dog chasing its tail, or that is has been reduced to little more than an autoerotic enterprise with no hope or desire for escape from bachelorhood. Quite the opposite, actually. As already mentioned in the prologue, there's been a dramatic refinement over time in the way people think and talk about the central issues. Post-war developments—such as the stream of new interpretations, the various no-go theorems, experiments at the quantum level, and more recently quantum information—have not only put a distinctly new spin on old debates, but have also given rise to a flurry of new questions (and even a few precious answers).

In fact, it is now far from obvious what a contemporary foundationalist would regard as the key issues awaiting resolution. There are no hard-and-fast rules. What one person may experience as a genuine and pivotal difficulty—to be disregard only at our peril—may be perceived by someone else as a petty concern or mere pseudo-issue. And even once you find two people settling on the same problem, you can bet that they'll hold divergent views of what the problem is really all about and what the best course of action might be.

To get a good sense, then, of a representative range of present-day foundational priorities, let's ask our interviewees to lay out the playing field for us.

GUIDO BACCIAGALUPPI · I think recent progress in various fields within foundations has brought up, or renewed, interest in a number of very important questions—although maybe none are so pressing as to impede further progress pending their resolution.

Hidden-variables programs, that is, pilot-wave theories of the de Broglie–Bohm type, have progressed enough in recent years that the question of direct experimental evidence that might decide between them and quantum mechanics has become meaningful. The central idea is the analogy between pilot-wave theories and classical statistical mechanics, in particular the possibility of observable nonequilibrium effects. The range of application of pilot-wave theories is now large enough that they can be applied to quite exotic phenomena that might reveal systematic violations of the Born rule. Antony Valentini in particular has been pioneering the exploration of these possibilities. Such violations would be the most direct evidence in favor of a revision of quantum mechanics.

Within collapse theories, recent work—especially by Pearle and by Nicrosini and Rimini in physics, and by Wayne Myrvold in philosophy—has brought us very close to finally deciding whether a satisfactory relativistic collapse theory is possible. That is a very big question, and it is surprising that so few researchers actively engage in it. (Maybe this is a side effect of an apparent shift in the preoccupations of the community, partially away from more traditional approaches and more toward the new field of quantum information. Indeed, at the Sixteenth U.K. Foundations Meeting just a few months ago, it was quite noticeable that only a handful of talks were in the subject areas of hidden variables, collapse theories, and Everett interpretations.) The experimental question of deciding between collapse theories and quantum mechanics has also made progress, but it is not quite as promising as in the case of pilot-wave theories. This is due to the fact that the appearance of spontaneous collapse can be always mimicked by decoherence induced by some appropriate environment (coupled with one's favorite no-collapse interpretation). What is particularly worrisome is the suspicion that a rival no-collapse theory might not even need to invoke some hitherto unobserved, mysterious environment to do the job, but that once gravitation is quantized, it might provide just the right kind of environment to reproduce some of the currently best candidates for collapse theories (which tend to be mass-density based). A paper by Bernard Kay some twelve years ago or so made this point in a particularly striking manner.

Everett interpretations have also made quite spectacular progress in recent years, principally thanks to work by Simon Saunders in the 1990s, and by David Deutsch, David Wallace, and others in the 2000s. They appear, in fact, to have solved—or to have convincing strategies for solving—all the classic questions that used to trouble them. There are still a few question marks, but I would not say there are very pressing questions for Everett. (Personally, I think there are some questions about the details of relativistic locality and of the various accounts of mentality, which I am exploring with Laura Felline, and some lingering issues about probabilities, as raised, for instance, by Peter Lewis.)

The development of the cluster of approaches around quantum information has brought renewed interest in axiomatic foundations of standard quantum mechanics, and the reconstruction problem of quantum mechanics has seen a sudden flood of very impressive and diverse results from a number of researchers (among others, Hardy, Goyal, and Chiribella–D'Ariano–Perinotti—quoting just the ones I happen to be most familiar with). Among these developments, one particular instance that never ceases to amaze me is Rob Spekkens's "toy theory," which reproduces qualitative analogues of scores of quantum effects (excepting computational speedup, Bell-inequality violation, and Kochen–Specker theorems), based purely on a notion of an epistemic limitation on the description of system states. These and similar results carry with them insights into what the truly crucial difference might be between classical and quantum theories, and decisive progress along these lines would be a truly splendid thing.

Some of the other questions I would be most intrigued to see resolved are those surrounding the relation between standard quantum field theory and the axiomatic approach of algebraic quantum field theory, but I am not sure I am competent enough to comment in detail.

Finally, if I may mention a particular interest of mine, I believe that the relation between quantum mechanics and the direction of time needs to be explored further and may yet have surprises in store. Part of this interest, of course, stems from my period at Huw Price's Centre for Time in Sydney, but part is rooted in my interest in decoherence, and is related to ideas I am exploring jointly with Max!

ČASLAV BRUKNER · Quantum theory makes the most accurate empirical predictions. Yet it lacks simple, comprehensible physical principles from which it could be uniquely derived. Without such principles, we can have no serious understanding of quantum theory and cannot hope to offer an honest answer—one that's different from a mere "The world just happens to be that way"—to students' penetrating questions of why there is indeterminism in quantum physics, or of where Schrödinger's equation comes from. The standard textbook axioms for the quantum formalism are of a highly abstract nature, involving terms such as "rays in Hilbert space" and "self-adjoint operators." And a vast majority of alternative approaches that attempt to find a set of physical principles behind quantum theory either fall short of uniquely deriving quantum theory from these principles, or are based on abstract mathematical assumptions that themselves call for a more conclusive physical motivation.

One strategy for progress on this front is to view quantum theory within the context of general theories that conform to reasonable axioms about probabilities, and then to contrast the alternatives. Surprisingly, in the last decade it was found that what one might have expected to be uniquely quantum features—such as probabilistic predictions for individual outcomes (indeterminism), the impossibility of copying unknown states (no cloning), or the violation of "local realism"—are actually highly generic for general probabilistic theories. So, is there any reason why we see phenomena obeying the laws of quantum theory rather than of any other possible probabilistic theory?

Most recently, there have been several approaches to reconstructing quantum theory on the basis of a small set of reasonable physical axioms that demarcate phenomena that are exclusively quantum from those that are common to more general probabilistic theories (see my answer to Question 3, page 66, for my own reconstruction attempt). Typically, however, the proposed axioms partially use abstract mathematical language. One should, in my opinion, insist on reducing this language as far as possible to a phenomenological meaning, and not be afraid to combine these simple elements of everybody's experience with abstract concepts such as "information" or "knowledge."

Modern reconstructions of quantum theory partially meet this demand by being entirely developed in terms of primitive laboratory operations, such as preparations, transformations, and measurements. Bohr's insistence on the usage of classical terms is respected insofar as these operations are classically describable, but they are not linked to the concepts of time, position, momentum, or energy of "traditional" physics. As a result, one derives a finite-dimensional, or countably infinite-dimensional, Hilbert space as an operationally testable, abstract formalism concerned with predictions of future experiments and frequency counts, which are ultimately based on clicks of detectors and nothing more. While I consider the quantum state to be a tool for calculating the probabilities of whatever future measurements we may choose to carry out, I want to make the point that we do appoint physical labels to the states in any particular orthonormal basis, and that we do deal with notions of position, momentum, fields, specific forms of Hamiltonians, and so forth. The abstract quantum formalism, however, tells us nothing about how we should go about building a useful instrument for measuring, say, position, as opposed to any other observable.

In my opinion, the clue for this will not be obtained without an understanding of the concept of distance—or of the more abstract idea of nearness—of points lying in ordinary real space. In the abstract quantum formalism, any two different eigenvalues of the position observable correspond to orthogonal quantum states, without any concept of closeness or distance. The terms "close" and "distant" make sense only in a *classical* context, where those eigenvalues are treated as close when they correspond to neighboring outcomes in real space. Is it possible to arrive at notions of nearness, distance, and space—and, furthermore, at the theories referring to these notions, such as the theory of relativity, quantum field theory, and elementary-particle theory—merely on the basis of clicks in detectors? Or is it necessary to presuppose these notions, prior to the construction of physical theories? To me, this is one of the most pressing contemporary questions in the foundations of quantum mechanics.

Preferred tensor factorizations, coarse-grained observables, and symmetries might help to indeed demonstrate that all known basic theories of physics are a consequence of abstract quantum theory. The most elementary system, or qubit, lives in an abstract state space with $SU(2)$ symmetry, which is locally isomorphic to the group $SO(3)$ of rotations in three-dimensional space. Thinking about directional degrees of freedom—i.e., about spin—this symmetry finds its operational justification in the symmetry of the configuration of macroscopic instruments by which the spin state is prepared and measured. But from where have the macroscopic instruments acquired this symmetry in the first place?

I would like to suggest that under the everyday conditions of coarse-grained measurements, the systems consisting of a large number of elementary systems, such as macroscopic instruments, acquire the symmetry of their elementary constituents. For example, in 2007 Johannes Kofler and I derived the following result. Suppose we mimic restricted measurement precision by bunching together eigenvalues of spin projections into slots. Then the spin coherence states—which are states of many identical elementary spins—acquire an effective description as a classical spin embedded in ordinary three-dimensional space. The orientation of this classical spin requires two angles to be defined, which gives rise, through the relative angle, to the notion of "neighboring" orientations. Thus, the reason for three-dimensional real space being *the* space of the inferred world is offered through a circular but *consistent* movement in the reconstruction, in which it is legitimate to recover the elements with which one started the reconstruction. Von Weizsäcker coined the name *Kreisgang* ("circle walk") for such movements. The epistemological framework of classical physics and three-dimensional ordinary space are required at the "beginning" of the *Kreisgang* to specify the configuration of macroscopic instruments by which the quantum state is prepared and measured. The *Kreisgang* is "closed" by showing that under the everyday conditions of coarse-grained measurements, a description of macroscopic instruments emerges in the terminology of classical physics, and three-dimensional ordinary space emerges from within quantum theory. I conclude by remarking that this program is not completed—and perhaps not completable.

JEFFREY BUB · We don't really understand the notion of a quantum state, in particular an entangled quantum state, and the peculiar role of measurement in taking the description of events from the quantum level, where you have interference and entanglement, to an effectively classical level where you don't. In a 1935 article responding to the EPR argument, Schrödinger characterized entanglement as "*the* characteristic trait of quantum mechanics, the one that enforces its entire departure from classical lines of thought." I would say that understanding the nonlocality associated with entangled quantum states, and understanding measurement, in a deep sense, are still the most pressing problems in the foundations of quantum mechanics today.

Having said that, I don't think we are going to get anywhere by sitting back and reflecting on the meaning of measurement or the notion of state in physics, or in trying to "solve the measurement problem." It's not that we don't know how to solve the measurement problem: Bohm's theory is a solution, so-called modal interpretations provide formal solutions, the Everett interpretation is another solution, the Ghirardi–Rimini–Weber theory is a rival theory that avoids the measurement problem. It's rather that there's nothing like a general consensus that any of these proposals are getting it right. Einstein commented in a letter to Max Born that Bohm's theory "seems too cheap to me." He was referring to the deterministic character of Bohm's theory. My feeling is that all these ways of thinking about quantum mechanics are "too cheap," because they all attempt to explain away the irreducible indeterminism of quantum mechanics—rather than providing a conceptual framework for thinking about a universe in which, to put it somewhat anthropomorphically, a particle is free to choose its own response to a measurement, subject only to probabilistic constraints, which might be nonlocal.

I think the way forward is to consider the sort of question raised by Wheeler: why the quantum? Or, the more focused question posed by Popescu and Rohrlich in their 1994 article, in which they introduced the notion of a nonlocal box: why is quantum theory not more nonlocal, given that you can have more nonlocality without thereby allowing the possibility of instantaneous signaling between the parties? This question has been extraordinarily fruitful in leading to new insights about quantum nonlocality and seems to me the most promising route to advancing our understanding of what is really involved in the transition from classical to quantum physics.

ARTHUR FINE · My general attitude toward science is pluralistic, in the sense that I regard every major theory in science as open to reasonable interpretations that differ from one another over some essentials. This is certainly true in the case of quantum theory, where interpretations differ over collapse and the need for an external observer, over determinism and indeterminism, over whether Lorentz invariance is merely phenomenological, over realism and instrumentalism, and so on. Faced with this array, one might experience a pressing need to sort things out so as to narrow the options, hopefully, to the one "correct" interpretation. I do not share that attitude. Rather, I see the interpretive array as part of a healthy freedom of choice whose payoff comes from the different heuristic paths suggested by the differing interpretations. So I don't think that finding the "right" interpretation of quantum mechanics is a pressing problem at all.

Still, there are problems that we would all like to understand better. One is the whole question of locality. Reflections that stem from the Bell theorem have suggested that quantum phenomena exemplify nonlocality: acting here can immediately influence happenings way over there. I have never seen an argument for this conclusion that does not involve assumptions that go well beyond reliable theory and data. Indeed, several generations now of excellent experimental investigations have not yet produced a conclusive verdict concerning the violation of the Bell inequalities themselves. The problem remains as to whether one can satisfy efficiency requirements (both on detection and on synchronization of coincidence) and, in the same experiment, manage to rule out communication between the two (or more) wings where the measurements are made. Although there are plans for experiments that claim to do this, none seem to work. It may be that none can work, since modern simulation techniques suggest that statistics in violation of the Bell inequalities can be generated classically in a wide range of circumstances, including the conditions proposed in most experimental designs. Thus, entanglement may turn out to be a significant resource in quantum information theory, but not of such significance foundationally as has been supposed.

One general issue raised by the debates over locality is to understand the connection between stochastic independence (probabilities multiply) and genuine physical independence (no mutual influence). It is the latter that is at issue in "locality," but it is the former that goes proxy for it in the Bell-like calculations. We need to press harder and deeper in our analysis here.

CHRISTOPHER FUCHS · John Wheeler would ask, "Why the quantum?" To him, that was the single most pressing question in all of physics. You can guess that with the high regard I have for him, it would be the most pressing question for me as well. And it is. But it's not a case of hero worship; it's a case of it just being the right question. The quantum stands up and says, "I am different!" If you really want to get to the depths of physics, then that's the place to look.

Where I see almost all the other interpretive efforts for quantum theory at an impasse is that despite all the posturing and grimacing over the "measurement problem" and the "mysteries of nonlocality" and what have you, none of them ask in any serious way, "Why do we have this theory in the first place?" They see the task as one of patching a leaking boat, not one of seeking the principle that has kept the boat floating this long (for at least this well). My guess is that if we can understand what has kept the theory afloat, we'll understand that it was never leaky to begin with. The only source of leaks was the strategy of trying to tack a preconception onto the theory that shouldn't have been there.

What is this preconception? It almost feels like cheating to say anything about it before Question 4 ... but I have to, or I can't answer the rest of Question 2! The preconception is that a quantum state is a *real thing*—that there were quantum states before there were observers; that quantum states will remain even if all observation is snuffed out by nuclear holocaust. It is that if quantum states are the currency of quantum theory, the world had better have some in the bank. Take the Everett interpretation(s)—the world as a whole has its wave function, darned be it if observership or probability is never actually reconstructed within the theory. The Bohmian interpretation(s)? The wave function is the particle's guiding field; observers never mentioned at all. GRW interpretation(s)? Collapse is what happens when wave functions get too big; of course they're real. Zurek's "let quantum be quantum"? It is, as far as I can tell, a view that starts and ends with the wave function. There is no possibility that two observers might have two distinct (contradicting) wave functions for a system, for the observers are already *in* a big, giant wave function themselves.

So when I say "Why the quantum?" is the most pressing question, I mean this specifically within an interpretive background in which quantum states aren't real in the first place. I mean it within a background where quantum states represent observers' *personal* information, expectations, degrees of belief.

"But that's *just* instrumentalism," the philosopher of science says snidely (see my answer to Question 14, page 253). "You give up the game before you start." Believe me, you've got to stand your ground with these guys when their label guns fly from their holsters! I say this because if one asks "Why the quantum?" in this context, it can only mean that one is being *realist* about the *reasons* for one's instrumentalities. In other words, even if quantum theory is purely a theory for apportioning and structuring degrees of belief, the question of "Why the quantum?" is nonetheless a question of what it is about the actual, real, objective character of the world that compels us to use this framework for reasoning rather than another. We observers are floating in the world, making decisions on all that we experience around us: why are we well-advised to use the formalism of quantum theory for that purpose and not some other formalism? Surely it connotes something about the general character of the world—something that is contingent, something that might have been otherwise, something that goes deeper than our decision-making itself.

With this one gets at the real flavor of this *most pressing problem in the foundations of quantum mechanics* from the point of view of QBism. It takes on two stages. The first is to find a crisp, convincing way to pose quantum theory in such a way that it gets rid of these trouble-making quantum states in the first place. What I mean by this is, if quantum theory is actually about how to structure one's degrees of belief, it should become conceptually the clearest when written in its own native terms. To give an example of how this might go, consider the Born probability rule as it is usually represented: one starts with a quantum state $\hat{\rho}$, say for some d-level system, and some orthogonal set of projection operators $\hat{D}_j$ representing the outcomes of some nondegenerate observable. The rule is that the classical value D_j registered by the measuring device (no hat this time) will occur with probability

$$p\left(D_j\right) = \mathrm{tr}\left(\hat{\rho}\hat{D}_j\right).$$

A recent result of QBism, however, is that if a certain mathematical structure always exists in Hilbert space (we know it does for $d = 2$ to 67 already), then in place of the operator $\hat{\rho}$ one can always identify a *single* probability distribution $p\left(H_i\right)$, and in place of the operators $\hat{D}_j$ one can always identify a set of conditional probability distributions $p\left(D_j \mid H_i\right)$, such that

$$p\left(D_j\right) = (d+1)\sum_i p\left(H_i\right)p\left(D_j \mid H_i\right) - 1.$$

The similarity between this formula and the usual Bayesian sum rule (law of total probability) is uncanny. It says that the Born rule is about degrees of belief going in, and degrees of belief coming out. The use of quantum states in the usual way of stating the rule (that is, rather than degrees of belief directly) would then simply be a relic of an initial bad choice in formalism.

If this program of rewriting quantum theory becomes fully successful (working for all d, for instance), thereafter there should be no room for the distracting debates on the substantiality of quantum states—they're not even in the theory now—nor the tired discussions of nonlocality and the "measurement problem" the faulty preconception inevitably engendered. At this point, a second stage of the pressing question would kick in: it will be time to take a hard look at the new equations expressing quantum theory and ask how it is that *they* are mounted onto the world. What about the world compels this kind of structuring for our beliefs? To get at that is to really get at "Why the quantum?" And my guess is, when the answer is in hand, physics will be ready to explore worlds the faulty preconception of quantum states couldn't dream of.

GianCarlo Ghirardi · I believe that the most pressing problems are still those that have been debated for more than eighty years by some of the brightest scientists and deepest thinkers of the past century: Niels Bohr, Werner Heisenberg, John von Neumann, Albert Einstein, Erwin Schrödinger, John Bell. To characterize these problems in a nutshell, I cannot do better than stressing the totally unsatisfactory conceptual status of our best theory by reporting the famous sentence by Bell: "Nobody knows what quantum mechanics says exactly about any situation, for nobody knows where the boundary really is between wavy quantum systems and the world of particular events."

I also share Bell's opinion that the fact that this wonderful and extremely successful theory is radically incapable of accounting for our definite perceptions does not matter in practice, at least not presently. But I cannot accept that the basic theoretical construction for our understanding of natural phenomena is internally inconsistent, and that it is not able to account for the way it postulates measuring processes to take place. I will repeatedly come back to this point in my subsequent comments. But from the very beginning, I want to emphasize with great strength that science, this wonderful and unbelievable creation of the human mind, finds its real reason of existence in its ability to allow for an objective and always-growing understanding of reality. As such, an internally inconsistent theoretical scheme—one that becomes acceptable only by resorting to vague, not well-defined, imprecise, and fundamentally contradictory verbal assertions—cannot be taken as real progress in our grasping *God's thoughts*.

In this spirit, and given that theoretical schemes exist that are logically consistent and predictively equivalent—or even identical—to standard quantum mechanics (here I have in mind particularly the spontaneous-collapse theories and Bohmian mechanics), I am naturally led to share another position of Bell's, which he expressed with great clarity in *Against Measurement* and in his Touschek Lectures. Namely, the great problem now is which one of the existing "exact" theories admits a fully satisfactory relativistic generalization. Here it is useful to recall that Bell used the term "exact" to denote a theory that "neither needs nor is embarrassed by an observer."

SHELLY GOLDSTEIN · If I were to take this question to be concerned only with the most pressing problems in the foundations of quantum mechanics *today*, then I suppose I would point to the tension between quantum nonlocality and relativity. Relativity is widely regarded both as a fundamental physical principle and as being incompatible with any sort of genuine action-at-a-distance. Quantum nonlocality is arguably (correctly, I believe) an experimentally verified consequence of quantum mechanics that would clearly seem to involve genuine action-at-a-distance. Does relativity then have to be abandoned, or can it be reconciled with quantum nonlocality, appearances to the contrary notwithstanding?

I think it would be better, however, to respond to the following question: what *have been* the most pressing problems in the foundations of quantum mechanics? And to this I suppose the standard answer is the measurement problem, or, more or less equivalently, Schrödinger's cat paradox.

The problem here is that the usual description of the state of a system in a quantum-mechanical universe is of a rather unusual sort. It is given by a rather abstract mathematical object, called the wave function or the quantum state vector (or maybe the density matrix) of the system, an object whose physical meaning is rather obscure in traditional presentations of quantum theory. Moreover, in these presentations we are usually rather emphatically discouraged from supplementing our description of a quantum system with further—possibly more familiar but maybe exotic and elusive—variables, or even from contemplating such a possibility.

If one accepts, however, that the usual quantum-mechanical description of the state of a quantum system is indeed the complete description of that system, it seems hard to avoid the conclusion that quantum measurements typically fail to have results: pointers on measurement devices typically fail to point, computer printouts typically fail to have anything definite written on them, and so on. More generally, macroscopic states of affairs tend to be grotesquely indefinite, with cats seemingly both dead and alive at the same time, and the like. This is not good!

These difficulties can be avoided by invoking the measurement axioms of quantum theory, in particular the collapse postulate. According to this postulate, the usual quantum-mechanical dynamics of the state vector of a system (given by Schrödinger's equation)—the fundamental dynamical equation of quantum theory—is abrogated whenever measurements are performed. The deterministic Schrödinger evolution of the state vector is then replaced by a random collapse to a state vector that can be regarded as corresponding to a definite macroscopic state of affairs: to a pointer pointing in a definite direction, to a cat that is definitely dead or definitely alive, and so on.

But doing so comes at a price: one then has to accept that quantum theory involves special rules for what happens during measurement, rules that are in addition to, and not derivable from, the quantum rules governing all other situations. One has to accept that the notions of measurement and observation play a fundamental role in the very formulation of quantum theory, in sharp conflict with the much more plausible view that what happens during measurement and observation in a quantum universe, like everything else that happens in such a universe, is a consequence of the laws governing the behavior of the constituents of that universe—say the elementary particles and fields. These laws apply directly to the microscopic level of description, and they say nothing directly about measurement and observation, notions that arise and make sense on an entirely different level of description, the macroscopic level.

I believe, however, that the measurement problem, as important as it is, is nonetheless but a symptom of a more basic difficulty with standard quantum mechanics: it is not at all clear what quantum theory is about. Indeed, it is not at all clear what quantum theory actually says. Is quantum mechanics fundamentally about measurement and observation? Is it about the behavior of macroscopic variables? Or is it about our mental states? Is it about the behavior of wave functions? Or is it about the behavior of suitable fundamental microscopic entities, elementary particles and/or fields? Quantum mechanics provides us with formulas for lots of probabilities. What are these the probabilities of? Of results of measurements? Or are they the probabilities for certain unknown details about the state of a system, details that exist and are meaningful prior to measurement?

It is often said that such questions are the concern of the foundations of quantum mechanics, or of the interpretation of quantum mechanics—but not, somehow, of quantum mechanics itself, of quantum mechanics simpliciter. I think this is wrong. I think these, and similar, questions are a reflection of the fact that quantum mechanics, in the words of John Bell, is "unprofessionally vague and ambiguous."

What is usually regarded as a fundamental problem in the *foundations* of quantum mechanics, a problem often described as that of *interpreting* quantum mechanics, is, I believe, better described as the problem of finding a sufficiently precise *formulation* of quantum mechanics: a *version* of quantum mechanics that, while expressed in precise mathematical terms, is also clear as physics.

And it is hard for me to imagine how this can be achieved, in any fundamental physical theory, unless that theory involves, as part of its description of the state of a system, an explicit space-time ontology (for a relativistic version, and a spatial ontology whose state changes with time for the nonrelativistic version). This ontology might be a particle ontology, involving world lines in space-time, or a field ontology, involving a field on space-time, or perhaps both, or perhaps neither but something else. In any case, the space-time ontology amounts to a certain kind of decoration of space-time, to the specification of what Bell has called the local beables of the theory.

Theories involving different local beables, or involving the same local beables but different laws for the local beables, would be different theories—for example, different *versions* rather than merely different *interpretations* of quantum theory.

DANIEL GREENBERGER · For reasons I'll explain in my answer to Question 7 (see page 152), I don't think the measurement problem will be solvable soon, or possibly ever. We will probably have to know more about nature for that. But there are other questions that are intriguing, such as whether a single particle has a wave function, or whether we have to talk about ensembles, and whether the wave function represents solidly observable probabilities, or just subjective information that we have about the system.

I myself have been worrying along different lines. I don't think we treat mass properly in quantum theory. It enters as a parameter, while energy enters as an operator. If $E = mc^2$, then I don't think that's consistent, and there is much evidence for that. In the same vein, the concept of proper time is much more subtle in quantum theory than it is in classical physics. For example, if you send a particle wave packet through a beam splitter, each part has its own proper time. If the two parts then get accelerated differently, their proper times run at different rates. If now the two parts get recombined, say at another beam splitter, what exactly is the proper time of the recombined particle? This is a practical question because the particle can be unstable, and its decay time will be controlled by the proper time that has elapsed. Surely the two parts cannot remember their separate histories. That would violate the essence of how quantum theory works.

Connected to this problem is the serious disconnect between quantum theory and general relativity. Quantum theory works with position and momentum, which intrinsically brings in the mass of the particle, while relativity works with particle trajectories, position and velocity, purely geometrical concepts, and independent of the mass. As a consequence, the weak equivalence principle breaks down in quantum mechanics. I think that these problems are the essence of why we don't have a theory of quantum gravity. It goes way beyond the mathematical complications of a non-linear theory. I think we don't understand gravity at the simple physical level of the equivalence principle. We don't know nearly enough to even begin to make a theory of quantum gravity. (If someone succeeded in making such a theory mathematically, which certainly could happen, I think it would be a serious step backward—everyone would believe it, and it would probably win a Nobel prize. Nobody could test it, and in my opinion, it would be almost guaranteed to be wrong, since it would be based on ideas that do not fit together on the simplest level.) I'll have more to say about this in my answer to Question 15 (see page 265).

LUCIEN HARDY · The most well-known problem in quantum foundations is the measurement problem—our basic conception of reality depends on how we resolve this. I will address this problem in my answer to Question 7 (see page 153). The measurement problem is tremendously important. But there is another problem that is even more important—and that may well lead to the solution of the measurement problem. This is to find a theory of quantum gravity. The problem of quantum gravity is easy to state: find a theory that reduces to quantum theory and to general relativity in appropriate limits. It is not so easy to solve. The two main approaches are string theory and loop quantum gravity. Both are deeply conservative, in the sense that they assume it will be possible to formulate a theory of quantum gravity within the quantum formalism as it stands. I do not believe this is the right approach. Quantum theory and general relativity are each deeply conservative, and deeply radical, but in complementary respects. Quantum theory is conservative in that it works on a fixed space-time background, but it is radical in that probabilities play an indispensable role. General relativity is conservative in that it is deterministic (probabilities are not necessary), but it is radical in that the space-time background is not fixed but rather depends on the distribution of matter. In my opinion, a theory of quantum gravity will have to take the radical road in each case. It will be probabilistic, and it will have nonfixed causal structure. In fact, we can expect it to be a bit more radical still. It will, most likely, have indefinite causal structure. The reason for this is that in quantum theory, when we have a physical quantity that can vary, we will typically have situations where there is fundamental indefiniteness as to the value of the quantity. Since causal structure is dynamical in general relativity, we therefore expect it to be subject to fundamental indefiniteness in quantum gravity. This means that it will sometimes be the case that there is no matter of fact as to whether a given interval is spacelike or timelike. The basic mathematical apparatus of quantum theory needs a fixed space-time background (at least it requires a background time with respect to which the state evolves), and the basic mathematical apparatus of general relativity is deterministic. Neither framework is likely to be capable of accommodating a theory of quantum gravity, since neither possesses the radical feature of the other, and neither has indefinite causal structure. Hence, we require a deeper framework with new conceptual and mathematical apparatuses.

It is instructive to look at the transition from Newton's theory of gravitation to Einstein's theory of general relativity. We can take a limit to get from Einstein's theory back to Newton's theory. The mathematical apparatus of general relativity, however, is very different from that of Newton's theory. Newtonian gravity suffers from a deep conceptual problem: the force of gravity is not local. In general relativity, locality is restored, because the gravitational force is propagated locally through the space-time continuum (through matter-induced curvature of this very continuum). Even though Newton's theory turned out not to be fundamental, it is interesting to ask what the best interpretation of it is. One reasonable answer is that it should be regarded as a theory of curved space rather than of curved space-time. Such an interpretation of Newton's theory (as formalized by Cartan) only became evident after Einstein had formulated his theory of general relativity in terms of the curvature of space-time. This point, which is due to Wayne Myrvold, raises the possibility that we will best understand quantum theory—which suffers from its own deep conceptual problems—in retrospect as a limiting case of a deeper theory, such as a theory of quantum gravity. If this is true, then we need to work on quantum gravity to have a hope of properly solving the measurement problem.

The problem of quantum gravity requires, in my opinion, the development of a new mathematical framework. This could be as radical a departure from the frameworks of quantum theory (Hilbert spaces) and general relativity (tensor calculus) as the tensor calculus for general relativity is from the mathematics of Newtonian mechanics. The problem of quantum gravity is, I believe, a foundational problem, and the tools and methods of foundational thinking need to be brought to bear on it.

ANTHONY LEGGETT · To my mind, within the boundaries of "foundations of quantum mechanics" strictly defined, there is really only one overarching problem: is quantum mechanics the whole truth about the physical world? That is, will the textbook application of the formalism—including the use of the measurement axiom, possibly at a very late stage—continue to describe experimental results adequately for the indefinite future? If the answer should turn out to be no, then, of course, there would be any number of further questions to be raised, but they would no longer be about quantum mechanics. If the answer is yes, then I believe there is really not much left to be asked (see also my answer to Question 3, page 79).

I think that there is, however, one question that—while in some sense more general than being about quantum mechanics as such—may be relevant to our future perceptions of the meaning of the formalism. This is the issue of the basis and status of the conventional viewpoint on the arrow of time. To be more specific, if it were to become accepted in a more general context that this arrow could, as it were, reverse itself locally and temporarily—as has in effect been suggested by a number of thinkers—then I believe this might recolor our thinking about the measurement problem and about other aspects of the formalism.

TIM MAUDLIN · The most pressing problem today is the same as ever it was: to clearly articulate the exact physical content of all proposed "interpretations" of the quantum formalism. This is commonly called the measurement problem, although, as Philip Pearle has rightly noted, it is rather a "reality problem." Physics should aspire to tell us what exists (John Bell's "beables"), and the laws that govern the behavior of what exists. "Observations," "measurements," "macroscopic objects," and "Alice" and "Bob" are all somehow constituted of beables, and the physical characteristics of all things should be determined by that constitution and the fundamental laws.

What are commonly called different "interpretations" of quantum theory are really different theories—or sometimes, no clear theory at all. Accounts that differ in the beables they postulate are different physical theories of the universe, and accounts that are vague or noncommittal about their beables are not precise physical theories at all. Until one understands exactly what is being proposed as the physical structure of the universe, no other foundational problem, however intriguing, can even be raised in a sharp way.

DAVID MERMIN · Here are three.

One: In the words of Chris Fuchs, "quantum states: what the hell are they?" Quantum states are not objective properties of the systems they describe, as mass is an objective property of a stone. Given a single stone, about which you know nothing, you can determine its mass to a high precision. Given a single photon, in a pure polarization state about which you know nothing, you can learn very little about what that polarization was. (I say "was," and not "is," because the effort to learn the polarization generally results in a new state, but that is not the point here.)

But I also find it implausible that (pure) quantum states are nothing more than provisional guesses for what is likely to happen when the system is appropriately probed. Surely they are constrained by known features of the past history of the system to which the state has been assigned, though I grant there is room for maneuver in deciding what it means to "know" a "feature."

Consistent historians (see also my answer to Question 16, page 279) maintain that the quantum state of a system *is* a real property of that system, though its reality is with respect to an appropriate "framework" of projectors that includes the projector on that state. Since the reality of most other physical properties is also only with respect to suitable frameworks, for consistent historians the quantum state of a system is on a similar conceptual footing to most of its other physical properties. Quantum cosmologists maintain that the entire universe has an objective pure quantum state. I do not share this view. Indeed, I do not believe it has a quantum state in any sense, since there is nothing (nobody) outside the entire universe to make that state assignment. Well, I suppose it could be God, but why would he want to make state assignments? Einstein has assured us that he doesn't place bets. (See also my answer to Question 4, page 102.)

Two: How clearly and convincingly to exorcise nonlocality from the foundations of physics in spite of the violations of Bell inequalities. Nonlocality has been egregiously oversold. On the other hand, those who briskly dismiss it as a naive error are evading a direct confrontation with one of the central peculiarities of quantum physics. I would put the issue like this: what can one legitimately require of an *explanation* of correlations between the outcomes of independently selected tests performed on systems that no longer interact? (See also my answer to Question 8, page 176.)

Three: Is the experience of personal consciousness beyond the reach of physical theory as a matter of principle? Is the scope of physics limited to constructing "relations between the manifold aspects of our experience," as Bohr maintained? While I believe that the answer to both question is yes, I list them as problems, because most physicists vehemently reject such views, and I am unable to explain to them why they are wrong in a way that satisfies me, let alone them.

I regard this last issue as a problem in the interpretation of quantum mechanics, even though I do not believe that consciousness (as a physical phenomenon) collapses (as a physical process) the wave packet (as an objective physical entity). But because I do believe that physics is a tool to help us find powerful and concise expressions of correlations among features of our experience, it makes no sense to apply quantum mechanics (or any other form of physics) to our very awareness of that experience. Adherents of the many-worlds interpretation make this mistake. So do those who believe that conscious awareness can ultimately be reduced to physics, unless they believe that the reduction will be to a novel form of physics that transcends our current understanding, in which case, as Rudolf Peierls remarked, whether such an explanation should count as "physical" is just a matter of terminology.

I am also intrigued by the view of Schrödinger (in *Nature and the Greeks*) that it was a mistake dating back to the birth of science to exclude us, the perceiving subjects, from our understanding of the external world. This does not mean that our perceptions must be parts of the world external to us, but that those perceptions underlie everything we can know about that world. (See also my answer to Question 14, page 256.) Until the arrival of quantum mechanics, physics made good sense in spite of this historic exclusion. Quantum mechanics has (or should have) forced us to rethink the importance of the relation between subject and object.

LEE SMOLIN · The measurement problem—that is to say, the fact that there are two evolution processes, and which one applies depends on whether a measurement is being made. Related to this is the fact that quantum mechanics does not give us a description of what happens in an individual experiment.

To put it differently, the only interpretations of quantum mechanics that make sense to me are those that treat quantum mechanics as a theory of the information that observers in one subsystem of the universe can have about another subsystem. This makes it seem likely that quantum mechanics is an approximation of another theory, which might apply to the whole universe and not just to subsystems of it. The most pressing problem is then to discover this deeper theory and level of description.

ANTONY VALENTINI · The interpretation of quantum mechanics is a wide open question, so we can't say in advance what the most pressing problems are. As the history of physics shows, it's only in hindsight that one can say who was looking in the right direction. What's important is that we leave the smoke screen of the Copenhagen interpretation well behind us, and that talented and knowledgeable people think hard about this subject from a realist perspective.

Instead of answering the question, I can offer a list of things I'd like to see done in the near future, as they seem important as far as I can tell.

It would be good if the ongoing controversy over the consistency of the Everett interpretation could be settled. It would be helpful to know if that theory really makes sense (on its own terms) or not. It would also be good to see further experiments searching for wave-function collapse. More generally, I'd like to see more experiments that test quantum theory in genuinely new domains—as in the recent three-slit experiment.

In modern theoretical physics, there are a number of important issues that deserve more attention from a foundations perspective, such as the question of Hawking information loss in black holes, and the problem of time in quantum gravity. The description of the quantum-to-classical transition in the early universe also deserves more foundational scrutiny.

As for my own current line of research—which focuses on the possibility of nonequilibrium violations of quantum theory, in de Broglie–Bohm theory and in deterministic hidden-variables theories generally—there are some outstanding issues that need a lot more work. One is the need for more detailed calculations and numerical simulations of relaxation to quantum equilibrium in the early universe, with the aim of obtaining precise predictions of where residual nonequilibrium violations of quantum theory might be found today—for example, in the cosmic microwave background or in relic cosmological particles. My work so far points in the direction of super-Hubble wavelengths as the area to look at, but much more remains to be done. I have also made some proposals to the effect that Hawking radiation could consist of nonequilibrium particles that violate the Born rule in a way that might avoid information loss, and there are a host of theoretical questions to be investigated to develop that proposal further.

Finally, there is the important general question of whether it's possible to construct a reasonable hidden-variables theory without an ontological wave function.

De Broglie–Bohm theory has several features that have been shown to be common to all hidden-variables theories (under some reasonable assumptions): nonlocality, contextuality, and nonequilibrium superluminal signaling. De Broglie–Bohm theory also has the feature of an ontological wave function, and it would be good to know if this is another common feature of hidden-variables theories or not. Alberto Montina has worked on this recently, but more needs to be done.

DAVID WALLACE · I think anyone's answer to this is going to depend above all on what they think of the quantum measurement problem. After all, the measurement problem threatens to make quantum mechanics incoherent as a scientific theory—to reduce it, at best, to a collection of algorithms to predict measurement results. So the only reason anyone could have not to put the measurement problem right at the top of the list would be if they think it's solvable within ordinary quantum mechanics. (Someone who thinks it's solvable in some modified version of quantum mechanics—in a dynamical-collapse or hidden-variables theory, say—ought to think that the most pressing problem is generalizing that modified version to account for all of quantum phenomena, including the phenomena of relativistic field theory.)

As it happens, though, I *do* think the measurement problem is solvable within ordinary quantum mechanics: I think the Everett ("many worlds") interpretation solves it in a fully satisfactory way, and while I think there are some philosophical puzzles thrown up by that solution—mostly concerned with probability and with emergence—that would benefit from more thought, I wouldn't call them *pressing*. Not from the point of view of physics, at any rate.

So from my point of view, the "most pressing problems" aren't going to be ultra-broad problems like, "What does quantum mechanics as a whole mean?" They're going to be a bit more detailed, a bit more concerned with particular puzzling features of the conceptual and mathematical structure of quantum mechanics. (The advantage of the Everett interpretation—the main *scientific* benefit it's brought, I'd say—is that it allows us to ask those questions without getting tangled up in worries about whether there are hidden variables or dynamical collapses or whatever not included in our equations, and without all sorts of doubletalk about "experimental contexts" and "the role of observers" and "subjective quantum states" and so on.)

All that said, here's the problem that leaps out for me. Just how are we to understand the apparently greater efficiency of quantum computers over classical ones? When I started as a physics grad student in the late 1990s, we had two really great quantum algorithms—Shor's algorithm, which factorizes large numbers, and Grover's algorithm, which finds the biggest number in a list—and both of them were dramatically more efficient than the best-known classical algorithms. Shor's algorithm in particular had had a huge impact, because the problem of factorizing large numbers *both* is one of the standard examples of a difficult computational problem, *and* is crucial in decoding a lot of codes that were and are thought to be basically undecodable by classical computers. So everyone who was working in quantum information—including me at the time—was very excited by this, and pretty much all of us thought that Shor's and Grover's algorithms were going to be the tip of the iceberg, that there were going to be dozens or hundreds of these amazing quantum algorithms. But actually, ten years and more later, and those algorithms are still pretty much all we've got. Even if you could solve the technical problems involved in making a quantum computer that would fit on your desktop, at the moment there's not much you could do with it that you can't do with your existing classical desktop.

Now that's embarrassing for people writing grant applications. But it's also bizarre from a foundational point of view. It's one thing to discover that quantum mechanics has a completely different computer-complexity theory from classical mechanics. It's quite another to discover that it's almost identical but *not quite*. My hunch is that we're missing something pretty profound here.

The second problem I'd identify is a bit easier to attack, and indeed we've got quite a long way with it already, but there's further to go. It's fairly clear now that the really big mysteries in quantum theory come not so much from superposition as from entanglement (after all, classical electromagnetism admits superpositions). But getting a detailed quantitative grasp of what's going on in multipartite entanglement is really hard. We've got a variety of tools, and a variety of results, but it feels as if we still haven't found the right way of thinking about it, or maybe the right mathematical framework to use, such that it all becomes less opaque and less mysterious. (I think the very graphical "language" that Bob Coecke and his coworkers are developing is really promising here, but it's early days.)

I'll mention one more thing, which might not normally be classified as "quantum foundations"—and which I guess isn't exactly "pressing," because we've been stuck with it for decades. The last twenty or thirty years have made it really clear that quantum mechanics is way, way different from classical mechanics, and that it's possible to understand why the world looks classical without having to keep classical concepts as basic. (I'm thinking, in particular, of the role of decoherence theory, and the way we've basically managed to wean ourselves of the correspondence principle.) But the way we construct quantum theories, particularly in quantum field theory, is still almost invariably to start with a classical theory and then "quantize" it. That really, really shouldn't be necessary, but it seems to be. We need to find some way of thinking about quantum fields that doesn't require this link to classical fields.

ANTON ZEILINGER · We have learned from quantum mechanics that naive realism is not tenable anymore. That is, it is not always possible to assume that the results of observation are always given prior to and independent of observation. To me, the most important question is to find out what exactly the limitations are. This can only be found out by carefully exploring quantum phenomena in more complex situations than we do today.

A deep reanalysis of the fundamental concepts underlying quantum mechanics is also necessary, analogous to the careful analysis of the notions of space and time by the Viennese philosopher–physicist Ernst Mach. Mach's analysis paved the way for the abandonment of the notions of absolute space and time, and for their replacement by the modern notions in special and general relativity.

WOJCIECH ZUREK · Understanding the role of information; or, to be more precise, clarifying the relation between information and existence. I think that this was always—that is to say, since about 1925—the key. It is the essence of the measurement problem.

When you read Bohr, von Neumann, Wigner, Everett, or Wheeler, it is clear that they were aware of this. Bohr may not have had information theory at hand when he was thinking about matters of interpretation, but his insistence on the communicability of the measurement outcomes in everyday language points in that direction. Von Neumann and Wigner worried about the role of the conscious observer in the process, and the precondition for (and maybe even the essence of) consciousness is information acquisition and processing. Everett has long passages on information and quantum theory in his thesis, and he even devises an information-theoretic version of Heisenberg's indeterminacy principle. Wheeler's "It from Bit" goes further, by turning tables on the usual understanding of information as representing what exists and proposing that it might be the material that reality—the "It"—is made out of.

In a sense, the interplay between information and existence—between what is known and what exists—is older than quantum theory: it was central to physics since at least Boltzmann and Maxwell. The origin of the second law and the threat posed by Maxwell's demon are a premonition of the problems that are central in quantum theory. Indeed, one may defend the thesis that the quantum discoveries of Planck and Einstein (for example, stimulated emission) that paved the way for modern quantum theory happened because thermodynamics "knew" that information plays a central role in physics. One of the best illustrations of this interdependence is the famous (classical and thermodynamic) discussion of Szilárd, who in effect deduced—years before Shannon—some of the key ideas of information theory. It also puts the observer (the demon) squarely in the center of the action. This theme of the physical significance of information persists in quantum measurements.

So, already thermodynamics made it clear that "information is physical." Newtonian mechanics, however, allowed for a separation of what *is*—what exists—from what *is known*: a point in phase space is a legal representation of the state of a classical system, and it need not be altered by the observation aimed at making its location precise.

This separation of information from states was tenable in classical physics, but it breaks down in quantum theory—it breaks down in our universe. I think that by now many people recognize how central information is to quantum physics. On a technical level, this started with Heisenberg and his indeterminacy principle. But even with all that we know now about the interplay of quantum physics and information (including Bell's theorem, the no-cloning theorem, quantum error correction, and so on), I sense that the real mystery is still barely touched.

QUESTION 3

MY FAVORITE INTERPRETATION

*What interpretive program can make the best
sense of quantum mechanics, and why?*

IT IS NO SECRET that a shut-up-and-calculate mentality pervades classrooms everywhere. How many physics students will ever hear their professor mention that there's such a queer thing as different interpretations of the very theory they're learning about? I have no representative data to answer this question, but I suspect the percentage of such students would hardly exceed the single-digit range.

I vividly remember my own first course on quantum mechanics, which happened to be a graduate-level class. During the entire academic year, there was never as much as a hint that this abstract axiomatic and mathematical edifice we were using might be calling for some kind of interpretation, or some kind of motivation from deeper physical principles—nor that whole generations had already been busying themselves with the creation, refinement, and debate of various interpretive programs.

You may now suspect that I zonked out and missed the crucial moment when the lecture finally turned to foundational matters. I doubt I did. After all, I had gone into physics in the first place because of all the fascinating stuff I had read in Heisenberg's and Schrödinger's books on the philosophy of quantum theory. My ears were tuned to even the slightest foundational vibration in the lecture hall.

Did our professor simply want us to focus on solving problems and passing exams, fearing that any talk of foundational matters could drag us into an abyss of confusion? Did he not share any interest in interpretive issues? Or did he perhaps not even *know* they existed? Sometimes I still ponder these questions.

It's a curiosity of quantum theory to call loudly for an interpretation. No other physical theory has ever done so with the same kind of persistence. And it's not as if the quantum-interpretation business is a later invention of nitpicking philosophers who were out to create artificial problems that no one else had cared about or thought of before. Quite the contrary: the genesis of quantum theory itself is intimately tied up with attempts to elucidate its meaning. After all, Bohr didn't invite the crème de la crème of physicists to his Copenhagen institute and take them on long strolls through the adjacent park because he needed help with polishing quantum theory's mathematics. It was matters of interpretation that kept these pioneers awake at night.

So why is quantum mechanics widely perceived to be in need of interpretation? Many would say it has to do with the degree to which the entities of the theory correspond to something in the world around us, and with the degree to which the theory is couched in terms of tangible physical statements.

Classical mechanics seems pretty much trouble-free in this regard. We're all intuitively familiar with material objects and their trajectories. And it requires only a small stretch of the imagination to go a little further and swallow, without interpretive hiccups, the idealization of mass points and the concept of a force. (Newtonian gravity's attraction-at-a-distance might be a little harder to swallow, but there's arguably still no danger of choking.)

Classical electrodynamics wasn't such a clear-cut case anymore. It stirred much debate at the time. People asked many questions. What exactly is the nature of those mysterious electric and magnetic fields? Are they something real? How do they propagate? Aether, anyone?

On the heels of electrodynamics followed Einstein's theory of special relativity. The birth of this theory was an intellectual achievement comparable to that of quantum mechanics, because it too uprooted some of our most cherished notions and deeply ingrained intuitions about the structure of the world. Yet next to the churning sea of quantum theory, relativity feels like an interpretive backwater. Where are the warring factions, each rallying around their particular reading of relativity? It seems they're few and far between.

How come relativity is blessed with such admirable interpretive peace and tranquility? Many people point (correctly, I think) to the fact that the theory of relativity can be stated in plain language, involving a couple of clear physical principles. The basic mathematical structure of relativity—the Lorentz transformations—was more or less in place before Einstein came along. But it took Einstein's genius to supply the physical explanation for why these curious equations happened to work so well in accounting for our observations. To be sure, Einstein's principles describe a behavior of the physical world that challenges our intuitions. But they still talk about things we can readily conceive of, such as the speed of light. The theory of relativity may have baffling implications, but its formulation isn't all that baffling.

With this interpretive nonchalance of relativity as the contrasting backdrop, quantum theory's arduous quest for its own meaning is generally seen as a consequence of the fact that the theory appears to be little more than some unmotivated mathematics together with a few axioms. Though one shouldn't take the comparison with special relativity too far, it's as if quantum theory is still stuck at the level of the Lorentz transformations, waiting for a new Einstein to furnish, in a few crisp statements, a compelling physical justification for the existence and success of the quantum formalism. (Question 10, *Reconstructions*, has more on this quantum-theory-from-first-principles idea.)

The quantum-mechanical entity that describes the state of a physical system—the wave function—lives in a high-dimensional complex vector space and is quite radically removed from ordinary experience. It also embodies a strange tension: on the one hand it is said to provide a complete description of the state of a single system, on the other hand it exhibits a perplexing statistical character (see Question 4, *Quantum States*). And once we've trudged our way through the thicket of wave functions, Hilbert spaces, Hermitian operators, and unitary evolutions, we finally get to the part of quantum theory that supposedly makes direct contact with the world: the measurement axiom. But some people argue that this axiom lives in uneasy coexistence with the implications of the rest of the formalism (see also Question 7, *The Measurement Problem*), and I wouldn't be surprised to see them turn up at a conference dressed in T-shirts sporting the slogan I WENT TO THE TEMPLE OF QUANTUM DIVINITY AND ALL I GOT WAS THIS LOUSY MEASUREMENT POSTULATE.

And so legions of foundationally-minded folks set out to tackle what each perceived as quantum theory's deficiencies. Dissatisfaction with the "Copenhagen orthodoxy," among other things, kick-started an ever-growing network of alternative denominations. (Note that contrary to popular belief, there's actually no such thing as a single, unitary Copenhagen interpretation. The image of such an interpretation is, as Don Howard put it, "a postwar myth" and "an invention of the mid-1950s, for which Heisenberg is chiefly responsible.")

Today, the soul-searching quantum foundationalist can choose from a great many faiths, each with their individual gospel of interpretive salvation. There's Everett's relative-state interpretation and its full-fledged many-worlds and many-minds versions. There's the de Broglie–Bohm hidden-variables picture with its particle trajectories. One has modal, consistent-histories, ensemble, and instrumentalist camps. There are neo-Bohrian, post-Copenhagen incarnations, rooted in the notion of information or in a Bayesian reading. And one can even leave the territory of interpretations proper and opt for collapse theories, which make predictions different from those of standard quantum mechanics.

Faced with such an overwhelming smorgasbord of options—and given that all interpretations are empirically equivalent (with the exception of new theories, like collapse models and nonequilibrium versions of de Broglie–Bohm)—how does any individual become attracted to a particular faith, let alone turn into a fervent advocate? This is the question we will explore in this chapter. We'll ask our interviewees to tell us about their favorite interpretation and the motivation for their choice—or, alternatively, to explain why they remain uncommitted, or why quantum theory may not even need interpretation in the first place.

It's only natural that interpretive preferences will, at least partially, have something to do with personal taste and philosophical temperament (see also Question 13, *Beliefs and Values*). For example, *you* may be keen on salvaging a classical worldview; *I* may want to take the plunge and pursue an interpretation that radically breaks with all classical intuition. Or, *you* may prefer to apply Occam's razor to the laws of the formalism; *I* may want to make the cut at the level of the ontology implied by this formalism. And on it goes. No wonder we're still waiting for any kind of consensus to be reached in matters of interpretation.

And there's no magic cure-it-all: with every interpretation, you win some, but you also lose some, and whether something is to be regarded as a gain or a loss in any given instance will depend on who you ask. Two people may see one and the same aspect of a particular interpretation in starkly different lights. Take Everett's scientific-realist reading of the wave-function formalism as an example. One person may celebrate this interpretive move as the one that lets the quantum do the talking; as the one that takes to heart the message of quantum theory in the most consistent and unadulterated manner; as the one that has no need for wasting and mincing words, for hiding behind philosophical and semantic smoke screens, for elevating man-made terms such as "irreducibly classical concepts" and "complementarity" to principles of nature. But another person may feel the exact opposite, judging the desire to promote a formal entity—the wave function—to the all-encompassing, objectively existing essence of the universe as symptomatic of a *classical* mindset. And they might see the Everett interpretation as possessed by a philosophical agenda of absolutism and monism—an agenda that William James, long before Everett's time, captured thus:

> So the universe has always appeared to the natural mind as a kind of enigma, of which the key must be sought in the shape of some illuminating or power-bringing word or name. That word names the universe's PRINCIPLE, and to possess it is, after a fashion, to possess the universe itself. "God," "Matter," "Reason," "the Absolute," "Energy," are so many solving names. You can rest when you have them. You are at the end of your metaphysical quest.

Needless to say, personal circumstances of education and academic environment also often play a role in the choice of interpretation. Perhaps the first book on quantum foundations you've read left such an indelible mark on your mind that from then on the author's interpretive position has become your own. And so you may sympathize with Bohr's thinking, say, or rather with John Bell's quest for a theory of "beables." It's also not uncommon to find younger foundationalists carry on the interpretive legacy of their thesis supervisors. When thinking about the influence of such personal or intellectual role models, I'm often reminded of my third-grade essay, "What I Want to Be When I Grow Up." I told my teacher, with all the earnestness of a nine-year-old, that I wanted to become a senior legal adviser with the human-resources department at the University of Munich—which just so happened to be my father's position at the time. I also recall that a classmate wrote that he wanted to become a terrorist with the Red Army Faction. He now works as an architect. (And no, his father had *not* been a terrorist.)

One bargaining point can be the utility of an interpretation. We may ask ourselves, "How can a particular interpretation, and thus a specific way of thinking about quantum mechanics, help me tackle a problem that I wouldn't have known how to get a handle on otherwise?" Or, "How can an interpretation guide me to new insights that I wouldn't happened upon without the distinctive vantage point provided by the approach?" Ceteris paribus, if an interpretation has enough cash value in this regard, it may win people over who would otherwise remain indifferent or undecided.

Now, there is indeed evidence that certain discoveries and ideas were spurred by particular interpretive angles. Let me mention a few examples. The decoherence program has its roots in Dieter Zeh's independent rediscovery of the Everett interpretation. This interpretation also seems to have fueled some of David Deutsch's ideas on quantum information and quantum computing. The rampant nonlocality of Bohmian mechanics led Bell to investigate whether this must be a generic feature of all hidden-variables theories, and thus Bell's theorem was born (see Question 8, *Bell's Inequalities*). Recently, I learned that some quantum chemists have begun using Bohmian trajectories to speed up their computations. The close interplay between an information-based foundational mindset and ongoing work in quantum information theory is self-evident, although it's difficult to judge which side has the inspirational lead. Suffice to say that by far not every practitioner of quantum information is also an adherent of an informational interpretation—there are also Everettians and Bohmians and other affiliations in the pack of quantum informationalists. Collapse models, to name one more example, keep motivating experimental proposals for testing the limits of quantum mechanics, which is an enterprise as laudable as it is arduous. And finally, a cynic might say that the shut-up-and-calculate "interpretation" could perhaps claim the largest prize, by liberating people from worrying about fundamentally unsettling but practically irrelevant issues, and by just letting these people get on with their hands-on business so they can make all those beautiful predictions and discoveries and carry out ingenious experiments and build useful things.

These examples all make for heart-warming anecdotes. But none of them has turned out to be the one big practical selling point that would make a visible difference in tilting the balance in favor of a particular interpretation. It will be exciting to see what the future holds in this regard.

This brings us to the next question. Will the forest of competing interpretations thin out over time? Will we see the day when we all come together in a show of solidarity and unity and rally around the same interpretation? It is debatable how probable—and even desirable—such a scenario would be. Perhaps what would be needed to help things along is some kind of seismic event that would change (or simply inform) our view of quantum theory in a universally agreed way.

One such event could be a grand future experiment—say, one that decisively demonstrates a clash with the predictions of quantum theory (see also Question 11, *The Experiment of My Dreams*). But it's far from clear, in fact, that such a finding would necessarily pick out a single new interpretation or make interpretation redundant. It's more likely to result in a general upheaval spawning an altogether fresh set of theories and problems.

On the theoretical side, progress in our understanding of the origins of the quantum formalism—say, in terms of physical or information-theoretical principles—may be exactly the breath of fresh air that's needed (see Question 10, *Reconstructions*). Of course, new and more general theories, such as quantum gravity and other attempts at unification, may also completely reshuffle the deck of cards at some point (see Question 15, *Unification*).

But let's suppose for a moment that no such revolutionary events will take place, that quantum theory in its current form is indeed here to stay, and that therefore rivaling (and empirically equivalent) interpretive readings will continue to thrive. Such an irreducible plurality of interpretations would tell us that we're free to embellish—some may say encumber—the formalism with entities of our choice, if such a maneuver helps us visualize what's going on, but that in doing so we'll be crossing into strictly metaphysical terrain. And if we follow such a reading to its logical (if radical) conclusion, then quantum theory might even contain a lesson about the task of physics: that the search for "what the world is made of," for a unique, definitive, and fundamental ontology at the heart of everything, may ultimately be misguided.

Guido Bacciagaluppi · I do not believe that there is a single interpretive program that makes better sense of quantum mechanics than all the rest. I believe that a number of programs are, indeed, successful, but they have different strengths and weaknesses.

There are different criteria by which one might judge how well an interpretive program makes sense of the theory, and these criteria will generally lend support to different programs. One can judge programs based on how well they fit with the rest of accepted physics (in particular with special relativity), or by their promise of novel physical discoveries (even superluminal signaling!); by how well they incorporate important aspects of the theory (e.g., decoherence), or by how they employ proven insights from other fields within physics (e.g., statistical mechanics); by how well they make sense of specifically physical questions, such as the analysis of measurements (or nonlocality, or the classical regime, and so on), or by how well they make sense of specifically philosophical questions, such as the nature of probability (or mentality, or personal identity, and so on); or, finally, by how well they fit with one's general views on science and knowledge.

The main programs I consider to be successful are the three traditional ones in the foundations literature: hidden-variables theories (in the form of de Broglie–Bohm pilot-wave theories and related incarnations), spontaneous-collapse theories, and Everett interpretations. This list omits the older Copenhagen interpretations and newer quantum-information approaches, but for specific reasons.

While I have come to realize that charges of downright incoherence against Copenhagen-style interpretations are exaggerated, I do believe that these interpretations do not make as good sense of quantum mechanics as other approaches do (essentially for the reason that they treat some bona fide physical problems as pseudoproblems, and are thus self-limiting). Incidentally, one of the approaches that I believe have made Copenhagen-style interpretations philosophically intelligible is the recent one by Caves, Fuchs, and Schack, seeking to characterize quantum mechanics as a theory of the beliefs of rational agents in a quantum world.

As regards more generally the cluster of approaches loosely collected under the label of quantum information, insofar as they have given rise to foundational work on entanglement, nonlocality, and other quantum phenomena, they have been extremely successful, and they have also achieved striking results within the reconstruction problem of quantum mechanics. But insofar as quantum information is regarded as a comprehensive interpretive program, I think it is fair to say it is still mostly work in progress. Indeed, as Alex Wilce aptly points out, the reconstruction problem and the measurement problem of quantum mechanics are complementary questions, and they call for different kinds of answers.

Returning to pilot-wave theories, collapse theories, and Everett interpretations, I believe they fare very differently on a number of the criteria I have gestured toward above. A full analysis of the pros and cons of these programs is obviously more than I could attempt here, but there are a few interesting points I would like to mention.

As regards the promise of new physics (and of new experimental results), pilot-wave theories rank highest; in collapse theories, quite surprisingly, the question is still open; and the idea is totally alien to Everett interpretations. As regards instead the compatibility with relativity, the opposite is true: Everett interpretations are bound to be relativistic by construction; the judgement is still open in the case of collapse theories; and pilot-wave theories stand in open opposition to relativity (except at the phenomenological level).

Decoherence is constitutive of contemporary Everett interpretations, where the "worlds" are none other than decoherent histories. But it plays a crucial role also in pilot-wave theories and, arguably, in collapse theories—essentially because in both theories the configuration-space trajectories and the collapse mechanism can build on the preparatory work of decoherence. On the other hand, this lends Everettians their most notable rhetorical weapon against the other two approaches, namely, the dynamically autonomous existence of empty waves (in pilot-wave theories) and of wave-function tails (in collapse theories).

All possible approaches to the philosophy of probability are reflected in the main approaches to the philosophy of quantum mechanics (and might correlate to certain sympathies and antipathies). The nature of probability is the same in statistical mechanics and in pilot-wave theories: in both theories, probabilities are purely epistemic. Collapse theories do not offer an analysis of probabilities, which are in fact primitives (possibly best thought of in terms of propensities?). Everett interpretations, following recent work in particular by David Deutsch and by David Wallace, have probabilities emerging at the level of worlds from the deterministic Schrödinger evolution, in a way that is the most perfect embodiment yet of the spirit, if not the letter, of David Lewis's "Principal Principle." And Caves, Fuchs, and Schack play the hitherto neglected card of taking quantum probabilities as subjective degrees of belief in the sense of de Finetti. (For some more details, see the last part of my answer to Question 6, page 129.)

If one considers that collapse theories solve the measurement problem by purely physical means, they ought to be the interpretation of choice for physicists. (Or maybe physicists ought to prefer Everett interpretations, which leave the edifice of quantum physics singularly untouched.) Instead, Everett interpretations can have a special attraction for philosophers, because they provide new perspectives on many traditional issues—from emergence to personal identity, from the philosophy of mind to the philosophy of probability. On their part, pilot-wave theories are built on the analogy with thermodynamics and classical statistical mechanics, allowing for the free flow of conceptual resources (as well as problems!) from the classical to the quantum setting. And, finally, another double-edged sword for pilot-wave theories is that they take seriously the thermodynamic analogy in special relativity, interpreting the latter as a phenomenological theory of principle. As I put it once in an examination question, "Is the Bohm theory the constructive version of special relativity (this is not a misprint)?"

ČASLAV BRUKNER · To a large extent, the various debates about the interpretation of quantum mechanics can be seen as debates about what quantum physics refers to. Does it directly refer to reality—or to our information, on the basis of which we construct reality? I find very suggestive the role information played in the early debates on the meaning of quantum mechanics, most notably in the Bohr–Einstein dialogue. No matter how sophisticated the claim was that it should be possible to *both* observe the interference fringes *and* identify which of the two slits the particle goes through, it was invariably found that the flawed mechanism lurking behind this claim can't, in fact, violate the following principle: any increase of partial information about the particle's path will always mean a corresponding loss in visibility of the interference pattern, and vice versa. Most importantly, it is not relevant whether we read out that information. All that is necessary is for the information to be present somewhere in the universe.

The evidently significant role information plays in the analysis of interference experiments has persuaded me that we should try to understand quantum mechanics by putting primacy on the concept of information (or on the concept of probability, which again can be seen as a way of quantifying information). Indeed, in 2009 Borivoje Dakic and I demonstrated that classical probability theory and quantum theory—the only two probability theories for which we have empirical evidence—are special, in that they fulfill three reasonable axioms on the systems' information-carrying capacity. There are two key ideas in this reconstruction:

(1)   An elementary system has the information-carrying capacity of at most one bit.
(2)   All systems of the same information-carrying capacity are equivalent.

These ideas build on Zeilinger's proposal from 1999 for a foundational principle for quantum mechanics, and on our joint work from 2001. Roughly, statement 1 specifies the structure of a single qubit as the simplest two-dimensional quantum system. Statement 2 then characterizes the structure of a higher-dimensional system in such a way that any two-dimensional subspace has again the information-carrying capacity of one bit. If one requires that between any two pure states there exists a *continuous* reversible transformation, one separates quantum theory from classical probability theory.

Above two guiding ideas lay down my expectations of providing scientifically sound resolutions to all the standard quantum puzzles. The randomness of an individual measurement outcome follows from the fundamental limitation on the information-carrying capacity as specified by statement 1. This capacity is simply not enough to determine the outcomes of all conceivable measurements: the outcomes of some measurements must necessarily contain an element of irreducible randomness.

From statement 2 it follows that there is logically no difference between an elementary system containing the bit locally, and the one containing it in correlations between measurements on two or more subsystems. Entanglement arises when the bit resides in the correlations—when it is actually a mistake to think of the subsystems, each containing a bit, separately. Since this feature is independent of the spatiotemporal arrangements of the measurements on the subsystems, one has a violation of Bell's inequalities. Finally, the "collapse of the wave function" is not a physical process but simply the acquisition of fresh knowledge about a physical system. (See also my other answers for more details.)

JEFFREY BUB · The program of interpreting quantum mechanics tends to treat the theory like a problem child in the family of theories and propose therapy. The aim is to get quantum mechanics to conform to some ideal of classical comprehensibility. If this is what it means to "make the best sense of quantum mechanics," then I think the exercise is misguided. Rather, we should be trying to make sense of quantum mechanics as an indeterministic theory where the probabilities are "uniquely given from the start," as von Neumann put it, in terms of a geometric feature of Hilbert space, namely, the angles between events represented by Hilbert-space subspaces. This means not just contrasting quantum mechanics with classical theories, where probabilistic correlations between two systems can be reduced uniquely to a shared probability distribution over joint deterministic states that are also product deterministic states (local states) for the systems separately, but considering quantum mechanics relative to other indeterministic nonclassical theories.

A natural class of theories to consider is the class of theories satisfying a "no signaling" constraint. We expect that if Alice and Bob are in separate regions of the universe, then no information should be available in the marginal probabilities of measurement outcomes in Alice's region about alternative choices made by Bob. In particular, Alice should not be able to tell what observable Bob measured in his region, or whether Bob performed any measurement at all, by looking at the statistics of her measurement outcomes, and conversely. Of course, if this no-signaling constraint is violated, then Alice and Bob could signal superluminally, but this is not primarily a relativistic condition. It is a much more elementary constraint, satisfied by classical and quantum theories as well as "superquantum" theories, where the correlations violate the Tsirelson bound.

One can represent the probabilistic structure of a theory in this class as a convex set—loosely, a set such that from any point in the interior you can see any point on the boundary. As a simple example, consider a classical theory with just two deterministic states, labeled 0 and 1. The convex set in this case is represented by the line segment between 0 and 1, with the two boundary points representing extremal, or pure, states. The points between the pure states represent mixed states—convex combinations of extremal states: $\mathbf{p} = p\mathbf{0} + (1-p)\mathbf{1}$, for $0 \leq p \leq 1$. If you have more deterministic states, the structure is a convex polytope (the analogue of a polygon in many dimensions), with the vertices representing the deterministic states. For a classical theory, the polytope is a simplex: a polytope generated by $n + 1$ vertices that are not confined to any $(n-1)$-dimensional subspace—for example, a tetrahedron as opposed to a rectangle.

Note that the lattice of subspaces of a simplex (the lattice of vertices, edges, and faces) is a Boolean algebra, with a one-to-one correspondence between the vertices, corresponding to the atoms of the Boolean algebra, and the facets—that is, the $(n-1)$-dimensional faces—corresponding to the co-atoms. The classical simplex represents the classical state space regarded as a space of classical (multipartite) probability distributions; the associated Boolean algebra represents the classical event structure.

The simplest quantum system is the qubit. The convex set in this case has the structure of a sphere (the Bloch sphere), with the points on the boundary representing the extremal, or pure, states. This is, of course, not a simplex, not even a polytope. For superquantum no-signaling theories, the convex set is a convex polytope that is not a simplex. Some of the vertices of the polytope represent deterministic states, other vertices represent indeterministic extremal states. For a bipartite system, these are Popescu–Rohrlich (PR) boxes.

A PR box is a hypothetical superquantum device with two inputs, $x \in \{0,1\}$ and $y \in \{0,1\}$, and two outputs, $a \in \{0,1\}$ and $b \in \{0,1\}$. We can think of the x-input as controlled by Alice, who monitors the a-output, and the y-input as controlled by Bob, who monitors the b-output. For a comparison with quantum mechanics, the inputs would correspond to measurement choices and the outputs to measurement outcomes. Alice's and Bob's inputs and outputs are required to be correlated according to:

$$a \oplus b = x \cdot y,$$

where $\oplus$ is addition mod 2, that is,

(1)   same outputs (i.e., 00 or 11) if the inputs are 00 or 01 or 10;
(2)   different outputs (i.e., 01 or 10) if the inputs are 11.

The box is nonlocal in the sense that the x-input and a-output can be separated from the y-input and b-output by any distance without altering the correlations, which are "more nonlocal" than quantum correlations in the precise sense discussed in my answer to Question 8 (see page 168). The no-signaling constraint is satisfied because the marginal probabilities of Alice's outputs are all one-half and do not depend on Bob's input; this means that Alice cannot tell what Bob's input was by looking at the statistics of her outputs, and conversely. Additional PR boxes are derivable from the standard PR box by relabeling the x-inputs and a-outputs conditionally on the x-inputs, and the y-inputs and b-outputs conditionally on the y-inputs.

A simplex has the rather special property that a mixed state can be represented *uniquely* as a mixture of extremal, or pure, states, the vertices of the simplex. No other convex set has this feature. So in the class of no-signaling theories, classical theories are rather special. For all nonclassical (= nonsimplex) theories, there is no unique decomposition of mixed states into pure states. For such theories, there can be no general cloning procedure for an arbitrary extremal state without violating the no-signaling constraint, and similarly there can be no measurement in the nondisturbing sense that one has in classical theories, where it is in principle possible, via measurement, to extract enough information about an extremal state to produce a copy of the state without irreversibly changing the state.

The quantum theory is a member of this class of nonsimplex theories. Hilbert space as a projective geometry (that is, the subspace structure of Hilbert space) represents a non-Boolean event space, in which there are built-in, structural probabilistic constraints on correlations between events (associated with the angles between events)—just as in special relativity the geometry of Minkowski space-time represents spatiotemporal constraints on events. These are kinematic, i.e., predynamic, objective probabilistic or information-theoretic constraints on events to which a quantum dynamics of matter and fields conforms through its symmetries, just as the structure of Minkowski space-time imposes spatiotemporal kinematic constraints on events to which a relativistic dynamics conforms.

In this sense, Hilbert space provides the kinematic framework for the physics of an indeterministic universe, just as Minkowski space-time provides the kinematic framework for the physics of a non-Newtonian, relativistic universe. From this perspective, there is no deeper explanation for the quantum phenomena of interference and entanglement than that provided by the structure of Hilbert space, just as there is no deeper explanation for the relativistic phenomena of Lorentz contraction and time dilation than that provided by the structure of Minkowski space-time.

This is, in broad outline, what I would call an information-theoretic interpretation of the nonclassical features of quantum probabilities, in the sense of Shannon's notion of information, which abstracts from semantic features of information and concerns probabilistic correlations between the physical outputs of an information source and a receiver. On this view, what is fundamental in the transition from classical to quantum physics is the recognition that *information in the physical sense has new structural features*, just as the transition from classical to relativistic physics rests on the recognition that space-time is structurally different than we thought. This seems to me the interpretive program that makes the best sense of quantum mechanics.

ARTHUR FINE · As I explained in answer to Question 2 (see page 44), I do not think the interpretive situation calls for, or even allows for, a "best sense." Nevertheless, some interpretations certainly make more sense than others. Surely at the bottom of anyone's list of what is sensible are the "manies"; that is, the many-worlds and many-minds interpretations. There is, I think, no sense at all to be made of the splitting of worlds-plus-agents in many worlds. Of course, one can repeat the words over and over until one becomes deaf to the nonsense, but it remains nonsense nevertheless. Curiously, those who favor this interpretation concentrate their defense on dealing with some obvious technical issues: preferred basis, getting the right probabilities via "measures of existence" (or the like), questions of identity and individuation across worlds, and so on. But the fundamental question is just to explain what it means to talk of splitting worlds, and why we should not just write it off, à la Wittgenstein, as language on holiday. (Einstein once described the writings of Hegel as "word-music." Perhaps that would be a gentler way of dismissing many worlds.) The same problem of making good sense is at the heart of many minds as well for, recall, that in many minds *all* our definite beliefs are an illusion.

By contrast with the manies, the several collapse approaches that introduce a stochastic element in the dynamics seem to me perfectly sensible. Still, they strike many as ad hoc on several counts. First, they postulate a mechanism of spontaneous collapse that is uncomfortably similar to the phenomenon they were presumably intended to explain. (Why does the state function collapse on measurement? Well, that results from a lot of mini-collapses along the way.) Second, they require certain free parameters that need to be set by hand in order to make the account accord with experiment. This may look like curve fitting rather than fundamental theory construction. Finally, they appear to be indistinguishable experimentally from ordinary decoherence. If one were to go the route of embracing collapse, then one might want to find some compelling, fundamental principles that determine the collapse dynamics. (Easy to say!)

At first glance, the de Broglie–Bohm approach looks both sensible and attractive. If we have wave motion (the Schrödinger equation), then currents may be generated and particles can be moved about by the currents. But, of course, this nice picture becomes less physically attractive once one realizes that the "waves" and their currents play out in multidimensional configuration space. Moreover, the particles moved by the currents do not actually push back in the manner of ordinary flotsam. That is very unphysical and suggests that the "waves" are just a fiction (or, if you prefer, a placeholder) for positing laws of motion of the particles. Thus, this account simply postulates new laws of motion, highly nonlocal, and carefully contrived not to conflict with the statistical predictions of quantum mechanics, provided we build in just the right ignorance of initial particle positions. Given the chaotic trajectories dictated by these laws, it might well seem a miracle that the regular statistics of quantum mechanics fall out. Apparently, that was Bohm's attitude, since he thought of the de Broglie approach as a stopgap to be understood from below by averaging over a subquantum level. Thus, Bohm acknowledges that positing these special laws of motion requires further grounding, or explanation.

This brings us to old-fashioned quantum mechanics, with collapse. Perhaps we can mitigate the clearly nonsensible aspect associated with superpositions (Schrödinger's cat) by leaning on decoherence to provide an account of how things should look to us from our collective point of view. Then, are we so badly off after all?

CHRISTOPHER FUCHS · Asher Peres was a master of creating controversy for the sake of making a point. For instance, in 1982 he was asked to make a nomination for the Nobel prize in physics. He nominated Israeli prime minister Menachem Begin! Asher reasoned that Begin's decision to invade Lebanon proved him as qualified for a Nobel physics prize as he was for his earlier peace prize.

It certainly wasn't of the same magnitude, but Asher intended to make trouble when we wrote our 2000 "opinion piece" for *Physics Today*. Previous to our writing, the magazine had published a series of articles whose essential point was that quantum mechanics was *inconsistent*—it tolerated the unacceptable "measurement problem," and what else could that mean but inconsistency? Quantum theory would need a patch to stay afloat, the wisdom ran—be it decoherence, consistent histories, Bohmian trajectories, or a paste of Everettian worlds.

To take a stand against the milieu, Asher had the idea that we should title our article, "Quantum Theory Needs No 'Interpretation.'" The point we wanted to make was that the structure of quantum theory pretty much carries its interpretation on its shirtsleeve—there is no choice really, at least not in broad outline. The title was a bit of a play on something Rudolf Peierls once said, and which Asher liked very much: "The Copenhagen interpretation *is* quantum mechanics!" Did that article create some controversy! Asher, in his mischievousness, certainly understood that *few* would read past the title, yet *most* would become incensed with what we said nonetheless. And I, in my naiveté, was surprised at how many times I had to explain, "Of course, the whole article is about an interpretation! Our interpretation!"

But that was just the beginning of my forays into the quantum foundations wars, and I have become a bit more seasoned since. What is the best interpretive program for making sense of quantum mechanics? Here is the way I would put it now. The question is completely backward. It acts as if there is this *thing* called quantum mechanics, displayed and available for everyone to see as they walk by it—kind of like a lump of something on a sidewalk. The job of interpretation is to find the right spray to cover up any offending smells. The usual game of interpretation is that *an interpretation is always something you add to* the preexisting, universally recognized quantum theory.

What has been lost sight of is that physics *as a subject of thought* is a dynamic interplay between storytelling and equation writing. Neither one stands alone, not even at the end of the day. But which has the more fatherly role? If you ask me, it's the storytelling. Bryce DeWitt once said, "We use mathematics in physics so that we won't have to think." In those cases when we need to think, we have to go back to the plot of the story and ask whether each proposed twist and turn really fits into it. An interpretation is powerful if it gives guidance, and I would say the very best interpretation is the one whose story is so powerful it gives rise to the mathematical formalism itself (the part where nonthinking can take over). The "interpretation" should come first; the mathematics (that is, the preexisting, universally recognized thing everyone thought they were talking about before an interpretation) should be secondary.

Take the nearly empty imagery of the many-worlds interpretation(s). Who could derive the specific structure of complex Hilbert space out of it if one didn't already know the formalism? Most present-day philosophers of science just don't seem to get this: if an interpretation is going to be part of physics, instead of a self-indulgent ritual to the local god, it had better have some cash value for physical practice itself. If, for instance, the Everettian interpretation could have gotten us to realize the possibility of graphene before the Scotch tape of Geim and Novoselov, it would have been a conversion experience for me—I would be an Everettian today. That is the kind of influence an interpretation should have.

Most quantum foundationalists, I suspect, would say that this is an impossibly high standard to hold, but it shouldn't be. In any case, let me give an example that has a bit more chance to make some effect on the intelligentsia. Some years ago, I was involved in a paper that explored various properties of a certain set of quantum states on two qutrits (i.e., two three-level quantum systems):

$$|0\rangle \otimes |0+1\rangle \qquad |0\rangle \otimes |0-1\rangle \qquad |2\rangle \otimes |1-2\rangle$$
$$|2\rangle \otimes |1+2\rangle \qquad |1\rangle \otimes |1\rangle \qquad |1+2\rangle \otimes |0\rangle$$
$$|1-2\rangle \otimes |0\rangle \qquad |0+1\rangle \otimes |2\rangle \qquad |0-1\rangle \otimes |2\rangle.$$

Here $|0\rangle$, $|1\rangle$, $|2\rangle$ represent an orthonormal basis for each system, and $|0+1\rangle$ stands for the state $2^{-1/2}(|0\rangle + |1\rangle)$, and so on. There are two things to notice about this set of states. (1) The states form a complete orthonormal basis for the bipartite Hilbert space. Thus, if someone were to prepare one of the states secretly, another observer privy to the identity of the set but not to the particular state would be able to perform a measurement that identifies it with complete accuracy. But, (2) there is no entanglement in any of these states—they are all products. This gives the appearance that everything about point 1 is actually intrinsically local. This provokes the following question. If the "observer" is really two separate observers, each localized at one of the qutrits, can the unknown preparation still be identified with complete accuracy, particularly if the observers are allowed the full repertoire of quantum measurements (POVMs, sequential measurements, weak measurements, and the like), along with any amount of classical communication between themselves?

What guidance would the many-worlds interpretation(s) give on this question? If you're an Everettian, and you don't know the answer, think hard at this point before reading on. By thinking in terms of the Everettian imagery, would we be able to see the answer at least in rough outline before doing any prolonged calculations? You can guess what I suspect.

In any case, the answer is that localized observers *cannot* give a perfectly accurate identification of an unknown state drawn from this set. We called this effect "non-locality without entanglement" and gave further examples—for instance, one based on three qubits, and so on. The reason I bring this phenomenon up is because it is a particularly *ugly* and *unexpected* one where an epistemic view of quantum states (that they are states of knowledge, information, or belief, as Peres and I held, rather than agent-independent states of nature) has some teeth. In fact, there is no better way to see this than through the "toy model" Rob Spekkens constructed in his paper "In Defense of the Epistemic View of Quantum States: A Toy Theory" just for the purpose of demonstrating the unifying and far-ranging power of an epistemic view of quantum states. The toy theory is not quantum theory itself, nor does it pretend to be more than a source of ideas for deriving the real thing. Mostly, it is a framework for making it obvious and incontestable that the states from which its phenomena arise are epistemic, not ontic—i.e., they are decidedly not states of nature.

Here are two paragraphs from Rob's paper that get to the heart of the matter:

> We shall argue for the superiority of the epistemic view over the ontic view by demonstrating how a great number of quantum phenomena that are mysterious from the ontic viewpoint, appear natural from the epistemic viewpoint. These phenomena include interference, non-commutativity, entanglement, no cloning, teleportation, and many others [including non-locality without entanglement]. Note that the distinction we are emphasizing is whether the phenomena can be understood conceptually, not whether they can be understood as mathematical consequences of the formalism .... The greater the number of phenomena that appear mysterious from an ontic perspective but natural from an epistemic perspective, the more convincing the latter viewpoint becomes. ...

Of course, a proponent of the ontic view might argue that the phenomena in question are not mysterious if one abandons certain preconceived notions about physical reality. The challenge we offer to such a person is to present a few simple physical principles by the light of which all of these phenomena become conceptually intuitive (and not merely mathematical consequences of the formalism) within a framework wherein the quantum state is an ontic state. Our impression is that this challenge cannot be met. By contrast, a single information-theoretic principle, which imposes a constraint on the amount of knowledge one can have about any system, is sufficient to derive all of these phenomena in the context of a simple toy theory, as we shall demonstrate.

An anecdote Rob tells, and which is surely true, is that when someone tells him of some phenomenon in quantum information theory that they think is surprising, he quickly checks to see if an analogue of it can be found in the toy model—the toy model is intuitive enough that he can usually do that in his head. And most often, he finds that the phenomenon is there as well, signifying that it is coming about from little more than the epistemic nature of quantum states.

In other words, he can pull a little conceptual model from his pocket and gain quick insight into any number of technical questions in quantum theory, just by having started with the right conception of quantum states! That is physical insight; that is power in physics! *That is physics.* The next time I'm at a Bohmian or Everettian conference, I'll pose some problem in quantum theory that has me flustered. We'll see which one's worldview and intuition helps it find the answer first.

GIANCARLO GHIRARDI · Before answering this question, let me say that I don't think it's formulated in fully appropriate terms. The question refers to potential "interpretive programs," but I believe that the alternative possibilities one may consider cannot all be classified as "interpretations" of the existing scheme. I find it difficult to regard as an interpretation even theories that are predictively identical to the standard theory, such as Bohmian mechanics. Moreover, let's not forget that among the alternative candidates there are some that qualify as rival theories of quantum mechanics, since they are, in principle, testable against quantum mechanics.

Having made this clear, I believe that it will not surprise anybody to hear—given the many years I have devoted to it—that I attach a prominent role to programs like the dynamical-reduction approach, a particular instance of which is the GRW theory. I do not want to be misunderstood: I do not consider the collapse theories in their present form as sufficiently general to be taken seriously as new fundamental theoretical schemes. Basically, to my mind, they are phenomenological proposals involving parameters that should play the role of new fundamental constants of nature. I am also fully aware of the fact that it will not be easy to work out fully satisfactory, and sufficiently general, relativistic generalizations of these theories. I attach great relevance, however, to the underlying motivating factors that have pushed me into trying to work out such models and to some of the achievements they have made possible. Let me clarify this point.

I am firmly convinced that the superposition principle cannot have universal validity. Actually, I have worked out a quite general proof of this fact in a paper with one of my students, Angelo Bassi. Accordingly, I believe that at an appropriate scale somewhere between the microscopic and macroscopic world, such a principle must begin to be violated.

Reduction models make clear—by taking up a line of thought that nobody would have considered viable before its formulation, and that Einstein himself, in his "Reply to Critics," had declared impossible—that one can work out a dynamical scheme different from standard quantum mechanics, a scheme that is based on a unique universal dynamical principle and that accomplishes, without leading to any predictions contradicting those of the standard theory concerning all microscopic systems, the goal of forbidding the occurrence of superpositions of macroscopically and perceptually different states.

With reference to my introductory remarks, let me also stress the following point. Even if the present form of the reduction theories cannot be taken as definitive, the theories must violate linearity in a way that is much more constrained than usually thought, if they are to overcome the difficulties arising from the reduction process. For this reason, such theories might help us identify appropriate phenomenological areas that deserve our particular attention when attempting to demonstrate the limited validity of the linear nature of quantum mechanics.

SHELLY GOLDSTEIN · As my answer to Question 2 (see page 47) suggests, I believe that a fundamental physical theory must involve local beables, that is, variables describing the configuration of matter in space-time. A variety of choices for the fundamental local beables may be possible, and each choice may admit a variety of laws to govern the behavior of the fundamental local beables. Each such choice, of local beables and laws governing them, corresponds to a different theory, even when these different theories yield the same predictions for the results of experiments—even, that is, when they are *empirically equivalent*. When these predictions are those of orthodox quantum theory, the different theories are different *versions of quantum theory*.

The way such theories yield experimental predictions is via the fundamental local beables, in terms of which macroscopic variables are defined. Some of these macroscopic variables describe the results of experiments, so that the laws governing the behavior of the fundamental local beables have empirical implications. The orientation of a pointer, for example, is determined by the configuration of the fundamental local beables associated with the pointer—the configuration, say, of its particles—and the behavior of the pointer is determined by that of its particles.

This may seem rather obvious. I think it is. What is not so obvious, perhaps, and what, given the history of quantum mechanics and the surrounding controversy, is perhaps surprising, is that a choice of fundamental local beables and law for them yielding a version of quantum mechanics—yielding a theory empirically equivalent to quantum mechanics—should be possible at all.

But such a choice is possible and, insofar as nonrelativistic quantum mechanics is concerned, rather obvious. It involves a simple particle ontology: the local beables are the positions of the particles, and these move according to an equation of motion, the guiding equation, that involves the wave function of standard quantum theory and is naturally suggested by the structure of the evolution equation for that wave function (Schrödinger's equation).

The resulting theory—called the de Broglie–Bohm theory, or Bohmian mechanics—was discovered by Louis de Broglie in 1927 and rediscovered by David Bohm in 1951. The state of a system in this theory is given by the positions of its particles together with its wave function; the latter evolves according to Schrödinger's equation, while the former are governed by the guiding equation. Given Schrödinger's equation, this theory thus involves the obvious ontology evolving the obvious way.

Another version of quantum mechanics, called stochastic mechanics and discovered by Edward Nelson around 1966, involves the same local beables—the positions of particles—as Bohmian mechanics. But in this theory the particles evolve randomly according to a diffusion process defined in terms of the wave function, whereas the particle evolution in Bohmian mechanics is deterministic.

In some other versions of quantum mechanics with a particle ontology, the evolution of the particles is again deterministic, but with evolution laws for the particles—for example, those proposed by Deotto and Ghirardi—different from the one given by the guiding equation of Bohmian mechanics.

For quantum field theory and beyond, the choice of local beables and law governing their evolution need not be at all obvious. But a variety of choices consistent with the predictions of quantum field theory are nonetheless possible, at least when suitable cutoffs are imposed.

Thus, I believe that the "interpretive program" that "can make the best sense of quantum mechanics" is that of formulating a version of quantum mechanics involving local beables whose behavior is governed by a law expressed in terms of the wave function. Such a theory has the resources that seem obviously necessary if quantum mechanics is to make any predictions at all without the invocation of fundamental rules governing the results of measurement.

Of course, what seems to be necessary need not be so. It may be that Everett's many-worlds version of quantum mechanics need not involve any local beables in its explicit formulation. But whether this is so remains controversial; my own feeling is that it is not so.

DANIEL GREENBERGER · I don't know what the "best" interpretation is, but I can tell you the one I tend to use, since I find it interpretatively powerful, and since it is very useful in helping to create new experiments. And that is the old "Copenhagen" interpretation. While it is easy to find all sorts of philosophical problems with this interpretation, so that one is very reluctant to call it "the best," nonetheless it has certain features that correspond to the physics of the situation in a very natural way.

Hilbert space is a rather incredible place, and it tends to answer the questions you ask of it according to the symmetry of the problem. So if you are doing atomic physics in a many-particle system, all the orbital angular momenta add up to a total L, and all the spin angular momenta add up to a total S. The individual particles don't mean anything. If you have two entangled particles in a singlet state, the individual particles don't mean anything. Only the oppositely correlated spins. In any given situation, the system is arranged according to the symmetries that are present, and other realities get sidelined. So sometimes the individual particles are important, and sometimes the total spin, or total J, the total angular momentum. That is part of what makes quantum theory so un-classical, but it corresponds to what you actually measure. And all this is very natural in the Copenhagen interpretation.

Sometimes even space and time are unimportant, which separates it from the Bohm interpretation, where space and time are paramount. All of this is accomplished by what is known as transformation theory, which says that you can subject the system to a unitary transformation, which rearranges the vectors in Hilbert space (the states of the system) so that they are referred to a new set of orthogonal axes along which measurements can be made. And these arrangements respect only the symmetry of the problem, and not the classical features, such as independent particles. It imposes its own structure upon the problem, which you then have to be sensitive to. It's a beautiful feature of the Copenhagen interpretation, and it directly incorporates the superposition principle, the essence of the subject. So there is something experimentally "just right" about the Copenhagen interpretation, which is why these other interpretations have not replaced it and are not about to, in spite of the hype that some of the proponents give out.

LUCIEN HARDY · I do not believe any of the currently available interpretive programs. But I think they play an important role in helping us to understand quantum theory and, hopefully, move beyond to a theory of quantum gravity. From this point of view, it is worth evaluating the main approaches on their merits. A crucial question to ask of these interpretations is *to what extent will those features of the quantum formalism upon which the given interpretation depends survive in a theory of quantum gravity.* If quantum gravity is as different from quantum theory as general relativity is from Newtonian mechanics, then there is a danger that interpreting quantum theory will amount to little more than an academic exercise. We may have a good (even the best) interpretation of the quantum formalism, but this may tell us little about the true nature of reality if quantum theory is not the correct fundamental theory. Of course, such a danger exists at any stage in the development of physics, unless we really have the final theory. But given the problems with quantum theory and the unresolved nature of the problem of quantum gravity, there is particular reason to be cautious at this point in history.

The pilot-wave model of de Broglie and Bohm solves the measurement problem in arguably the simplest possible way. Real particles are guided in their motions by an equation involving the wave function. A single outcome is picked out by these particles. The model can be regarded as a resource for identifying candidates for possible conceptual properties of quantum theory in general (such as nonlocality). Further, it acts as a counterexample to the various misconceived claims that hidden-variables theories are impossible. Anybody who has not studied this approach cannot really say they have seriously attempted to understand quantum theory. It suffers, however, from two serious problems. First, the essential part of the pilot-wave model (the guidance equation for the trajectories) appears to be a rather ad hoc bolt-on to standard quantum theory (the Schrödinger equation). If it could be shown that the guidance equation *and* the Schrödinger equation both followed from some deeper considerations, then the case for the pilot-wave model would be much stronger. The second problem is that it is subject to the "many worlds in denial" attack (due to Harvey Brown and David Wallace). The wave function itself has to be regarded as a real existing thing in this model. This wave function contains a superposition of many terms, each of which contain the structure of a possible world evolving as the wave function evolves. Since the wave function is a real thing itself, it is difficult to resist the claim that each of these worlds must themselves be real and existing. From this point of view, the de Broglie–Bohm trajectory in configuration space that singles out a particular world is largely irrelevant—this world along with all the others exist. One way to counter this attack is to show that the wave function is a nonfundamental object. The idea that there are actual trajectories is not particularly dependent on the quantum formalism and could, in principle, be true in a theory of quantum gravity also. The real test, then, for the pilot-wave approach is whether, by thinking about trajectories and such, we could construct a theory of quantum gravity in a way that is not ad hoc and sees the wave function emerging only as a nonfundamental object.

The many-worlds interpretation asserts that reality is fully described by the wave function and that the latter evolves according to the Schrödinger equation. This innocent starting point leads to the rather crazy idea of a branching structure in which every term of the wave function (in an appropriate basis) corresponds to a different world. It has been argued that this interpretation follows from the equations as surely as it follows from the equations of classical physics that the earth moves. There remains the technical problem of showing that we can account for the usefulness of the notion of probability in a theory that is completely deterministic. In spite of impressive recent efforts by David Deutsch, David Wallace, and others, I do not believe this problem has been solved. If we assert, however, that the wave function corresponds directly to a real existing thing and it evolves unitarily, then it is hard to avoid the conclusion that there are many worlds (as has been argued very effectively by Simon Saunders and David Wallace), whether or not we are able to address the problem of accounting for probability. Nevertheless, I think this assertion expresses just a little too much faith in the mathematical formalism. First, it is not completely clear we should assert that the wave function is a real existing thing (see my answer to Question 4, page 99). Second, and even more significantly, the idea of a wave function across space evolving unitarily in time is, I believe, particularly susceptible to the strains put on the formalism by the need to incorporate general relativity. A theory of quantum gravity is unlikely to have a notion of evolution in time at a fundamental level, because we would expect causal structure to be indefinite in such a theory. Further, even at an effective level, unitary evolution is likely to break down when we have macroscopic superpositions involved (as has been argued for by Lajos Diósi and Roger Penrose). Hence, the very features of the quantum formalism on which the craziness of the many-worlds interpretation hinges are unlikely to survive in the next theory. We should not, then, lose too much sleep worrying about how we and our loved ones are faring in other worlds. I am confident there is only one world—this one.

A third line of approaches to interpreting quantum theory are the collapse models developed by Philip Pearle, GianCarlo Ghirardi, and others. In these approaches the Schrödinger equation is modified such that for small systems (such as a few electrons), the wave function evolves unitarily with high probability, while for big systems (such as measurement apparatuses) the wave function undergoes collapse to a particular outcome. My opinion is that this is, technically, the most correct of the approaches mentioned so far. I expect that a collapse model of some sort will be the effective theory that results from taking the limit of quantum gravity in the direction of applicability of quantum theory. We are likely to lose, however, a lot of structure in taking this limit, and so a collapse model provides a bad starting point for any attempt to construct a theory of quantum gravity from fundamental principles. Similarly, I expect that collapse is not the correct resolution to the measurement problem, because this has to be resolved at the level of the fundamental theory rather than any effective theory.

These three approaches are the main straightforward realist interpretations of quantum theory. There has always been another strand of interpretations, originating with the Copenhagen approach of Bohr. The most interesting modern-day descendant of that approach is the Quantum Bayesian interpretation promoted by Christopher Fuchs and his collaborators. Here, quantum mechanics is regarded as a theory for determining the optimal betting strategies of agents. By giving agents a central role, it adopts the much-debated split of the world into observer and observed. The agent, along with his apparatuses, makes measurements on the quantum world. The approach takes a Bayesian interpretation of probabilities—they are subjective degrees of belief that are updated when new data is collected. Probability is a deeply troublesome notion from a philosophical point of view (see my answer to Question 6, page 135). By giving the interpretation of probability a central role, the Quantum Bayesians are addressing this important issue head-on. In the Quantum Bayesian interpretation, the quantum state is simply a list of subjective probabilities. The quantum state therefore exists in the mind of the agent, rather than being a real existing thing in the world. Under this shift in understanding of what the quantum state is, the Quantum Bayesians can make a claim similar to that of the many-worlds believers to have adopted the correct interpretation of the equations of the formalism—quantum theory is simply a calculus for saying how to update subjective probabilistic beliefs of agents about the world. To be an interpretation of quantum theory in the standard sense, however, it would also have to say something about ontology. In response to this, Fuchs has proposed identifying commonalities that systems can possess that are independent of the subjective probabilities we have for them. Hilbert-space dimension is identified as one such commonality. This is promising, but it is not enough. We need to identify enough properties of the ontology to do the work of accounting for appearances (tables, chairs, and so on). Asserting the reality of Hilbert-space dimension alone does not do this. The approach is currently regarded, however, as an unfinished project. It is possible that it will eventually lead to a sufficient ontological picture. One further point in its favor is that the approach is not dependent on the particular formalism of quantum theory. It may, then, be feasible to apply this approach to help in the construction of a theory of quantum gravity.

Aside from those courageous people actively involved in developing one or the other of these interpretations, I think the best attitude for the rest of us is to be opportunistic with respect to the possible insights they can offer. The pilot-wave interpretation led John Bell to come up with his famous theorem showing that quantum theory is nonlocal. The many-worlds interpretation inspired David Deutsch to do his pioneering work on quantum computation.

ANTHONY LEGGETT · If, for the sake of argument, we assume that quantum mechanics is indeed going to continue to be the "whole truth" about the physical world for the indefinite future, and if, moreover, we exclude the consideration mentioned in the second paragraph of my response to Question 2 (see page 51), then I believe that the only interpretive program (if one can call it that) that makes any sense is what I call the "extreme statistical" approach. By this I mean the approach outlined, as regards all its essential features, by Leslie Ballentine in his well-known 1970 article in *Reviews of Modern Physics*. Here, the whole formalism of quantum mechanics is, in effect, *nothing but a recipe*—to be constructed and, if necessary, updated in the light of one's current knowledge—for the prediction of the probabilities of various alternative *directly observed* macroscopic outcomes; to seek any further "meaning" in the formalism is pointless and can only generate pseudoquestions. (This formulation might be regarded as a natural extension of Bohr's version of the Copenhagen "interpretation"—or, better, noninterpretation—in the light of recent experiments on SQUIDs and the like.) If one takes this view to its logical conclusion, then any further elaborations of it, such as "Ithaca," consistent histories, and (neo-) Bohmian mechanics, are little more than verbal window dressing—or, to put it more charitably, a matter of personal taste without any observable consequences.

TIM MAUDLIN · The question is better stated: which theories, based on the quantum formalism, can be used to make sense of the world? A very generic characterization of our naive, immediate experience of the world is as of a collection of macroscopic objects disposed in space and time. Bohr insisted that our description of experimental situations and the outcomes of experiments be couched in a language of macroscopic objects in space and time, and ultimately the data we use to test our theories is presented in such a form. So the obvious way for a physical theory to make contact with the observed world is to postulate some macroscopic objects in space and time.

But "macroscopic" is a vague term: it ought not to appear in the foundations of a theory. So a straightforwardly comprehensible physical theory ought to postulate some objects in space and time (Bell's "local beables"), and provide a clear account of what determines their motions. "Macroscopic" objects will then just be large collections of these fundamental local entities.

The basic innovation of quantum theory is the postulation of a "wave function" or "quantum state" that plays some role in determining or constraining how localized objects behave. So a comprehensible quantum theory should be clear about the wave function: what it is, how it evolves (if it does), and the role it plays in influencing the local beables. A theory that postulates this sort of structure can make clear sense of the physical world. And any theory that denies the existence of local beables, or of a wave function, or of any influence of the wave function on the local beables, must surmount many substantial obstacles just to be understood.

There are several interpretive programs that meet the criteria: Bohmian mechanics and the GRW spontaneous-collapse program (with a clear account of the local beables) are the most highly developed candidates. The many-worlds tradition has yet to precisely articulate an account of local beables. And, of course, the absence of what we think of as unique outcomes to experiments raises a host of conceptual puzzles. The main problem at present is not to pick the best of the approaches, but to appreciate those approaches that present a clear physical account at all.

DAVID MERMIN · My sympathies are with those, going all the way back to Heisenberg and Peierls, who maintain that quantum mechanics is a set of rules for organizing our knowledge with a view to improving our ability to anticipate subsequently acquired knowledge. By "our knowledge," I mean my own knowledge combined with whatever other people are able to communicate to me of their own knowledge. I take this commonality of scientific knowledge to be one of the reasons why Bohr placed such emphasis on what can be expressed in ordinary language.

To John Bell's "Knowledge about what?" I would say knowledge about our perceptions—ultimately our direct, irreducible mental perceptions, which can, of course, be refined by the use of instruments devised for that purpose. To his "Whose knowledge?" I would say knowledge of whoever is making use of quantum mechanics. Different users with different perceptions may well assign different quantum states to the same physical system. What consistency requirements, if any, can be imposed on such descriptions, is an entertaining question. I have had some disagreements with some of my friends about this, as described in "Compatibility of state assignments," which I cite here because it cannot be found in the primary repository, arXiv, but only in the *Journal of Mathematical Physics* (volume 43, page 4560, 2002).

My answer to "Why?" has to be inferred from my answers to most of the other sixteen questions.

LEE SMOLIN · I believe that quantum mechanics must be the statistical mechanics of a more fundamental theory. That theory must be fundamentally nonlocal; indeed, it seems likely that it describes a world in which space has not yet manifested itself. In such a theory, space and locality would emerge from a more fundamental description, and the nonlocality in quantum mechanics would arise from statistical fluctuations of the fundamental nonlocal degrees of freedom.

I thus believe that the hidden variables that Einstein called for are relational: they do not describe a more detailed description of an individual elementary particle, but rather they describe a more detailed description of the network of relationships that space emerges from.

I have developed these hypotheses in a number of papers between the early 1980s and the present. I have been able to realize different versions of the idea, but I am not satisfied with the results, for two reasons. One is that a fine-tuning of the parameters of the theory is necessary to get quantum mechanics to emerge. This turns out to be connected with a need to impose a symmetry relating the theory to its time reverse. The second is that no predictions have come to light that could be tested, nor have any implications emerged for other issues in physics.

ANTONY VALENTINI · In my view, the de Broglie–Bohm interpretation—or pilot-wave theory, as de Broglie originally called it—presents several deeply attractive features, in addition to the obvious merits of being realistic and deterministic.

First and foremost, as I said in my answer to Question 1 (see page 33), before I studied de Broglie–Bohm theory properly, I was very puzzled by why we can't use quantum nonlocality for signaling. It was as if there was some sort of conspiracy at work in the laws of physics. To explain this, I had drawn the conclusion that we were stuck in some sort of statistical equilibrium state whereby uncertainty noise happens to mask the underlying nonlocal effects (a kind of equilibrium "balancing" condition, as I had vaguely thought of it). When I studied pilot-wave theory and saw that it was a consistent theory, I was amazed to see that it provided a precise realization of what I had been looking for: it was a nonlocal theory for which the nonlocality was washed out, or averaged to zero, in the state of "quantum equilibrium"—the state in which the hidden configurations have a Born-rule distribution. Other people working on the theory usually took the Born probability rule as an axiom, alongside the equations of motion, but to me it seemed obvious that the theory should be considered for arbitrary distributions. I was able to show that such distributions give rise to nonlocal signals at the statistical level. I also proved an analogue of the classical coarse-graining H-theorem, which gave a general understanding of how evolution toward equilibrium occurs for an isolated system, as has since been confirmed by numerical simulations. It seemed natural to me to suppose that relaxation to equilibrium had taken place in the remote past, presumably soon after the big bang. The nonlocality associated with early nonequilibrium might then explain why the early universe was so homogeneous. More importantly, the puzzle of why we can't use quantum nonlocality for signaling could be given a simple answer: it's a peculiarity of the fact that we happen to be stuck in an equilibrium state. There's no conspiracy in the laws of physics; we are simply trapped in a special state with special properties. Both signal-locality and the uncertainty principle could be shown to be contingencies of equilibrium.

This viewpoint opens up the possibility of a new and wider "nonequilibrium" physics, in which superluminal signaling is possible and the uncertainty principle can be circumvented. Relativity, too, is violated in this new physics, which contains a notion of absolute simultaneity associated with a preferred state of rest. Nonequilibrium particles that violate the Born rule could exist today, perhaps in the form of relic particles from the very early universe—particles that decoupled before they had time to relax completely to equilibrium. I've also speculated that nonequilibrium particles might be generated by evaporating black holes, on the grounds that their states could then carry more information and possibly avoid the Hawking information loss.

I find pilot-wave theory attractive in another respect. In de Broglie's original formulation, it is a radically non-Newtonian theory, with the extraordinary feature of a wave in configuration space, which determines the *velocities* of systems instead of their accelerations. It's an Aristotelian dynamics, with a natural Aristotelian kinematics. Nonlocality in ordinary space is explained as an effect of the configuration-space dynamics. This seems to me much more like the radical conceptual shift I had expected when I first studied quantum theory (see my answer to Question 1, page 33). We throw out classical forces and classical dynamics, introduce a new and radical entity (the pilot wave in configuration space), and construct a new theory of motion. That's the sort of thing that Faraday and Maxwell did with the electromagnetic field in the nineteenth century, and that Einstein did with gravity in 1915. And that's what de Broglie did in the 1920s.

In 1923 de Broglie concluded that when a particle is diffracted by an obstacle without touching it, the nonrectilinear motion violates Newton's first law. De Broglie proposed a new form of dynamics based on velocities. Particle motions were guided by waves, in a way that unified the variational principles of Maupertuis and Fermat. But de Broglie's achievement went unnoticed. He is remembered chiefly for the relation between momentum and wavelength, but that was merely a by-product of his new dynamics. Even those who work on his theory often fail to credit de Broglie—who in 1927 had the full many-body dynamics in configuration space, not just the one-body theory as is often claimed.

Unfortunately, in 1952, Bohm presented the theory in a pseudo-Newtonian form, based on acceleration and the quantum potential, which made it look much more like classical physics than it really was. Bohm's important contribution was to show how the theory accounts for the general quantum theory of measurement. But I never took Bohm's version of the dynamics seriously. It looks artificial, like writing classical general relativity in terms of flat space-time with a tensor field that distorts rods and clocks. It can be done, but it's not a natural language to use. De Broglie's original dynamics, which Bell used and advertised, seems much more appropriate. In recent work with Samuel Colin and Ward Struyve, we have shown that Bohm's dynamics is actually unstable, in the sense that nonstandard momentum distributions (which are allowed in Bohm's dynamics but not in de Broglie's) do not relax to quantum equilibrium. So my preference for de Broglie's dynamics is no longer merely a question of taste. I think Bohm's dynamics is actually untenable.

De Broglie's remarkable work in the 1920s remains mostly unknown, even among historians. In my view, in certain key respects, he understood the fundamental dynamics better than Bohm did. Matters are further confused by some who refer to de Broglie's dynamics as "Bohmian mechanics." For a proper understanding of the theory, it helps to know how and why de Broglie constructed it in the 1920s—instead of thinking of it anachronistically in terms of a "completion" of modern quantum theory. In my view, the theory is still widely misunderstood, even by some of its most fervent supporters, partly because de Broglie's original work is still being ignored. Associating the theory primarily with Bohm is not only wrong as regards credit, it also deprives us of de Broglie's insights.

While de Broglie's dynamics is attractive in itself, for me it has always been first and foremost a means to provide a concrete model of the idea I had about physicists being trapped in an equilibrium state that hides nonlocality. I thought from the outset that the essence of this idea would hold in *any* reasonable deterministic hidden-variables theory—as I eventually showed explicitly. Now, at present, we don't know what the true hidden-variables theory is. Pilot-wave theory might be right or approximately right; of course, it could also be quite wrong. I think it's a worthy guess, as it contains a number of features, such as nonlocality, that are known to be generally true for hidden-variables theories. But to find out what the correct theory is, we'll need an empirical window.

I tend to compare pilot-wave theory with the early models of the kinetic theory of gases, in which molecules were hard spheres. That was the simplest assumption to make, and it was a good strategy to develop the resulting theory as far as possible, until things like the explanation of Brownian motion gave us an empirical window onto the world of atoms. Similarly, I hope that developing pilot-wave theory to its logical conclusions will lead to an empirical window onto the world of hidden variables.

I wouldn't be willing to bet a huge sum that the details of pilot-wave dynamics are correct. I *would*, however, be willing to bet a considerable sum that there is a non-local hidden-variables theory behind quantum mechanics, and that the only reason we can't send superluminal signals today is because we're trapped in a state in which the hidden variables have an equilibrium distribution. Locality and the uncertainty principle are not laws, they are merely peculiarities of equilibrium. Quantum theory is a special case of a much wider, nonequilibrium physics, in which nonlocal signaling is possible and the uncertainty principle can be beaten. I think this is likely to be true. And it's a good scientific rule of thumb to say that if the laws of physics permit something to happen, then it will happen somewhere. So I expect that nonequilibrium violations of quantum theory will eventually be found. When they are found, and we are able to see our way through the fog of quantum noise, what will we find? Will we see trajectories obeying pilot-wave dynamics? Maybe, maybe not. But we will, I think, see a nonlocal world radically different from the world we're familiar with. And we will realize how misled we've been all this time, wrongly thinking that the Born rule and its associated features are fundamental when they are not.

DAVID WALLACE · What interpretation? Everett's "many-worlds" interpretation.

Why? Here's the short version. Normally, we don't get worried about "interpreting" physical theories—we don't really need to interpret general relativity, or classical electromagnetism. We just take the theory as representing (part of) the structure of the world, so that states of the theory correspond to states of the world according to the theory. In quantum mechanics, things like Schrödinger's cat made us think that couldn't be the case—what kind of a state of the world is it in which a cat is a superposition of alive and dead? So we thought we had to give up on the usual story and find some alternative, clever way to think of the theory, or else change it into one which didn't have the same problem. What Everett did was tell us what kind of state of the world it is: it's a state of the world in which there are two cats, and one's alive and the other's dead. (Or really, there are two lots of cats, and one lot are alive and the other lot are dead.) And given the way quantum entanglement works, that pretty quickly means there are two (lots of) copies of the solar system, one with a live cat and one dead cat. Two worlds, in other words—at least locally.

What's the advantage of the Everett interpretation in particular? Here's one way to put it. In trying to interpret quantum mechanics, you've got two yes/no choices to make. Choice one: are you going to change the physics? Are you going to stick with the Schrödinger equation and the quantum state, or are you going to add dynamical-collapse processes or hidden variables or backward-in-time interactions or something? Choice two: are you going to change the philosophy? Are you going to stick with the straightforward way of reading a scientific theory as just telling us what the world is like, or are you going to start saying "a scientific theory is just a predictive algorithm for experiments" or "observers can't just be modeled as physical systems" or "ordinary logic is wrong" or something? If you answer no to both questions, you're stuck with the Everett interpretation, because the Everett interpretation is just the "take quantum mechanics completely literally" interpretation.

Now, for each choice, there are a lot of people who have said yes to each choice. And both those answers have led to interesting insights. But, basically, the task they're setting themselves is pretty challenging. Answering yes to the first question essentially commits you to redoing the last seventy-five years of developments in quantum theory—not just coming up with alternatives to nonrelativistic quantum mechanics, but coming up with alternatives to QED, to the Standard Model, to neutrino-mass developments of it, and so on. Answering yes to the second question essentially commits you to overturning a really pretty solid consensus in philosophy of science that scientific theories really do have to be understood as making claims about what the world is like, and aren't just shorthands for claims about how experimental devices work. (And it's a consensus that I think pretty much all scientists share when they're not actively philosophizing. Are there really astrophysicists who think that the reason for talking about stars is to model patterns of detections on photoplates, not vice versa?)

But here's the crucial point. You can make those yes/no choices for any scientific theory you care to name. You can do it for *paleontology* if you want to! Spend time coming up with alternative theories for fossil formation, or decide that dinosaurs are just theoretical constructs used in theorizing about fossils. It's a free country. But the only motivation for answering yes to either question in the particular case of quantum mechanics and not in general is that you think there's some special problem with the no/no answer in quantum mechanics: that is, you think the "take quantum mechanics completely literally" interpretation—the Everett interpretation—*doesn't make sense*.

Okay, so does it make sense? Well, the main worries people have raised are: what justifies the "many worlds" description? And what about probability? I don't have space here to do more than comment briefly, but I think these are both resolvable. The "many worlds" language follows from decoherence theory—from the various processes that dynamically suppress quantum interference—once we realize that they're not supposed to be part of the fundamental ontology of the theory, but just something approximate, something emergent. And the probability issue turns out to be a strength, not a weakness, because (1) once you start thinking hard about probability *in Everettian quantum mechanics*, you realize probability is philosophically really mysterious *in general*, but (2) it turns out that there are ways of understanding probability in quantum mechanics that don't work in classical mechanics. (I'm thinking of the so-called decision-theoretic approach to quantum probability that goes back to David Deutsch, though really it's more about symmetries of the quantum state, and the decision-theoretic gloss is just there to operationalize the concept of probability and dodge some philosophical worries.) Actually, probability is just one of several places where what looks like an Everett-specific philosophical problem turns out to be an old problem in a new guise.

Philosophers who come across the Everett interpretation tend to get worried about whether it makes sense, but physicists are more likely to ask, "What's the point of it," or "Can it be tested." The answer to both questions is basically that there isn't any point using the Everett interpretation instead of ordinary quantum mechanics, and that you can't test the Everett interpretation against ordinary quantum mechanics, but that's because really, the Everett interpretation just *is* ordinary quantum mechanics—maybe not the quantum mechanics of the undergrad textbooks, which explicitly invoke wave-function collapse, but the quantum mechanics we mostly use in practice, where we model measurement processes physically and apply nonunitary evolution only because we're tracing out—that is, deciding to neglect—some environmental degrees of freedom. Viewed from this angle, the point of the Everett interpretation is to allow us to do quantum theory without either taking measurement as some kind of primitive or having to change the formalism. And the right way to test the Everett interpretation is to test the universality of the superposition principle and the unitary dynamics.

ANTON ZEILINGER · To me, the Copenhagen interpretation, and its new developments based on the notion of information, are the best. The reason is that Copenhagen is the most austere interpretation; it doesn't assume anything unnecessary. I feel that in all the other interpretations, one tacitly makes assumptions akin to preserving some kind of prequantum realism.

WOJCIECH ZUREK · I think the relative-state view of Everett (and Wheeler!), in the form in which it does not prejudice the interpretation by making it into "many worlds," is the best *framework* for interpretation. It is the most flexible (and most quantum!) way to think about quantum theory and our universe. The relative-state approach also has the virtue of making it clear why it is not the last word on the subject. This is important, because it is crucial to expose open questions, rather than to cover them up with superficial answers. The relative-state view makes it clear that

Bohr's concern about Everett's program is still valid. This concern focused on what defines observables when everything is quantum—for instance, the question of what makes a certain direction in space "z" when there is no classical Stern–Gerlach apparatus with its magnetic field.

It should be clear from the above that my vote for relative states is not a vote for a final solution. Rather, it is based on the fact that the stripped-down (pre-many-worlds) nature of relative states makes it obvious that the emperor has no clothes: that quantum theory has no interpretation. And this helps. For example, the relative-states point of view made it easier to develop the theory of decoherence.

Decoherence explains how the classical realm can emerge from within the quantum universe—something that seemed so puzzling three decades ago. Decoherence therefore constitutes very significant progress (although it is *not* an interpretation per se). It can fit within both Bohr's and Everett's points of view. Perhaps it is even a bridge between them. A natural consequence of decoherence is environment-induced superselection, or "einselection" for short. It defines what states persist in spite of decoherence. The so-called pointer states, which can survive decoherence intact, are then the obvious candidates for classical states. This suggests an "existential interpretation" that builds a bridge between the malleable quantum world and the more resilient classical world of our everyday experience. But these results follow directly from quantum theory and can fit within either Bohr's point of view (where they delineate the quantum–classical border) or Everett's relative states (where they constrain "relativity" and help define branches of the universal state vector).

I believe the way decoherence and its consequences—pointer states, einselection, and the like—are being investigated is a model for how to make progress on the issue of interpretation: take quantum theory in an interpretation-free form, and use it—its formalism—as a guide.

It is also important to be realistic about one's expectations. Some physicists seem to expect that a *single* good idea should solve any big problem. So when a new idea leads to significant progress but does not settle every question to everyone's satisfaction, it tends to be dismissed.

I think decoherence was a good idea. It has shed new light on problems that were over half a century old. It has led to new experiments. It does explain why we do not see flagrant manifestations of the quantum underpinnings of our universe in our everyday life. But it rests on assumptions that—like Born's rule for probabilities—deserve to be reexamined, as they date back to Bohr's view of the universe and do not fit a universe that is quantum to the core.

To sum up, the only interpretational program I would wholeheartedly advocate is to take the quantum postulates of quantum theory very seriously and see where they lead. There are still good questions to be studied. Exploring the dual "epiontic" nature of quantum states (which tell us about what we know, but also help define what is) is one of them. And any pressure to come up with an interpretation that attempts to bypass all of the hard work essential to answering such intermediate questions is at best counterproductive. Haste makes waste.

QUESTION 4

QUANTUM STATES

What are quantum states?

Once upon a time, there was a quantum state. He considered himself someone who was leading a purposeful life, because he knew that physicists around the world relied on him for doing their fancy calculations and experiments. And he helped them get along as best as he could, and this was something he did indeed extremely well, and he was proud of his role. Yet in his more introspective and philosophical moments, he couldn't quite shake the feeling that he had been languishing in a sort of perennial identity crisis.

"What am I, deep down?" he kept asking himself. "Am I really nothing but an abstract mathematical entity, a tool that people can just pick up to get their daily business done, without having to waste a thought about why I work so well? Don't I stand for something more? But what? Have I been underestimated? Overestimated? Misunderstood?"

Knowing that he wouldn't find answers to his questions in the company of practical-minded scientists, he went to see some renowned specialists in the foundations of quantum mechanics to learn more about his true self. After all, those foundationalists were the people who had cared about and debated his identity ever since quantum theory had been brought into existence.

When he was done making his round of calls, he came away bewildered. Everyone had told him something different! He could be so many things! One foundationalist had suggested that he, the quantum state, should think of himself as someone much larger and more important, perhaps even as the stuff the universe was made of. Someone else had agreed that the quantum state was a physical being all right, but that his duty was to push a bunch of particles around. The next few people had rather crushed his ego, but their diagnosis electrified him at the same time. They had claimed that he wasn't a thing of the physical world at all, but that instead he represented information, knowledge, or beliefs. And they had said that this recognition would liberate him from the burdens others had placed on his shoulders, and that he would no longer be haunted by all the foundational dilemmas in which he had found himself caught up for as long as he could remember.

The quantum state remembered that Barack Obama had once described himself as a kind of Rorschach test, and he started to feel like a Rorschach test himself.

People around him seemed to see in him what they chose to see, using him as a representative and vehicle of their very own personal ideas and dreams, as if he was nothing but a blank screen for everyone's projections. And somehow he managed to accommodate all these disparate visions without stumbling.

What makes the interpretation of quantum states such a recurrent theme of foundational discussions? How can quantum states be such chameleons of meaning? There are many aspects to this matter. The first issue is simply one of abstraction. In classical physics, what we call "the state of a system" is merely a shorthand, a catalogue of numerical values of physical quantities whose meaning is crystal-clear: position x equals such and such value, velocity v equals such and such value, and so on. And thus the correspondence between physical properties and their formal representation is plain, direct, and intuitive.

Now enter the quantum age, and this correspondence crumbles before our eyes. How are we expected to find a reflection of our world in something as radically abstract as a normalized vector in a high-dimensional Hilbert space? Not directly anyway, the standard mantra of quantum theory tells us. Instead, we ought to bring in measurement to make the link between formalism and physical reality. The measured physical quantity is formally represented by a Hermitian operator (or, more generally, by a positive operator-valued measure, though this generalization makes no conceptual difference here). After the measurement, the quantum state becomes an eigenstate of this operator, and the corresponding eigenvalue represents the value of the measured physical quantity now possessed by the system. Voilà!

If we could just keep going in this manner, eventually we'd be able to build up a catalogue of actually possessed values of physical quantities, not too different from the situation in classical physics. But, of course, we can't: since measurements, in general, alter the quantum state, we'll lose our previously measured values along the way. Quantum states are utterly fragile. For every bit of information gained, an inevitable disturbance will be introduced. Gaining a maximum amount of information about the system is much more subtle business in quantum mechanics than in classical physics. This is another aspect of the idiosyncratic nature of quantum states (and of quantum measurement, for that matter), and it has some intriguing consequences.

For example, let me put a system in front of you. Can you determine its state for me? In classical physics, you'd get to work right away, measuring a set of properties that would appear appropriate to the nature of the object. Your friends may even offer their help and redo some of these measurements, just to confirm you didn't mess up. These measurements won't interfere with each other, and at the end of the day, you and your friends will agree on the results (provided, of course, no one did a slipshod job). Given that our senses have delivered such experiences since childhood, it's no wonder that the image of a preexisting, observer-independent physical reality exerts such a powerful hold on our intuition.

In quantum theory, however, your task will be doomed: whatever you do, you're almost guaranteed to change the state with your first measurement, and you will have learned little if anything about the state of the system before the measurement.

The no-cloning principle—the impossibility of duplicating an unknown quantum state—is yet another reflection of this fact.

There are further aspects to the peculiar nature of quantum states that get foundationalists wound up. I said that the degree of abstraction inherent in the quantum-state description tends to induce a general sense of disconnect between formalism and reality. But there's something else that contributes to this feeling. When parts of the quantum state appear to reflect structures of our world, then the state as a whole is often a superposition of these parts and would therefore seem to represent a grotesque caricature of reality, as made vivid by the example of Schrödinger's cat. To add insult to injury, the dynamical laws of quantum theory imply that quantum states should morph into such creatures everywhere and all the time, with quantum systems becoming rapidly entrapped in an ever-expanding web of entanglement.

And here's yet another issue. Quantum states are supposed to be all there is to say about the physical state of the system. But even if we know the system's quantum state, we're fundamentally unable to predict the result of almost any measurement we may care to perform next. Does this mean that nature is intrinsically random (see Question 5, *Randomness*)? Does this imply that quantum states are "incomplete"—that quantum mechanics cannot be the whole story, and that we must look for another layer beneath it (see Question 8, *Bell's Inequalities*)? As with many foundational matters, it all depends on who you ask.

The abstraction of quantum states, their tenuous link with the physical world, their curious properties especially under measurement, their occasionally perceived incompleteness: it should come as no surprise that quantum states have been a focus of foundational inquiries since day one. Thus, in many ways, the problem of interpreting quantum mechanics boils down to (or, at least, starts from) the problem of interpreting quantum states. As a rule of thumb, your view of what quantum states stand for will be a telltale sign of your interpretive preferences. It may not uniquely single out a particular interpretation of quantum mechanics, but with just a couple of extra constraints it will.

Politics shows us how handy it is to pigeonhole the great variety of views into two main contrasting movements. In the case of quantum states, these are the ontological ("ontic") and epistemic camps. The ontic camp reads quantum states in a scientific-realist way, as real physical properties of systems and as objective things of the world. (Incidentally, some of Schrödinger's first attempts at interpreting quantum theory were along those lines.) But once you join this camp, you quickly run up against the problem of Schrödinger's cat, because a cat state would represent some form of physical reality just as any other quantum state. And how you choose to deal with this problem will shuttle you toward a certain interpretation of quantum mechanics. If you think that quantum mechanics shouldn't be burdened with additions or modifications, and if the idea of residing in a giant-Schrödinger-cat-like universe doesn't scare you off (knowing that decoherence will arguably ensure that you wouldn't necessarily notice anything particularly strange), you may be drawn to Everett's purism about wave functions. If you can't live with the image of yourself holding on to merely one branch in a gargantuan tree of parallel worlds, you'll probably be busy tickling quantum theory to see if it lets you sneak in a collapse mechanism. And if hidden variables are no anathema to you, then perhaps you'll go for one or another flavor of Bohmian mechanics and reinterpret the wave function as a physically real guiding field for particles. The interview responses below amply illustrate how an ontic interpretation of quantum states can sustain rather different interpretive attitudes toward quantum theory as a whole.

Let's turn to the epistemic approach. Its basic idea is as simple as it is appealing. It is to let go of the notion that a quantum state is a physical entity, and to view it instead as a representation of what an observer knows or believes about the system in front of him. Collapse of the wave function? Not to worry, it's just updating of knowledge. Schrödinger's cat, Wigner's friend? Merely an illustration of different levels of knowledge. And so on.

But if the matter was this straightforward, everyone would have signed up long ago for membership in the epistemic club. Let's have the scenario of Wigner's friend illustrate the catch. Wigner's friend, inside the laboratory, carries out a measurement on a quantum system and assigns to the system a pure state corresponding to the outcome she's seen. Wigner himself, waiting outside, treats his friend as just another quantum system and describes the measurement interaction in terms of a von Neumann measurement, which means that the final state he will assign to friend-plus-system will be an entangled one. This state assignment is not simply grounded in Wigner's ignorance of an objective measurement outcome inside the lab, but it reflects the possibility that Wigner could, in principle, apply a suitable unitary countertransformation to friend-plus-system (treated, for the purpose of this gedankenexperiment, as an isolated system) and thereby undo the entire measurement process that his friend has supposedly carried out.

Now, whose quantum-state assignment is correct? Wigner's, or his friend's? Who "knows more," who has the "correct" information? No answer seems possible. The crucial question, then, seems to boil down to this: to what extent may we regard a quantum state, epistemically interpreted, as a representation of knowledge or information about something objective out there in the world—as something that facts of the world can make right or wrong, better or worse, more or less complete? That is, how much residual objectivity, if any, can and should an epistemic interpretation grant the quantum state, without sacrificing the purity and power of the epistemic mode of thinking?

The cleanest and most forceful way of cutting through the Gordian knot tied by Wigner's friend is to answer "None!" (This is what the Quantum Bayesian interpretation does.) Because as long as we insist that the epistemic content of quantum states retains even a hint of objectivity, the quantum state is just a metalevel that mirrors what's happening in the world. This means that we're still left with the task of accounting for how and when those events can be said to have actually occurred. It means that there is, after all, a right and a wrong quantum state, and this suggests we're back to square one when it comes to the conundrum of Wigner's friend, and now some of us will probably argue that the whole talk of "knowledge" and "information" appears like mere semantic window dressing.

But if we completely give up on the objectivity of quantum states, where does this leave us? Aren't we in danger of losing all foothold on reality, of being unable to give a good reason for the empirical success of quantum theory? Or is this move the only consistent way to go about an epistemic viewpoint—and will it perhaps even open our eyes to new opportunities once we've committed ourselves to taking the plunge? The answers that epistemic interpretations give to these questions vary. Accordingly, these interpretations differ in how far they're willing to go: we can tell them apart by their degree of radicalism about the subjectivity of quantum states.

If this introduction has given the impression that quantum states merely fill our short lives with undue puzzlement and obstacles, then I should end with a disclaimer. This can't-do view, held by generations raised on Heisenberg's uncertainty principle, is now being displaced by the intoxicating realization that in a quantum world we can do so much *more* than in a classical setting. From exponentially faster algorithms to completely secure cryptography, quantum information theory has paved the way toward a new appreciation of quantum states, seeing them as enablers rather than obstructions. (See Question 9, *Quantum Information*, for more on this subject.) And so the future is bright for the quantum state: although his identity crisis might not be resolved any time soon, admiration for his subtle skills will only grow.

GUIDO BACCIAGALUPPI · The answer to this—and to many other foundational questions about quantum mechanics—depends on the overall approach one adopts toward the theory. I think it is most helpful to distinguish between four kinds of approaches: pilot-wave theories, spontaneous-collapse theories, Everett interpretations, and post-Copenhagen approaches that use concepts from quantum information. (I think the latter have not yet developed into a full-blown approach to interpreting quantum mechanics, but they provide a new perspective on the traditional questions.) Once one makes these distinctions, it becomes much easier to sketch the possible answers to the question of what quantum states are.

According to pilot-wave theories, quantum states are objectively real, namely, they are pilot waves (or, synonymously, guiding fields). These determine the evolution of the configurations of the system, be those particle configurations, field configurations, or what not. Strictly speaking, there is only one quantum state, namely, the quantum state of the universe. But one can also talk about quantum states of subsystems, particularly in those cases in which standard quantum mechanics talks about collapsed quantum states. These effective states are components of the full quantum state that happen to be the only ones responsible for guiding the configurations of the system, because different components have been separated in configuration space, and due to decoherence, they remain separated. Quantum states in pilot-wave theories take on also a second role, that of defining the (time-dependent) equilibrium distribution over the configurations, but that role is inherited from their dynamical role as pilot waves and is not fundamental; indeed, on most readings, quantum equilibrium is a merely contingent feature of the theories.

For collapse theories—whether the well-developed stochastic ones, or the putative nonlinear ones—as well as for Everett interpretations, quantum states are the stuff that makes up the universe. The main difference between the two approaches is that in collapse theories, the quantum state gives rise to a unique subjectively experienceable world (apart from the presence of tails that one needs to argue are irrelevant), while in Everett interpretations, subjectively experienceable worlds arise from suitable components of the quantum state (which are dynamically independent due to decoherence), so that unlike in the case of collapse theories, there is a one-to-many correspondence between the quantum state and such worlds. There are some interpretational variants within collapse theories and within Everett interpretations—in particular, as to whether to interpret the quantum state by adopting an identification of physical properties with subspaces of Hilbert space in the style of quantum logic, or through the associated (emergent?) three-dimensional mass density; and as to whether to interpret it in terms of global worlds or of local neurophysical correlates of consciousness. But I believe the main point is independent of these further differences.

Among researchers who look to quantum information for the interpretation of quantum mechanics, the working hypothesis is that quantum states are somehow bearers of information, although there appears to be considerable disagreement about what such information is about (or is). The approach that I find most congenial—or, simply, the one I understand best—is that championed by Caves, Fuchs, and Schack, for whom quantum states are compendia of subjective degrees of belief. I think the approach still needs to acknowledge the issues related to Wigner's friend, but the application of subjective probabilities to quantum mechanics—very much in the sense of de Finetti—is a fascinating philosophical option, which these authors (I am glossing over some differences between them) are exploring very thoroughly and fruitfully.

Finally, I should add that there are theories, namely, Nelson's mechanics and related theories, that might be thought of as variants of pilot-wave theories without a fundamental pilot wave, but seeking to reconstruct the quantum state from more fundamental features of the dynamics. So far, this program has not succeeded, for subtle but important technical reasons (one can derive Madelung-style hydrodynamic equations for a pair of functions R and S, but these equations are not equivalent to the Schrödinger equation for $\psi = R e^{iS/\hbar}$, because of the specific multivaluedness of the complex function ψ). That may not be the last word on the subject, however, and apart from work within Nelson's mechanics, some recent work within quantum-information approaches on whether the quantum state is ontic or epistemic may shed further light on it. Incidentally, it seems to me that the early Heisenberg held qualitatively similar views (as expressed, for instance, in his joint report with Born at the 1927 Solvay conference), namely, that while transition probabilities in quantum mechanics were objectively real and could be efficiently calculated using Schrödinger's formalism of wave functions, the wave functions themselves were only an effective concept.

Č A S L AV   B R U K N E R   ·   The argument over the physical meaning of quantum states is typically pictured as one between two fundamentally opposed views: the ontic view and the epistemic view. The ontic view first assumes the existence of, and then refers to, the behavior of systems as they are, independent of any empirical access. The epistemic view refers to what we can know and infer from observations. The arguments against the ontic view have a long history. Already in 1927, at the fifth Solvay conference, Einstein pointed out that within the ontic view of the wave function (also called "ψ-ontic"), the "collapse of the wave function" in a measurement would imply an instantaneous propagation of the change in the value of the wave function everywhere. The problem is dissolved—that is to say, it does not appear in the first place—within the epistemic view. When the wave function has a nonzero value at some position in space at some particular time, it does not mean that the system is physically present at that point, but only that our knowledge allows the system the possibility to be found at that point at that instant if measured. Upon measurement, our knowledge is updated, and consequently its representation in form of the wave function instantaneously changes all its components, including those that describe our knowledge in regions of space quite distant from the site of the measurement.

We should be careful, however, about adopting the epistemic view too fast. The epistemic view should be differentiated. This differentiation has to do with the question "Knowledge about what?" Insofar as the quantum state is understood as the observer's incomplete knowledge about a presupposed ontic state in which the system is at any given time, the epistemic view (also called "ψ-epistemic") falls into the category of those views claiming observation-independent reality and is essentially not different from the ontic view (I find the name "crypto-ontic" more appropriate). It is then not surprising that the ψ-epistemic interpretation of the quantum state also leads to a failure of locality as demonstrated by Bell's theorem.

There is an alternative. In a "full-fledged" epistemic view, the quantum state can be understood as a mathematical representation of the observer's knowledge necessary to compute probabilities for outcomes of measurements following specified preparations. This operationalist's view also seems objectionable. Shimon Malin put it nicely: "What if the knower is a physicist who had a martini before trying to 'know'? What if a person who knows just a little physics learns of the result? What if he had a martini? Somehow we feel that such questions are irrelevant." To avoid difficulties of this kind regarding the epistemic interpretation, we can consider a quantum state as representing not the knowledge of actual but fictitious observers. Nobody would claim that quantum theory is not applicable when observers are not there. But when a quantum state is calculated, it is useful to think of a fictitious observer for whom the quantum state symbolizes her knowledge. Peres is right in stating that there is no reason to apologize for that. Fictitious observers are not restricted to quantum theory. They are also used in thermodynamics when we say that a perpetual-motion machine of the second kind cannot be built, or in the theory of special relativity when we say that no signal can be transferred faster than the speed of light. To conclude, the quantum state is a representation of knowledge necessary for a fictitious observer—respecting her experimental limitations—to compute probabilities of outcomes of all possible future experiments.

JEFFREY BUB · In a classical simplex theory, the extremal states are deterministic and can be interpreted as the "truthmakers" for propositions about the occurrence and nonoccurrence of events—that is, as representations of physical reality. A mixed state then represents ignorance about the deterministic state, about what is the case with respect to events. In a nonclassical theory with indeterministic extremal states, this interpretation is no longer appropriate and the state is simply a credence function, a bookkeeping device for keeping track of probabilities. See my answer to Question 6, page 130, for further amplification.

ARTHUR FINE · Pauli thought that using the word "state" (*Zustand*) in quantum mechanics was not a good idea, since it conveyed misleading expectations from classical dynamics. I think the word is just fine. A state is represented by a vector (OK, a ray) in the state space, and that vector provides a way of assigning probabilities to measurement outcomes for all the measurable quantities in the theory. Thus, it answers all the counterfactual "What if?" questions that seem to be physically realizable. In this sense, the state vector codes up maximum information about the system. Born's rule for probabilities tells us how to decode. A feature peculiar to quantum theory is that when systems interact, in general they lose their individual states in favor of a state for the combined system. Some are inclined to see deep metaphysical significance in this formal feature, equating loss of state vector with loss of individuality. I don't see why, since in the most typical cases we can easily define interactions with the component subsystems via operators defined only on the state space of that subsystem. The possibility of doing this is essential to standard ways of demonstrating no-signaling in the case of spatially separated entangled systems. Indeed, maintaining the integrity of the component subsystems is essential to understanding the very meaning of signaling from one subsystem to another.

Originally, Schrödinger thought that his states described a fuzzy bit of reality (something like charge density). But, in correspondence in 1935, Einstein persuaded him that this was not correct, by directing his attention to the case of an unstable pile of gunpowder. After a while, the gunpowder will either explode or not, but the state vector of the whole system at that time would be a superposition involving both exploded and not-exploded component terms. As Einstein wrote, "Through no art of interpretation can this be understood as a description of reality, since there is no intermediary between exploded and not-exploded." Einstein's gunpowder is the forerunner of Schrödinger's cat, whose description Schrödinger provided in the next round of their correspondence. These examples show cleanly (no issue of locality here) that if we want to think of the quantum state as describing some bit of reality, then—as in the case of the gunpowder or the cat—the quantum description is incomplete. Maximum information, quantum-style, leaves out quite a lot.

Standard moves to compensate for this information gap include versions of the Everett interpretation, hidden variables (or modal interpretations), stochastic collapse theories, and decoherence. Only the latter stays within the bounds of quantum theory as it is generally practiced and understood. The problem with decoherence, however, is that it is perspectival. We would like to have information as to whether the gunpowder has actually exploded (or not). Decoherence, at best, tells us what the gunpowder situation would look like from a (collective) point of view. That is not the same. For in reality, even if—let us say—it appears to be an explosion, the real quantum state remains superposed.

CHRISTOPHER FUCHS · I remember a conference banquet once in which a discussion arose over how quantum states should be classified linguistically: should they be nouns, verbs, or adjectives? I said that they're exclamations, sometimes even expletives! I still like that answer; maybe I should stop here. OK, I relent.

In my answer to Question 2 (see page 45), I cheated the jurisdiction of Question 4 by declaring already that quantum states are not real things from the Quantum Bayesian view. But what can that mean, and doesn't it contradict my answer to Question 3 (page 70) in any case? Aren't epistemic states real things? Well ... yes, in a way. They are as real as the people who hold them. But no one would consider a person to be a *property* of the quantum system he happens to be contemplating. And one shouldn't think of a quantum state in that way either—one shouldn't think of it as a property of the quantum system to which it is assigned. What I mean more particularly is that there is nothing external to the observer's or agent's history (intrinsic to the quantum system and its surroundings) to enforce the quantum state he *should assign* to it. For the QBist, a quantum state is of a cloth with *belief*—in the end, it is a personal judgment, a quantified degree of belief. A quantum state is a set of numbers an agent uses to guide the gambles he might take on the consequences of his potential interactions with a quantum system. It has no more substantiality than that.

This way of looking at quantum states is what comes about when one starts to think of quantum theory as a physically influenced *addition* to logic. Think first of formal logic: it is a set of criteria for testing the consistency between truth values of propositions. Logic itself, however, does not have the power to set truth values. It only says of any given set whether it is consistent or inconsistent; the actual values come from another source. In cases where logic reveals a set of truth values to be inconsistent, one must return to the original source, whatever it may be, to find a way to alleviate the discord. But which way to alleviate the discord—which truth values to change, which ones to leave the same—logic itself gives no guidance for.

The path back to quantum states from this starting point comes about from a personalist Bayesian take on probability theory. By this understanding, probability theory should be viewed as an extension of formal logic; it is the extended calculus decision-making agents ought to use when they hold uncertainties rather than truth values. The key idea is that like with logic, probability theory is a calculus of consistency—this time, however, for degrees of belief (quantified as statements of action or gambling commitments). Particularly, probability theory has the power to declare various degrees of belief as consistent with each other or not, but there its power stops: the particular beliefs it exercises its check on come from a source outside of probability theory itself.

What is the source? When it comes to formal logic, one is tempted to think of it as the facts of the world. The facts of the world set truth values. But it is not the world that is using the calculus of formal logic for any real-world problem (like the ones encountered by practicing physicists). The "source" is rather a finite subscriber to the service, one with limited abilities and resources; the source is always one of us—flesh and blood and fallible through and through—the kind of thing IBM Corporation is taking its first baby steps toward with its *Jeopardy!*-playing supercomputer Watson. The source of truth values in any application of logic are our *guesses*. Thus, it would be better to be completely honest with ourselves: applications of formal logic get their truth values from an *agent*, pencil and paper in hand, playing with logic tables not so differently than crossword puzzles. The facts of the world only later let the agent know whether his guesses were acceptable or unacceptable judgments.

The story remains the same, not one ounce different, with probability theory. Particular probability assignments have nothing on which to fall back but the very agent using the calculus—it is the agent's degrees of belief that he is checking for consistency. If they turn out to be inconsistent, he had better think harder, search his soul, until he sees a way forward. The external world he interacts with tosses him *hard facts* "at the end of the day," not the beliefs he begins with. The beliefs he starts with and bases his actions upon are his own contributions to the world.

Now, our path back to quantum theory is complete because I want to say this: a quantum state just *is* a probability assignment. The particular character of the quantum world places new, physically-influenced consistency requirements on our mesh of beliefs (like the second equation in my answer to Question 2, see page 46), but in the end, even quantum probabilities must port into probability theory more generally. A quantum state assignment is only one element in a much larger Bayesian mesh of beliefs each agent inevitably uses for his calculations. It is a numerical *commitment* to how he will gamble and make his decisions when he plans to interact with a quantum system. And everyone knows that many an expletive entails its own commitments as well!

GianCarlo Ghirardi · In its standard form, the theory assumes that quantum states are the mathematical entities that characterize, in the most accurate way possible *in principle*, the situation of an individual physical system. In a precise sense, knowledge of the state vector—the quantity which formally specifies the quantum state of a system—allows us to make definite predictions concerning the outcomes of all conceivable prospective measurements on the system.

The characteristic feature to be stressed is that in general, such predictions are only probabilistic. Only in particular instances—only for a small set of specific measurement procedures, which depend on the state—do such probabilities take the values one or zero, such that one can be certain that the result will or will not occur. Taking this perspective, one might state that quantum mechanics has taught us that we cannot, in general, consider too many "properties" as objectively possessed by a system: while some of them are associated to probability one and can be considered as actual, the vast majority have the ontological status of potentialities. This is a synthetic way of rephrasing the indeterminacy principle, one of the basic aspects of the theory.

When one takes perspectives different from the standard one, one is led to attribute a new status to the very concept of state. For example, within Bohmian mechanics, a state is typically identified via the simultaneous assignment of the state vector (the one of the standard theory) and of the precise positions of all elementary constituents (the particles) of the system under consideration.

Theories that dynamically induce wave-packet reduction give rise to yet another situation. In the GRW theory, the evolution of the state vector is determined by general dynamical entities—for example, by the Hamiltonian appearing in the evolution equation and by the stochastic localization processes, which are randomly distributed according to precise statistical laws. When it comes to the ontological characterization of the GRW theory and its variants, different positions have been taken. In a talk at the Centenary celebration of Schrödinger, Bell expounded the GRW approach and emphasized that he regarded the wave function as the characteristic quantity of the proposal: "There is nothing in this theory but the wave function. It is in the wave function that we must find an image of the physical world, and in particular of the arrangement of things in ordinary three-dimensional space." He also stressed that the wave function governs the places and times at which localization processes of elementary particles take place: "These are the mathematical counterparts in the theory to real events at definite places and times in the real world. A piece of matter then is a galaxy of such events."

Subsequently Bell seems to have slightly changed his mind—and I have discussed this point with him on various occasions—by claiming that the wave function (in $3N$-dimensional configuration space) by itself, that is, without further reference to explicit physical quantities, must be taken as the ontologically fundamental element. In a letter he sent me on October 3, 1989, he stated: "As regards Ψ and the density of stuff, I think it is important that this density is in the $3N$-dimensional configuration space."

I disagree with Bell on this crucial point. I have suggested—and this position has been considered as appropriate by many physicists and philosophers of science—that "what the theory is about" is essentially the mass density of the whole universe, averaged over spatial regions of the order of 10^{-15} cm^3. Recently, an analogous ontology has been adopted by Roderich Tumulka in the context of his development of relativistic generalizations of the GRW theory. In accordance with Bell's original suggestion, Tumulka chose the "flashes" (the times and positions at which localizations occur) as the basic beables of the theory. Such flashes strictly mirror the mass-density distribution.

SHELLY GOLDSTEIN · The most puzzling and controversial entity in quantum mechanics is the wave function, or quantum state. There seems to be little agreement as to what sort of physical entity the wave function is, or even as to whether it is a physical entity at all. Is it subjective or objective? Does it merely express our information about a system, or is it genuinely physical? And if it is objective and genuinely physical, does it reflect a concrete sort of physical reality, or something more elusive?

It is easy to understand why someone would resist the idea that the wave function is physical. The thought is that what is physical lives on physical space (be it three-dimensional, or of a somewhat higher dimension as is required for string theory) or on space-time, but not, like the wave function, on a space of hypothetical configurations (the *configuration space* of the system), whose dimension for a many-particle system is rather large, and is enormous for macroscopic systems.

At the same time, it is very hard to see how to make sense of quantum mechanics without regarding the wave function as objective. For example, it is difficult to see how the pieces of the wave function in the double-slit experiment could lead to quantum interference effects if these different pieces reflected not something physical but rather merely the present state of our knowledge of the system. And despite the controversy about the status of the wave function in quantum mechanics, I'm aware of no well-developed version of quantum mechanics that does not seem to require an objective, physical wave function as part of its formulation.

Thus, I'm strongly inclined to the opinion that the wave function of quantum mechanics is real, that is to say, that it is objective and physical and not subjective and merely an expression of our knowledge. This is certainly so for the wave function in Bohmian mechanics, in which it plays the role of a guiding field that governs the motion of the objectively existing particles.

More generally, the wave function plays a similar role in any of the fundamental physical theories involving local beables governed by a law of evolution involving the wave function: those theories that belong to the "interpretive program" that "can make the best sense of quantum mechanics" that I discussed in my answer to Question 3 (see page 74). And the wave function plays, it would seem, a similar role even in orthodox quantum theory, for which it governs the behavior of certain macroscopic variables (those describing results of quantum experiments).

A related question we might ask concerning the wave function is whether it is everything, or something, or nothing. If the wave function is not real, it is nothing; if it is real, it might be everything or it might be merely something. If it is everything, then the wave function of a system provides, contra Einstein, the complete description of that system. If it is something but not everything, then the description provided by the wave function should be supplemented by the values of additional variables, which—as I argued in my response to Question 2 (see page 47)—should be the fundamental local beables of the theory.

Now, it seems clear that the role of the wave function in a theory in which it is everything must be different from its role when it is merely something—different, that is, from its role in Bohmian mechanics and similar theories. The wave function can't play the role of governing the behavior of local beables or of any other additional variables if there are no such variables in the theory to begin with. For Everett's many-worlds version of quantum mechanics, which involves only the wave function, that wave function doesn't *govern* the behavior of some reality; rather it *is* the reality.

It has, however, been argued that even for an Everett-like version of quantum mechanics—for a theory in which all facts about a system are completely determined by a wave function obeying Schrödinger's equation—it is best to regard that wave function as governing the behavior of local beables. One such theory is Schrödinger's first quantum theory, formulated in 1927, a theory whose local beables are given by a matter density determined by the wave function. Valia Allori, Roderich Tumulka, Nino Zanghì, and I have argued that this theory is, in fact, a many-worlds theory.

By using a stochastic modification of Schrödinger's equation that naturally leads to wave-function collapse in situations in which, according to textbook quantum theory, the wave function should have collapsed, GianCarlo Ghirardi, Alberto Rimini, and Tulio Weber have formulated a version of quantum mechanics (the GRW theory) in which the state of a quantum system is completely determined by its wave function and for which there is nonetheless no measurement problem—and only one world. (Strictly speaking, the GRW theory is not a version of quantum mechanics, since it makes predictions for certain experiments—that have not yet been performed—that are different from those of quantum mechanics. The differences are very slight, however, so I will nevertheless speak of the GRW theory as a version of quantum mechanics.)

Allori, Tumulka, Zanghì, and I have, with Ghirardi, argued that the GRW theory too is best regarded as having local beables that are completely determined by the wave function. In this theory, so understood, as with Schrödinger's first quantum theory, the role of the wave function is to govern the behavior of the local beables. But unlike the situation with Bohmian mechanics, it does so not so much by being an ingredient in the evolution equation for these beables as by completely determining them.

For versions of quantum mechanics like Bohmian mechanics in which the wave function governs the behavior of local beables via an evolution equation for the local beables, it seems natural to regard the wave function as nomological—that is, not as a concrete physical reality but as a compact representation of the law of motion. For most systems in a quantum universe, however, it is problematical to regard the wave function in this way. Most, but not all. For the wave function of the universe, the proposal that it is nomological seems quite promising. And once one understands to his satisfaction the nature and status of the wave function of the universe, there should remain no further issue about the status of all the other wave functions of all subsystems of the universe, since these should be definable in terms of the wave function of the entire universe. (How this works is particularly clear in Bohmian mechanics.)

DANIEL GREENBERGER · Some of my ideas on this question are discussed in my reply to Question 6 (see page 133).

LUCIEN HARDY · A central starting question is whether quantum states are ontological. Does there exist, in reality, an actual ψ-field evolving in time? The pilot-wave, many-worlds, and collapse interpretations all assert that there does. Progress has been made recently on determining whether this must be a feature of any realist interpretation.

To set this up, we can define three notions of state. Imagine a system is emitted from some preparation device. First, we can define the *operational state* associated with the preparation. This state is given by any mathematical object that can be used to calculate the probability for any outcome of any measurement that may be performed on the system. In general, we can associate the operational state with a list of probabilities that are just sufficient to determine the probabilities for any measurement outcome. In the case of quantum theory, for a system associated with a Hilbert space of dimension N, the operational state can be specified by giving the probabilities for N^2 specially chosen measurement outcomes. In the case of a spin-half system, these could be $(p_{+z}, p_{-z}, p_{+x}, p_{+y})$, where p_{z+} is the probability of seeing spin up for a measurement of spin along the z direction.

Second, we define the *ontic state*. This is the actual underlying state of reality corresponding to the given instance of the preparation of the system. In general, we will not know the underlying ontic state and so must give a probability. Hence, we define our third notion, the *epistemic state* associated with a preparation. This is a list of probabilities, one for each possible ontic state of the system.

In 2004 I proved what I called the "ontological excess baggage theorem." This states that for systems corresponding to a finite-dimensional Hilbert space, the number of probabilities that must be listed for the epistemic state is infinitely more than the number that must be listed for the operational state. In other words, there are necessarily an infinite number of ontic states even for finite N. In 2007 Alberto Montina proved a significantly stronger result. He showed that the ontic state must have at least $(2N-2)$ continuous degrees of freedom. This is the same number as the wave function ψ has (after normalization and overall phase have been fixed). Further, in 2010 he showed that the ontic state must contain an object that evolves according to the Schrödinger equation (as does the wave function). This does not quite prove that ψ itself is ontic, but it does provide very strong evidence in this direction. If it is proven that the wave function ψ is necessarily ontic, then the case for one of the above-mentioned realist interpretations becomes stronger. There is, however, a compelling alternative.

The above proofs of Montina make the assumption that the state at time $t + \delta t$ is determined by the state at time t. This amounts to assuming that the evolution is local in time. If we have indefinite causal structure, as we'd expect in a theory of quantum gravity, then such locality in time makes less sense. First, we will not have a background time parameter with respect to which to define such evolution of the state. Second, any effective state may depend on properties that are nonlocal in time. Further support for this comes from Bell's theorem, which suggests nonlocality in space. If we have indefinite causal structure, then we cannot single out space in this way. We would have to have nonlocality in space-time. Montina has recently shown how, by relaxing the assumption of locality in time, it is possible to construct an ontic model for a qubit requiring only one real parameter (rather than two, which is the number of free real parameters in ψ for a qubit). He calls this "ontological shrinkage."

This indicates that the wave function need not be part of the ontological description in models that are not local in time.

Ultimately, we need an ontological understanding of physics. I anticipate that from the viewpoint of the correct ontological theory, the quantum state will be seen to be an effective object not having fundamental status in itself. In particular, I expect that the notion of a wave function across space evolving in time will turn out not to be fundamental.

What methodology should we use to actually get our hands on the correct ontological theory? In the absence of any clear idea of what the correct ontology is, I suggest adopting an operational methodology for the time being. I hope that this will ultimately lead to the correct ontological theory. Operationally, a state corresponds to a list of probabilities that can be used to calculate the probability for any outcome of any measurement that may follow the preparation with which the state is associated (as defined above). The notion of state used here is that it is something that is given at a time t and is used to predict the probabilities for the future (so that it pertains to a semi-infinite region of space-time). To construct a theory of quantum gravity, a good first step is to generalize this notion of state to being something that is given for an arbitrary region of space-time (instead of a semi-infinite one) and is used to predict probabilities for that region. Interestingly, it turns out that all the essential objects in quantum theory (states, effects, and transformations) can be understood as corresponding to lists of probabilities for appropriate arbitrary regions of space-time (as I have shown in my causaloid and duotensor formulations for operational theories). In such a formulation of quantum theory, we do not have to think of the state as something that evolves in time. States pertain to arbitrary regions, and there are rules for obtaining the state for a composite region from the states for the components. This enables us to calculate everything we want to calculate, without ever having to evoke the notion of an evolving state.

ANTHONY LEGGETT · According to the view expressed in my response to Question 3 (see page 79), a "quantum state" of a given system, as assigned by a given agent at a given time, is simply that part of the recipe that encodes the agent's current information about the history and present configuration of the system. This is distinct from the part of the recipe that specifies the evolution of the state in time (for example, in the case of standard nonrelativistic physics, the time-dependent Schrödinger equation together with the appropriate form of Hamiltonian), and distinct from the part that tells us how to predict the probabilities of the various possible outcomes of macroscopic measurement procedures (that is, the measurement axiom). Note that according to the above definition, one cannot in general speak of "*the* quantum state" of a given system at a given time, since different agents may possess different information and thus assign different states. The exception would be when at least one agent has the most complete information possible and can thus assign a pure state to the system. In this case, any other agent would have either the same or less information, so that one may legitimately treat the pure state as "*the* quantum state of the system," with the less-favored agents, if any, being, as it were, "classically" ignorant of some or all of the system's features.

TIM MAUDLIN · This is a difficult question, since it is not clear what a positive answer would look like. There are clearly real, objective physical degrees of freedom reflected in the quantum state of a system. Quantum states (wave functions) are not merely reflections of someone's information or knowledge of a system. As David Mermin once remarked, one can't explain the interference effects visible in a two-slit experiment by holding that *our knowledge* of the position of the particle went through both slits. Nor does it make sense to regard Schrödinger's equation as a claim about epistemology, that is, about how *our knowledge* of a particle evolves in time. The interference effects show that something physical is sensitive to the fact that both slits are open, so the "spread" of the wave function is something more objective than our mere lack of knowledge concerning the location of the particle.

But what kind of thing is the quantum state? The temptation is to try to assimilate it to something we are already familiar with in classical physics: a field, perhaps, or a law. There is no good reason, however, to expect classical physics or common sense to provide us with any good analogues. Better to try to characterize the features the quantum state must have to do the physical work we require of it.

The quantum state appears to be something intrinsically holistic. The state ascribed to a pair of particles, for example, contains more physical information than the states ascribed to the particles individually (i.e., the reduced states). For example, a pair of electrons in the singlet state have the same reduced states as a pair in the $m = 0$ triplet state, but these are clearly physically different situations. What is interesting is that the differences only appear in correlations between the particles, not in the behavior of either particle individually. It is from correlations like these that violations of Bell's inequality are constructed, so the holism of the quantum state appears to be the source of the nonlocality in quantum theory.

Ultimately, there should be but one quantum state: that of the whole universe. Our inquiries should be directed at it. It is theoretically possible that the universal quantum state is static—it never changes. At least, this is possible in a theory such as Bohmian mechanics, where all the visible changes in the world are changes in additional variables that are not determined by the quantum state. If a case could be made out for an unchanging universal quantum state, it would reinforce the idea that the quantum state has no good analogue in any classical ontology. Then we should just make a new ontological category for it.

DAVID MERMIN · The first of my answers to Question 2 (see page 52) primarily says what quantum states are not. It is harder to say what they are. I am intrigued by the fact that if quantum mechanics applied only to digital quantum computers, then the answer would be entirely straightforward. Quantum states are mathematical symbols. The symbols enable us to calculate, from the (explicit, unproblematic) prior history of a collection of Qbits—I commend to the reader this attractive abbreviation of "qubit"—the probabilities of the readings (0 or 1) of a collection of one-Qbit measurement gates to which the Qbits are then subjected. This procedure is made unambiguous by the rule that a Qbit emerging from a one-Qbit measurement gate reading 0 (or 1) is assigned the state $|0\rangle$ (or $|1\rangle$). This makes it possible to assign initial states with the help of one-Qbit measurement gates. Additional rules associate specific unitary transformations of the states of the Qbit(s) with the action of the other subsequent gates that appear in a computation.

Quantum states, in other words, are bookkeeping tools that enable one to calculate, from a knowledge of the initial preparation and the fields acting on a system, the probability of the outcomes of measurements on that system. This is what I take to be the Copenhagen interpretation of quantum mechanics. (I hereby renounce my earlier summary of Copenhagen, widely misattributed to Richard Feynman, as "shut up and calculate.") If the only application of quantum mechanics were to the operation of digital quantum computers, there would be no ambiguity or controversy about Copenhagen.

The Copenhagen view fits quantum computation so well that I am persuaded that quantum states are, even in broader physical contexts, calculational tools, invented and used by physicists to enable them to predict correlations among their perceptions. I realize that others have used their experience with quantum computation to make similar arguments on behalf of many worlds (David Deutsch) and consistent histories (Bob Griffiths). I would challenge them to make their preferred points of view the basis for a quick practical pedagogical approach to quantum computation for computer scientists who know no physics, as I have done with Copenhagen in my quantum-computation book. The approach to quantum mechanics via consistent histories in Griffiths's book, while something of a tour de force, does not strike me as either quick or practical.

LEE SMOLIN · In light of my view of quantum mechanics described in my answer to Question 3 (see page 80), quantum states are to be regarded as statistical states, which result from averaging over more fundamental nonlocal degrees of freedom.

ANTONY VALENTINI · In my view, de Broglie's pilot wave is a new kind of causal agent, a radically new kind of physical entity grounded in configuration space. To understand it, it's helpful to examine the historical parallel with two other physical entities that seemed mysterious when first introduced: Newton's concept of gravitational attraction-at-a-distance, and Faraday's concept of field.

Before the acceptance of Newtonian gravity, it seemed to many that scientific explanation should be reduced to Cartesian action-by-contact. If one body appeared to act on another at a distance, there must be an intervening medium that transmits the force through local action by contact. In Newtonian gravity, instead, a massive body can act directly on another at a distance, through empty space. Even Newton himself had difficulty with the idea, and continued to seek a deeper explanation for gravity in terms of an aetherial medium filling space. The concept of gravitational attraction arose by a process of abstraction, in which the conceptual scaffolding of a Cartesian medium was thrown away.

A similar step occurred in the nineteenth century when Faraday introduced the concept of field. Faraday looked at the pattern of iron filings around a bar magnet, and started to think that the pattern would exist even if the filings were taken away. It's hard for us today to appreciate what a conceptual leap that was. Remember, at that time, a force was understood to be present when a mass accelerates. In empty space, where there are no masses, how could a force exist all by itself? But Faraday believed that the magnetic "lines of force" seen in patterns of iron filings existed in their own right, even when the filings were absent. We eventually got used to the idea of forces existing and even propagating in empty space, where there are no masses or charges. Again, the concept arose by a process of abstraction. By abstracting away the iron filings, we're left with the new concept of fields.

Now, in my view, history repeated itself in 1927 when de Broglie introduced the concept of a pilot wave in configuration space. Again, it arose by a process of abstraction. In particular, early in 1927 he was struggling with a model he had of a system of particles as singularities of coupled fields in three-space. He was trying to show that the singularities would follow the guidance equation, which was supposed to emerge as an effective description of the motions, which were ultimately generated by the complicated coupled field equations. But he saw that as a provisional theory, he could simply take the guidance equation with the pilot wave in configuration space, and forget about the underlying model—just as Newton did with gravity, and just as Faraday and others did with electromagnetism. The difference, though, is that while we all recognize Newton and Faraday for their achievements, most physicists and historians simply don't know what de Broglie really did in the 1920s. I believe that, in 1927, de Broglie introduced a fundamentally new entity into physics, but to this day the world hasn't really noticed, and even those who are interested in his theory do not properly understand what de Broglie did. And to complete the analogy, particularly with Newton, de Broglie himself was uncomfortable with the idea and thought it should emerge as an effective theory along the lines he had been considering—a view he returned to in later life. He never really believed the pilot wave in configuration space was fundamental, just as Newton never believed that his theory of gravity was fundamental. Still, Newton's concept lasted for more than two hundred years.

In my view, de Broglie's pilot-wave concept deserves to be taken more seriously. It might turn out to be useful-but-wrong, like Newton's concept of gravitational attraction-at-a-distance. Or it might turn out to be an essential new concept—as happened with Faraday's concept of field, which survives even in quantum field theory. In any case, at present, I would suggest that the wave function is a new kind of causal agent, as new and radical as was Faraday's concept of field, but which, for historical reasons, has not been recognized.

To see how radical it is, consider the contrast with the idea of a field in space. An ordinary field can be probed using a test particle. An electric field, for example, can be measured by introducing an infinitesimal test charge and watching it accelerate. But if we try to do this for the pilot wave, we find that introducing a test particle actually increases the dimension of the configuration space on which the wave is defined. There is no such thing as a test particle for the pilot wave. So it's not comparable to ordinary fields, something that Bohm didn't really appreciate in 1952 but which de Broglie understood in 1927.

We may need to learn to think in terms of pilot waves, just as earlier generations learned to think in terms of fields, without a conceptual scaffolding based on more primitive notions. In particular, the pilot wave in configuration space provides a natural understanding of nonlocality in three-space.

Finally, I should comment on the notion of pilot wave for the whole universe. Aside from gaps in our understanding of quantum gravity, I see nothing problematic there. Some people claim that because there is "only one universe," the universal pilot wave cannot be contingent and must instead be lawlike. But that argument is spurious. In our current understanding of cosmology, the intergalactic magnetic field is not determined by physical laws—it is contingent, in the sense that, for all we know, the configuration of that field could have been different. The same goes for the space-time geometry of the universe as a whole. And the same can be said of the universal pilot wave.

DAVID WALLACE · At least if you mean pure states, they're states of the world. It's a bit misleading to take that as saying that they're real physical things, though. After all, in classical mechanics, the classical state—the phase-space point, that is—is a state of the world, but that doesn't mean that the world is a point in a really-high-dimensional space. What we mean by saying, of the state in a physical theory, that it's a state of the world, is that it represents, not facts about our knowledge of the world, but facts about the world itself.

Now, we might get worried about just what those facts are. In classical mechanics it's not so hard to answer: they're facts about where the particles are in space and how fast they're moving, or else in field theory they're facts about what the field strengths are in various spatial locations. In quantum mechanics too we can talk about the quantum state of a given space-time region (of course, it's normally a mixed state). I'm not sure how much point there is trying to get an intuitive grip on what the state of that region really represents, beyond "certain features of the structure of that region."

Of course, in saying this I'm rejecting the alternative view, that the state is somehow a codification of our ignorance, somewhat like the statistical-mechanical state. But I don't really think this is viable. In (classical) statistical mechanics, it's pretty easy to see what we're ignorant *of*: we're ignorant of what the real classical microstate is. But we know that making a strategy like that work in quantum theory is going to be incredibly difficult, because of the Bell–Kochen–Specker theorem. The alternative that's most frequently discussed is that the quantum state represents our ignorance of the possible results of measurements, but that forces us to take measurement as some primitive thing that can't be analyzed. I don't know how that can be squared with the fact that experimental physicists blatantly do analyse measurement processes all the time. (I actually think this is one place where the very abstract flavor of quantum information can get in the way; I say more about that in my answer to Question 9, page 193.)

What about mixed states? By and large, I think they represent states of the world too, but not necessarily states of this particular branch of the world (in Everettian—that is, many-worlds—terms). Say we prepare an EPR pair and throw one element of the pair away: the only quantum state available to represent the other element is a mixed state, and I don't see anything particularly wrong with saying that that really is the state of the qubit. Then if we let the qubit get decohered—say, if we measure it but don't look at the result—then it gets entangled with the macroscopic degrees of freedom of its vicinity, but *it* is still in a mixed state. Of course, relative to us, it's in some unknown pure state, so in that more limited sense the mixed state represents ignorance.

ANTON ZEILINGER · Quantum states are representations of knowledge—or, in modern language, of information. It is information about the apparatus and about the complete experimental situation, which are, in the end, always classical pieces of equipment. Quantum states allow us to make predictions—in general, probabilistic ones—about future measurement results. These results are, strictly speaking, again classical features of the apparatus. I should emphasize that in my opinion, individual systems can be described by quantum states. We can certainly prepare an individual system to be in an eigenstate of a specific apparatus. As a consequence, for example, each individual particle "knows" that it has to avoid the minima of the interference pattern in the two-slit experiment. I should remark that experiments with individual particles or quantum systems are at the core of quantum teleportation, to give just one example from quantum information protocols.

WOJCIECH ZUREK · Exactly.
   I'm tempted to turn this question around and first ask (1) how classical states differ from quantum states, and (2) whether and how quantum states can replicate basic properties of classical states. (Note that I do not believe that there is any point in asking the complementary question, namely, whether classical states can replicate quantum states; I think experimental evidence—violations of Bell's inequalities and the like—have already provided us with a convincing negative answer.)
   The key difference between quantum and classical states is in their accessibility. An unknown classical state can be found out without being changed in the process. In turn, this shows that classical states exist objectively. An unknown quantum state cannot be found out, as a measurement will usually reprepare it.
   As a consequence of this fragility, one is tempted to altogether deny the existence of quantum states and to reduce them to mere information in possession of the observer. But this is not completely fair: one can find out an unknown eigenstate of the measured observable without perturbing it. Moreover, one can confirm the existence of a known quantum state through repeated measurements. Indeed, one can turn this repeatability of measurements (along with the unitarity of quantum theory) into a proof of the inevitability of quantum jumps: one can show that the combination of repeatability and unitarity makes it impossible to find out an unknown quantum state. So in a sense, the same quantum postulates that make limited symptoms of existence (such as repeatability) possible also preclude the objective existence of unknown quantum states, by putting full-fledged objectivity off-limits.

This interdependence between the "objective existence" and "mere information" roles of quantum states makes it difficult for me to buy into programs that go all the way in either of these two directions. In particular, a literal many-worlds reading of Everett is just too *classical* an interpretation of states in quantum theory, as it asserts objective existence of all the branches and treats the state vector of the universe as an objectively existing entity—in effect, as a classical state. I tend to think that it should be possible to think of the "other" branches as information; after all, we cannot directly confirm their existence. In this case, the other branches are not real—at least not as real as "our" branch.

At the opposite end of the spectrum are attempts to derive "all of the quantum" from a subjective, observer-centered point of view. Naive subjectivist approaches fail in one obvious way: the observer has to be outside of the quantum realm, so that his subjective view of the universe can be based on something firm and nonquantum. How to construct an observer who is outside of the quantum realm—so that his subjective information can be a basis for the quantum world out there—from subjective quantum pieces is difficult to imagine.

On the other hand, I firmly believe that pushing even such extreme points of view as many worlds or the subjectivist approach to "the quantum" is a valuable exercise. We have certainly learned a lot from Everett and DeWitt, and we have definitely learned a great deal from Bohr, who at least some of those pursuing the subjectivist approach cite as their intellectual forefather. I believe the truth lies between these two extremes: I take from Everett the lesson that quantum theory is the best tool for explaining its own workings, but I take from Bohr (and Wheeler) the firm conviction that when we find out how it works, we will realize that information was an integral part of its machinery. (One might say that this attempt to have the best of both points of view is complementary.)

This brings me to the second question I posed at the beginning: can quantum states replicate basic properties of classical states? I believe the answer is a resounding yes: the essence of this view of the emergence of the classical lies in "quantum Darwinism"—in the selective proliferation of information about certain preferred states throughout the environment. Once this happens, such information becomes effectively objective: by trying out different possible measurements on subsets of states, the observer can find out the underlying state that has spawned such a progeny. To be sure, states of the measured environment subsystems will be destroyed, but there are still plenty more copies of the original in the environment, so one can find out what that state is, by trial and error, without erasing the information that is shared by the whole set of them.

So in a sense, as a consequence of quantum Darwinism one can kill a messenger without endangering the message. Moreover, as there are many copies, many observers can do this independently. It is not difficult to see that they will always agree about their findings. Thus, quantum Darwinism explains how robust objective reality—collective states that can be found out without being destroyed—can be built out of fragile quantum states. This process has to select a preferred set of states of the system, states that (as one can show) should be distinguishable and have to coincide with the pointer states that can survive decoherence.

Quantum Darwinism happens all the time: as you study this text, multiple photons carry the image of this page. Some of them reach your eyes. They suffice to deliver the information, but there are plenty more that go in other directions, so that many observers can look at the image and agree on what it is. This consensus is the essence of objectivity. So, "more is different": collections of quantum states allow one to account for all the symptoms of objective classical reality.

To sum up, we may not know what quantum states exactly are, but we do know how they differ from classical states: they are so fragile that an unknown state of a single quantum system cannot be found out. On the other hand, natural processes responsible for information transfer can correlate a multitude of quantum systems in the environment with a decohering quantum system. This system is then cocooned in, and shielded by, the collective state, which is robust enough to exhibit all the symptoms of classical reality. In particular, the state of the central quantum system can be found out indirectly from the imprint it has left on the environment. This allows for the underlying state to be revealed to many observers without endangering its existence. (By necessity, the state will be a pointer state, as it has to survive interaction with the environment in order to leave an imprint.) This is how fragile quantum states are organized by decoherence into an effectively classical, robust, and objective "reality."

RANDOMNESS

*Does quantum mechanics imply
irreducible randomness in nature?*

"F OR EVERY STEP, the footprint was already there." This sentence from Roberto Calasso's novel *The Marriage of Cadmus and Harmony* could well be a statement about the spirit of determinism. Determinism satisfies our longing for cosmic order and predictability. It is in this sense that determinism probably still underlies the mindset of many if not most scientists. But if we take a hard look at what determinism offers, we're immediately faced with a profound question. Why would the universe take any steps to begin with if determinism really was true and reality thus ultimately nothing but an immutable and static lump? In other words, what's the deeper significance—assuming any is needed, of course—of anything happening at all, beyond the initial *bang!* that one and for all hard-coded every future event?

Indeterminism, on the other hand, buys the universe creative freedom. The passing of time does not just amount to ticking off items on a blueprint. Instead, it represents a process of constant birthing and spontaneous (re)invention. But ought there not be *some* ultimate reason for an event to go one way or the other? How else could the world be analyzable and rationalizable in the last instance? Doesn't indeterminism fly in the face of our intuition about what science does and can do and should do?

And then there are those age-old questions concerning our sense and the possibility of free will. We get trapped in a damned-if-you-do-damned-if-you-don't situation: neither determinism nor indeterminism seem to quite capture the subtle combination of freedom and intentionality that we associate with our personal choices. Or are we perhaps misguided in seeking the basis of such a subjective notion as free will at the bare-bones level of fundamental physical laws? On this note, it's worth keeping in mind Einstein's words:

> I do not believe in free will. Schopenhauer's words, "Man can do what he wants, but he cannot will what he wills," accompany me in all situations throughout my life and reconcile me with the actions of others, even if they are rather painful to me. This awareness of the lack of free will keeps me from taking myself and my fellow men too seriously as acting and deciding individuals, and from losing my temper.

It is often said that quantum theory has helped loosen the iron grip exerted by the strict determinism of classical physics. But just how much has it done so? To what extent does quantum theory force us to relinquish the cushy pillow of a deterministic weltanschauung? Is the much-touted "quantum randomness" sufficient and suitable to inspire a sense of freedom, be it about the course of the universe as a whole, or about the course of our personal actions? Is quantum mechanics going as far as granting us—or forcing upon us, depending on one's point of view—the picture of a fundamentally unpredictable universe engaged in random acts of continual creation? A universe in which nothing is ever set in stone, including, perhaps, the very laws of physics, à la John Wheeler's "law without law" or Nancy Cartwright's "dappled world"? Conversely, could quantum randomness be merely a surface phenomenon, an artefact that can be exorcised with impunity, so that quantum theory can live in peaceful coexistence with a deterministic worldview?

Could quantum theory even point to a possible synthesis of determinism and indeterminism? For example, the Everett interpretation juxtaposes a notion of global, objective determinism and local, subjective indeterminism. On the one hand, it satiates classical, absolutistic urges by positing a completely deterministic, ready-made block universe (the "outside view"). On the other hand, from the subjective point of view of a local observer within this universe (the "inside view"), it is suggested that we can nonetheless recover the appearance of indeterminism as a consequence of the branching process. And so the Everettian tempts us with intuitive, classical imagery: an unrolling tree-like structure of objectively existing and deterministically forking paths, one of which "I" find myself scooting along—although without ever being able to influence the particular route I'm taking at each junction. Does that mean we get to have the cake and eat it too?

On a final note, it is likely that your individual view of the implications of quantum theory for the possibility (or even necessity) of a fundamental indeterminism in nature will be tied up with your other interpretive commitments—say, with your take on quantum states (see Question 4, *Quantum States*) and with your attitude toward hidden variables. Ultimately, personal sentiment might also have a say, if only subconsciously (see also Question 13, *Beliefs and Values*). For instance, do you feel unmoored and anxious at the thought of living in an indeterministic world, or rather liberated and empowered?

GUIDO BACCIAGALUPPI · As I have mentioned already (and will probably mention often again), the answer to many questions in the foundations of quantum mechanics depends crucially on the interpretational approach one adopts. They all have their pros and cons, and I, for one, am perfectly agnostic as to which approach might be the "best." As regards the question of irreducible randomness, the answer is yes in (most) spontaneous-collapse theories, no in (many) pilot-wave theories, and yes and no in (all) Everett interpretations.

Indeed, spontaneous-collapse theories, unless one includes also putative ones based on nonlinear Schrödinger equations, are explicitly stochastic, in that they postulate stochastic modifications of the Schrödinger equation.

In pilot-wave theories, the situation is slightly more complex: while the Schrödinger equation is unaltered, the guidance equation for the configurations might itself be deterministic or stochastic. If we take the classic de Broglie–Bohm theory as the main representative of pilot-wave theories, it includes no irreducible randomness. The origin of quantum uncertainty lies in our ignorance of initial configurations, just as in classical statistical mechanics, and in the case of quantum equilibrium (i.e., the standard $|\psi|^2$ distribution), this ignorance cannot be improved on, essentially because the theory contains a detailed story about how measurements systematically disturb the system.

Everett interpretations present the most unfamiliar picture, since here the Schrödinger equation is retained at the level of the universal wave function, but due to decoherence, the wave function splits into dynamically autonomous components, the different "branches" or "worlds," so that randomness—indeed, arguably perfectly genuine randomness—is an emergent notion from the internal point of view of each branch. (Different many-worlds or many-minds variants of Everett provide different ways of spelling out this fundamental point.)

ČASLAV BRUKNER · Quantum randomness manifests itself in two different scenarios: the impossibility of giving a causal explanation for the occurrence of a single measurement outcome, and the impossibility of algorithmically compressing sequences of random bits formed by concatenating such single outcomes. That these two ways of encountering randomness are different is apparent in the interpretational framework of de Broglie–Bohm mechanics. This theory is causal and deterministic, as each particle has a well-defined trajectory at every instant of time, just like in classical mechanics. And yet the appearance of pattern detection can be algorithmically incompressible, because the initial distributions of positions of particles (for example, at the beginning of the universe) might be so.

Among the strongest evidence for the irreducible randomness of an individual measurement outcome—such as the decay of an atom—has been the factual inability to introduce "causes" for individual outcomes in a satisfactory way. Classical realism assumes that there exist causes for outcomes of all potential measurements. These outcomes could have been obtained if the experimenter had chosen to perform these measurements—independently of whether any, and which, measurement has actually been chosen. The causes for the outcomes are said to be local if they are in the backward light cone of the measurements. Now, the theorems of Bell and of Greenberger, Horne, and Zeilinger demonstrate that the mere assumption of a coexistence of local causes for outcomes of a set of quantum experiments results in a contradiction.

Confronted with the impossibility of consistently assigning such local causes, one may assume, in an attempt to maintain a causal explanation, that the outcome of an experiment in one laboratory depends instantaneously on the choice of measurements in arbitrarily faraway laboratories—in other words, that it depends on the entire "experimental context." The de Broglie–Bohm theory expresses this in an explicit manner: the velocity of any particle depends on the value of the wave function, which depends on the whole configuration of the universe. To me, this instantaneous dependence on everything and everybody is a bit like denying the very purpose of time, which, as Wheeler once remarked, is "to prevent everything from happening at once."

Alternatively, one can depart from classical realism and accept that quantum probabilities are irreducible. This may be seen as an inevitable consequence of a simple and elementary idea that I developed with Zeilinger and follow in my own research: *the information-carrying capacity of an individual quantum system is fundamentally limited.* Specifically, the most elementary two-state system, or qubit, can carry at most one bit of information, and nothing more. Now, if a qubit can be prepared to specify the bit in only a single experimental context, then measurement outcomes in all other contexts *must* necessarily contain an element of randomness. There cannot be any "hidden" causes for the results, because otherwise the system would carry more information than is in principle available.

This should be contrasted to the required infinite information-carrying capacity in any causal theory. Even reproducing measurements on a single qubit requires infinitely many orthogonal hidden-variables states. It might be a matter of taste whether one is ready to shoulder this "ontological excess baggage" (in the words of Hardy) that's not doing any explanatory work at the operational level. But it is certainly conceptually distinctly different from my proposal that the information capacity of the most elementary systems—those that are by definition not further reducible—is fundamentally limited. A question of Feynman's makes this point decisively: "Why should it take an infinite amount of logic to figure out what one tiny piece of space-time is going to do?"

Finally, let me return to the interview question. When speculating about whether quantum randomness is "true," it is advisable to keep all options open. Just imagine that one day your experimentalist friends show up and report to you on their exciting experiment in which single photons, coming from a star seven-and-a-half-million light-years away, are overlapped on a beam splitter and detected behind either of the two outputs. On that day, the detection pattern they observe is the following binary sequence: 101010 (pause) 101010 (pause) ... Would you grin like a Cheshire cat?

JEFFREY BUB · I don't think quantum mechanics *implies* irreducible randomness in nature. Any nonclassical theory can be reinterpreted deterministically if one is willing to pay the price.

Consider, for example, the problem of reinterpreting a PR box deterministically—that is to say, consider the hidden-variable problem for a PR box. Basically, a PR box converts the conjunction of the inputs to the parity of the outputs. So one could consider a "Bohm box" with an internal mechanism involving a hidden parameter $\lambda \in [0,1]$. Suppose that when $\lambda < \frac{1}{2}$ the inputs are passed through an AND gate and the output of the gate is transferred to Alice's output a, while Bob's output b is set to 0. And suppose that when $\lambda \geq \frac{1}{2}$, 1 is added (mod 2) to the output of the AND gate before it is transferred to Alice's output a, while Bob's output b is set to 1. Symmetry between Alice and Bob can be restored by adding a second hidden parameter associated with a mechanism that flips the above map randomly between Alice and Bob. Evidently, if we assume a uniform distribution over λ, a Bohm box is phenomenally indistinguishable from a PR box.

It's clear that if there is no access in principle to the internal mechanism, then a Bohm box and a PR box are empirically indistinguishable, and nothing can rule out the possibility that a PR box is really a Bohm box with an intrinsically hidden internal mechanism. Note that we would have to allow the internal mechanism to function instantaneously, since the correlations of a PR box are assumed to be maintained nonlocally when Alice's part of the box is remote from Bob's part, and a similar observation applies to Bohm's theory. Of course, the correlations of a PR box are a lot simpler than the full probabilistic structure of quantum mechanics, and Bohm's theory is very ingenious in showing how interference and entanglement can be simulated with a deterministic theory, where the theory itself explains why the hidden variables are inaccessible once an equilibrium distribution has been achieved. Conceptually, though, the point is the same.

For all we know, Bohm's theory might be true. But one might say the same for Lorentz's theory in relation to special relativity, insofar as it "saves the appearances." Lorentz's theory provides a dynamical explanation for phenomena, such as length contraction, that are explained kinematically in special relativity in terms of the structure of Minkowski space-time. The theory does this at the expense of introducing motions relative to the aether that are in principle unmeasurable, given the equations of motion of the theory. Similarly, Bohm's theory provides a dynamical explanation of quantum phenomena—such as the loss of information on measurement, which is explained kinematically in quantum mechanics in terms of the structure of Hilbert space—at the expense of introducing the positions of the Bohmian particles, which are in principle unmeasurable more precisely than the Born distribution in the equilibrium theory, given the equations of motion of the particles.

Ultimately, the question is whether it is more fruitful in terms of advancing our understanding to consider quantum mechanics as a member of the class of nonclassical, i.e., nonsimplex, no-signaling theories that describe alternative irreducibly random universes, or whether we should think of quantum mechanics as a classical simplex theory that violates the no-signaling constraint (which would involve a preferred foliation in space-time), but where instantaneous signaling is impossible because we have no ability in principle to manipulate the hidden variables.

ARTHUR FINE · As usually understood, the answer is yes: quantum mechanics does imply irreducible randomness in nature. Unlike the case of a fair coin, where the probabilities for heads or tails are a function of the whole setup for tossing the coin (probability comes from variation among the tosses), in the case of a spinning electron the probability for finding spin-up or spin-down in fixed directions is inherent in the prepared state of the electron. For the coin, one can build a heads-over-tails tossing machine whose outcome is perfectly predictable. Start with heads, then the coin spins in a graceful arc—and winds up heads. Or vice versa. There is no such standard measuring device where the outcomes can be controlled for the x-spin of an electron in a state superposed over the x-spin-up and x-spin-down eigenstates. At least, this is the usual view.

Nevertheless, there are hidden-variables versions of the theory that are deterministic, like the de Broglie–Bohm version, and there are even versions, like that of Everett and many worlds, that are truly fatalistic, in the sense that everything is determined once and for all. In these versions, probabilities are epistemic, due—as one says—to ignorance. This is especially problematic in the Everett-related versions where, despite clever information-theoretic arguments, I would say that the probabilities really have to be put in by hand to agree with the usual density-operator trace formalism. In de Broglie–Bohm there are heuristic, physically robust lines of argument that can be used to justify the spread of initial positions to exactly what standard quantum theory requires. But these arguments, in both cases, have an air of artificiality, since they all stem from the fact that without these exact probabilistic limitations, the proposed version of the theory would not duplicate the results of the standard theory.

In any case, it is clear that in facing nature, deterministic versions of quantum theory require a probabilistic escort. So even in these nonstandard versions, nature will still appear to us *as though* there were irreducible probabilistic or random events. To be sure, one could still distinguish between randomness in our models of nature and true randomness in nature herself. But that would raise the general issue of how far to project features of our models (or theories) onto the world, an issue that has to be faced concerning science more generally, but not here.

CHRISTOPHER FUCHS · It strikes me that a question like this defeats the purpose of this volume. The point was to pose the same seventeen questions to all the contributors to see how their answers compared and contrasted. But if there are seventeen participants in this volume, they are surely reading seventeen *different* questions for this one. What does it mean?

For my own reading of it, here is the way I would make a start toward an answer. I would rather say that quantum mechanics on a QBist reading appears to imply an *irreducible pluralism* to nature. Nature is composed of entities, each with a fire of its own—something not fueled or determined by any of nature's other parts. The philosopher William James coined the terms "multiverse" and "pluriverse" to capture this idea and put it into contrast with the idea of a single, monistic, block universe. Unfortunately, the Everettians have co-opted "multiverse" in a grand act of Orwellian doublespeak for their monistic vision (what else is their universal wave function?), but "pluriverse" so far seems to have remained safe from these anti-Jamesian shanghais. I will thus use that term hereafter.

But what is a pluriverse more precisely, and what does it have to do with the specific issues of quantum mechanics? I will let James speak for himself on the first issue before returning myself to the second.

> [Chance] is a purely negative and relative term, giving us no information about that of which it is predicated, except that it happens to be disconnected with something else—not controlled, secured, or necessitated by other things in advance of its own actual presence. ... What I say is that it tells us nothing about what a thing may be in itself to call it "chance." All you mean by calling it "chance" is that this is not guaranteed, that it may also fall out otherwise. For the system of other things has no positive hold on the chance-thing. Its origin is in a certain fashion negative: it escapes, and says, Hands off! coming, when it comes, as a free gift, or not at all.
>
> This negativeness, however, and this opacity of the chance-thing when thus considered *ab extra*, or from the point of view of previous things or distant things, do not preclude its having any amount of positiveness and luminosity from within, and at its own place and moment. All that its chance-character asserts about it is that there is something in it really of its own, something that is not the unconditional property of the whole. If the whole wants this property, the whole must wait till it can get it, if it be a matter of chance. That the universe may actually be a sort of joint-stock society of this sort, in which the sharers have both limited liabilities and limited powers, is of course a simple and conceivable notion.

Additionally,

> Why may not the world be a sort of republican banquet of this sort, where all the qualities of being respect one another's personal sacredness, yet sit at the common table of space and time? ... Things cohere, but the act of cohesion itself implies but few conditions, and leaves the rest of their qualifications indeterminate. As the first three notes of a tune comport many endings, all melodious, but the tune is not named till a particular ending has actually come,—so the parts actually known of the universe may comport many ideally possible complements. But as the facts are not the complements, so the knowledge of the one is not the knowledge of the other in anything but the few necessary elements of which all must partake in order to be together at all. Why, if one act of knowledge could from one point take in the total perspective, with all mere possibilities abolished, should there ever have been anything more than that act? Why duplicate it by the tedious unrolling, inch by inch, of the foredone reality? No answer seems possible. On the other hand, if we stipulate only a partial community of partially independent powers, we see perfectly why no one part controls the whole view, but each detail must come and be actually given, before, in any special sense, it can be said to be determined at all. This is the moral view, the view that gives to other powers the same freedom it would have itself.

With James, this is QBism's notion of chance—*objective* chance, if you will. It is the residue of the Quantum Bayesian analysis of what the theory's probabilities are all *about*, along with a further analysis of the Wigner's-friend paradox.

QBism says that quantum theory should not be thought of as a picture of the world itself, but as a "user's manual" *any* agent can pick up and use to make wiser decisions in the world enveloping him—a world in which the consequences of his actions upon it are inherently uncertain. To make the point: in my case, it is a world in which *I* am forced to be uncertain about the consequences of most of *my* actions; and in your case, it is a world in which *you* are forced to be uncertain about the consequences of most of *your* actions. Yet both of us may use quantum theory as an addition to logic and probability theory when we contemplate our personal uncertainties about these very personal things for each of us.

This is where the Wigner's-friend question comes into play. This is a story of two agents with a different physical system in front of each: (1) the *friend*, with (say) an **electron** in front of himself, and (2) *Wigner*, with the **friend + electron** in front of himself. (Agents are italicized; systems are boldfaced.) Which agent's quantum-state assignment for his own system is the correct one? Quantum Bayesianism knows of no agent-independent notion of "correct" here—and this is why we say there is no paradox. The source of each assignment is the agent who makes it, and the *concern* of each assignment is not of what is going on out in the world, but of the uncertain consequences each agent might experience if he takes any actions upon his system. The only glaringly mutual world there is for Wigner and his friend in a QBist analysis is the partial one that might come about if these two bodies were to later take actions upon each other ("interact")—the rest of the story is deep inside each agent's private mesh of experiences, with those having no necessary connection to anything else.

But what a limited story this is: for its concern is only of agents and the systems they take actions upon. What we learn from Wigner and his friend is that we all have truly private worlds in addition to our public worlds. But QBists are not reductionists, and there are many sources of learning to take into account for a total worldview—one such comes from Nicolas Copernicus: that man should not be the center of all things (only some things). Thus, QBism is compelled as well: what we have learned of agents and systems ought to be projected onto all that is external to them too. The key lesson is that each part of the universe has plenty that the rest of the universe can say *nothing* about. That which surrounds each of us is more truly a pluriverse.

GianCarlo Ghirardi · In its standard formulation, there is no doubt that quantum mechanics assumes that there is an irreducible randomness in nature. The same holds true for the GRW theory, in which the randomness is, in a certain sense, increased. This is so because one has, in addition to the standard randomness related to the Hilbert-space description of physical systems, further stochastic processes occurring in the universe and affecting all its elementary constituents, with important consequences for the behavior of macroscopic objects.

Bohmian mechanics might be looked at as an attempt to introduce strict determinism into all natural processes. It must add, however, to the wave function the specification of the positions of all particles of a system, and it must assume that one cannot have any access to them (since otherwise one could prove quantum theory wrong). Accordingly, with respect to this approach, one may state that there is not an intrinsically irreducible randomness in nature—the random features of the theory have a merely epistemic status—but that one must accept the existence of a fundamentally uncontrollable set of variables whose knowledge is necessary in order to derive deterministic statements about natural processes.

Shelly Goldstein · It has often been argued that this is so. But this conclusion was rejected by Einstein. He expressed the belief that quantum randomness is like the randomness that sometimes arises in deterministic classical mechanics, most prominently in statistical mechanics.

Einstein believed that quantum randomness arises from averaging over our ignorance of detailed complicated initial conditions (involving the values of what are often called hidden variables), which, if only we could know them, would determine for us what appears to be a random result. The main point of von Neumann's no-hidden-variables theorem—the first of many, by many different physicists and mathematicians—was to mathematically demonstrate that Einstein was wrong: that quantum mechanics does indeed imply irreducible randomness in nature.

All such theorems are wrong. Upon examination, the proof of each is found to involve assumptions that by no means are implied by the experimental predictions of quantum mechanics (which I take "quantum mechanics" in this question to mean).

But, of course, one doesn't have to go to the trouble of examining the proofs. Bohmian mechanics is a simple counterexample to all of them. It is a deterministic version of quantum mechanics, a theory with no irreducible randomness that yields all of the quantum predictions.

And that it is a counterexample is very easy to check. The fact that many physicists continue to cite such no-hidden-variables theorems as an argument against possibilities such as Bohmian mechanics is really quite astonishing. There is simply no room for genuine controversy on this particular question.

While quantum mechanics does not imply that there is irreducible randomness in nature, it certainly does not imply that there is not. And, of course, it remains possible that there is. Some versions of quantum mechanics, like textbook quantum theory, stochastic mechanics, and the GRW theory, involve irreducible randomness, while others, like Bohmian mechanics and Everett's many-worlds, don't. As a matter of fact, what I regard as the most natural version of a Bohmian quantum field theory does involve irreducible randomness, arising in connection with particle creation and annihilation.

DANIEL GREENBERGER · Once you accept the idea of randomness as a basic feature of nature, there are things that you can do naturally that are very artificial otherwise. For example, how can a classical particle decay? The daughter particle must choose a direction to escape into, but there are no preferred directions in space. Classically, you can make a model with a built-in direction, and the particle comes out in that direction. But then you have to average over all particles, each of which has a different direction built in, in order to make the whole process isotropic (that is, independent of direction). But once you accept randomness, an individual particle can decay. It just has a certain probability to decay into each direction. So randomness gives you a certain freedom of action. The ease of having events taking place randomly is so great that I think it is here to stay. It frees one from the necessity of inventing arbitrary, nonobservable hidden-variables models.

Of course, a lot of baggage comes along with this, such as the breakdown of classical causality. This is what bothered Einstein so. It depends on what one is psychologically attuned to. Personally, I can live perfectly easily without causality in this sense. After all, things rarely turn out the way I expect them to, anyway. Why not make that a law of nature—Murphy's law of causality.

LUCIEN HARDY · There are two ways of understanding what we mean by irreducible randomness: (1) it could mean that we have no ability to predict what is going to happen, or (2) it could mean that at a fundamental level, nature is indeterministic. It is reasonably uncontroversial that quantum theory does predict the first kind of irreducible randomness. It does not, however, predict the second kind. Indeed, the pilot-wave model of de Broglie and Bohm proves that we cannot assert that it absolutely follows from the equations of quantum theory that nature is indeterministic at a fundamental level. The pilot-wave model is deterministic and provides a counterexample to such a claim.

We may, however, still be interested in the question as to whether nature is fundamentally deterministic, independently of what we can prove from quantum theory. I expect the answer will turn out to be no in an even more radical way than might be imagined. A prerequisite for a theory to be deterministic is that we have a time parameter so that we can predict the future from the past. This implies having something like a foliation of space-time into spacelike hypersurfaces. Only then can we evolve a state deterministically. In a theory of quantum gravity, however, we expect to have indefinite causal structure, and so we cannot have such a foliation. That is, in a theory of quantum gravity, we do not expect to have the necessary structure with respect to which we can even ask the question as to whether a theory is deterministic in the first place.

In fact, even probabilistic theories as usually formulated depend on having a notion of background time. A state (corresponding to a list of probabilities) at time t is evolved to time $t + \delta t$ by some transformation matrix. If we have indefinite causal structure, then we cannot adopt this point of view. In the absence of a background time, we must instead proceed by considering things such as $\mathrm{Prob}(A \mid B)$, where A and B are propositions pertaining to different arbitrary regions of space-time (where we do not assert that B is in the past of A). To deal with this type of situation (without evoking an evolving state), we need to adopt a two-stage approach. First, we must have a way of calculating whether the probability $\mathrm{Prob}(A \mid B)$ is *well-conditioned*—that is, whether it is equal to $\mathrm{Prob}(A \mid B \ \& \ C)$ for any independent condition C, so that the condition B is sufficient to determine $\mathrm{Prob}(A \mid B)$. If the probability is not well-conditioned, then we cannot sensibly talk about its actually having a value. In those cases where the probability is well-conditioned, we then go onto the second stage and calculate what the probability is equal to. I have shown how this approach can be implemented in my causaloid and duotensor frameworks. In the case that we have indefinite causal structure, I see no way round this two-stage approach. The generic situation is likely to be that the probability is revealed to be not well-conditioned at the first stage. When this happens, we might say that nature is irreducibly random beyond just being probabilistic.

ANTHONY LEGGETT · I think one has to distinguish between two questions:

(1)   If quantum mechanics should turn out to be the "whole truth," would this imply irreducible randomness?

(2)   Do the experiments that presently confirm the predictions of quantum mechanics imply irreducible randomness?

Regarding question 1, the premise is, of course, in some sense unrealistic, since we shall never be in a position to know for sure that experiments will conform to the quantum-mechanical predictions for the indefinite future. If we nevertheless make that assumption an explicit hypothesis, then I think that this would, by definition, make the apparent randomness we currently see irreducible, since even if an underlying deterministic level of reality in some sense "exists," by hypothesis we will never be able to access it.

As to question 2, my answer would be no. It seems entirely conceivable that a future revolution might restore microscopic determinism in some form or other, although we already know (see my answer to Question 8, page 175) that the resulting theory would have to have some other bizarre properties.

TIM MAUDLIN · No, as Bohmian mechanics illustrates. But this is not such an important feature one way or the other. Quantum mechanics does seem to imply some principled physical restriction on what we can predict—and perhaps what we can know—about the world. Logical positivists would have found the idea of unknowable aspects of the world empty: what you cannot reduce to observation, they would say, you cannot meaningfully discuss. But that view of meaning is untenable. Once we accept that the only way to find out about the world is to interact with it, it becomes obvious that the nature of the interaction may imply constraints about how much we can know. The miracle is that we are physically constructed in such a way as to know anything at all about the rest of the universe, not that we are not physically constructed to know, or predict, everything.

DAVID MERMIN · Yes. But "in nature" requires expansion. A more precise formulation would be that quantum mechanics implies irreducible randomness in the answers to most of the questions that we can put to nature. The probability of a photon that has emerged from a vertically oriented sheet of polaroid getting through one oriented at forty-five degrees from the vertical is irreducibly one-half, as is the probability of a slow-moving mu meson turning into an electron and a pair of neutrinos in the next microsecond-and-a-half. "Irreducible" means there is nothing we can condition the probabilities on that would sharpen them up.

Can you exploit quantum physics to make an ideal random-number generator? A distinguished Cornell computer scientist once made the long trek from the Engineering Quad to my physics-department office in the heart of the Arts College to ask me this question. He had been told this by a student, and didn't believe him. I said the student was right. I don't think he believed me either.

LEE SMOLIN · I believe that the randomness in quantum mechanics comes from our lack of control over relational nonlocal degrees of freedom. In the corresponding theories that I have developed, these nonlocal degrees of freedom connect each local degree of freedom, such as the coordinates of a particle, to many other local degrees of freedom, scattered across the universe. Whether nature ultimately has randomness in it or not (or whatever the interview question could ultimately mean), I believe that the quantum randomness represents ignorance of a deeper level of description.

ANTONY VALENTINI · Certainly not. There is at least one formulation of quantum mechanics—the pilot-wave theory of de Broglie and Bohm—that has no such randomness; therefore the conclusion cannot be drawn.

Pilot-wave theory has been extended to cover high-energy physics, with different approaches taken by different authors. In the best approach, in my view, bosons are described in terms of c-number fields, while fermions are particles with a pilot wave obeying the many-body Dirac equation. For fermions, we have to take the Dirac sea seriously. Even the vacuum is full of particles. This model was proposed by Bohm and Hiley, and its relation to quantum field theory has been clarified by Colin and Struyve. This, together with the bosonic field theory, provides a completely deterministic theory of high-energy physics, including processes such as pair creation. Some workers have proposed models of fermions in which pair creation has a fundamentally stochastic element, but those models are unnecessarily cumbersome. It would be odd if pair creation forced indeterminism upon us. In fact, it is straightforward to construct a completely deterministic theory of such processes.

Within pilot-wave theory, it has been claimed that the Born rule has a fundamental status as a preferred measure of "typicality" for the initial configuration of the universe. If this were so, in practice we would always be stuck with randomness for subsystems. But that argument inserts the Born rule by hand at the initial time. As I've said (in my reply to Question 3, see page 81), the theory certainly allows for "nonequilibrium" violations of the Born rule. Such violations for subsystems are "untypical" with respect to the global Born-rule measure. But to claim that they are therefore intrinsically unlikely is circular, because such violations are readily shown to be "typical" with respect to non-Born-rule measures. Also, I don't see a difference between "typicality" and "probability." To say that we will always have Born-rule randomness in practice, as some have argued, is in my view mistaken. There is no good reason to believe that. On the contrary, if one takes the theory seriously, it suggests that nonequilibrium will eventually be found somewhere, as I urged in my reply to Question 3 (see page 81).

I have shown that quantum nonequilibrium systems could be used to perform "subquantum measurements" on ordinary systems. These are measurements that violate the uncertainty principle and other standard quantum constraints. An extreme nonequilibrium ensemble, with arbitrarily small dispersion, could be used to perform analogues of the ideal, nondisturbing measurements familiar from classical physics. These would allow us to track the trajectories without disturbing the wave function, and to predict the future in ways that are not allowed by quantum theory. In other words, quantum randomness could be circumvented in the laboratory, if we possessed such nonequilibrium systems. And it's conceivable that such systems could exist today in the form of relic particles from the very early universe (see my answer to Question 11, page 223).

DAVID WALLACE · Okay, so that can't be answered without saying something about the measurement problem. Hidden-variables theories typically (not always) reject irreducible randomness; so do interpretations that take the wave function as just a measure of our ignorance. Dynamical collapse theories typically build in randomness explicitly.

My own view is that the only interpretive strategy that currently makes sense of quantum mechanics is the Everett (many-worlds) interpretation, for reasons I spell out in my answer to Question 2 (see page 55). And probability is really interesting from a many-worlds perspective, because there's clearly a sense in which nature is not random at all: the Everett interpretation says that the Schrödinger equation always holds, and the Schrödinger equation is deterministic. And yet there's clearly a sense in which the world at least *looks* random: when we do an experiment, we can't predict the outcome. And, in fact, the Everett interpretation guarantees that we can't predict the outcome, because it tells us that different outcomes happen in different branches.

Now, there's a line of argument that says that this just points to something incomprehensible, something unacceptable, about the Everett interpretation—that it tells us that probability in Everettian quantum mechanics doesn't make sense. And that's a serious line of argument and deserves a serious response, which I'm not going to give here in detail, but the short answer (here I'm repeating part of my answer to Question 2, see page 55) is that (1) once you start thinking hard about probability *in Everettian quantum mechanics*, you realize probability is philosophically really mysterious *in general*, but (2) it turns out that there are ways of understanding probability in quantum mechanics that don't work in classical mechanics.

So if that's right, whether there's irreducible randomness in nature according to quantum mechanics depends on your vantage point. From the third-person vantage point—put metaphorically, from God's perspective—there's no randomness in nature, everything just plays out according to the Schrödinger equation. But whether or not there's a God, *we* can't achieve that perspective. From our point of view, the randomness is irreducible.

ANTON ZEILINGER · Certainly the individual quantum event is irreducibly random (with the only exception being a situation in which the system is in an eigenstate of the measurement apparatus). It is not just that we do not know the reason why, for instance, a specific radioactive atom decays at the specific time we observe it to decay. Rather, there *is* no reason for this individual atom to decay at that specific time. The same holds for the Stern–Gerlach experiment. When we send in an x-polarized particle, it will trigger the spin-up (along z) and the spin-down detectors with equal probability. But which detector is triggered by a specific single particle is irreducibly random. To me, this irreducible randomness is one of the most important findings of physics ever. In my eyes, it is a consequence of the finiteness of information represented by the quantum state.

WOJCIECH ZUREK · The indeterminacy principle of quantum mechanics means that even if observers know the present state of the system perfectly, they cannot predict outcomes of all the measurements they are able to perform. Whether there is a state of the universe that includes all the possibilities (that is, all the branches) but evolves the superposition unitarily and therefore deterministically, and whether we are just a part of it and cannot perceive it all, is a very good question. But for us, who are stuck inside the universe, this overall deterministic evolution would anyway be irrelevant.

This conflict between global determinism and local interest of the observers—in other words, between the unitary evolution of the whole and the need to find out answers about the fragments (the local systems relevant to observers)—is responsible for quantum randomness. To see why, imagine a known perfectly entangled state of the form $|\heartsuit\rangle_S|\diamondsuit\rangle_E + |\spadesuit\rangle_S|\clubsuit\rangle_E$. One can use its symmetries to prove that an observer who knows this state must be completely ignorant about the two subsystems. The crux of the proof is straightforward: the correlations between the possible outcomes at the two ends (which we may call "system" and "environment," here denoted by S and E, respectively) can be manipulated with local interactions. Thus, one can swap $|\heartsuit\rangle_S$ and $|\spadesuit\rangle_S$ in the state $|\heartsuit\rangle_S|\diamondsuit\rangle_E + |\spadesuit\rangle_S|\clubsuit\rangle_E$ by acting only on the system S:

$$|\heartsuit\rangle_S\,|\diamondsuit\rangle_E + |\spadesuit\rangle_S\,|\clubsuit\rangle_E \quad \longrightarrow \quad |\spadesuit\rangle_S|\diamondsuit\rangle_E + |\heartsuit\rangle_S|\clubsuit\rangle_E. \qquad (*)$$

Such a swap operation exchanges the probabilities of the two possible results, $\heartsuit$ and $\spadesuit$. This is obvious, as E, the other subsystem, remained untouched. Therefore, the "new" probabilities of $\heartsuit$ and $\spadesuit$ (which before matched the probabilities of $\diamondsuit$ and $\clubsuit$, respectively) must now match the (unchanged) probabilities of $\clubsuit$ and $\diamondsuit$ instead.

The initial state of the whole composite SE, $|\heartsuit\rangle_S|\diamondsuit\rangle_E + |\spadesuit\rangle_S|\clubsuit\rangle_E$, can be restored, however, by taking the state $|\spadesuit\rangle_S|\diamondsuit\rangle_E + |\heartsuit\rangle_S|\clubsuit\rangle_E$ on the right-hand side of the expression $(*)$ above and swapping states in E:

$$|\spadesuit\rangle_S\,|\diamondsuit\rangle_E + |\heartsuit\rangle_S\,|\clubsuit\rangle_E \quad \longrightarrow \quad |\spadesuit\rangle_S|\clubsuit\rangle_E| + |\heartsuit\rangle_S|\diamondsuit\rangle_E.$$

This means that the probabilities of $\heartsuit$ and $\spadesuit$ are at the same time exchanged (by the swap on S) and unchanged (because one can restore the whole entangled state without touching S). This "exchanged and unchanged" requirement can be satisfied only in the case of perfect randomness—when the two probabilities are equal, $p(\heartsuit) = p(\spadesuit)$. If we say that certainty (for example, about the global state of SE) corresponds to probability of one (this is a normalization condition), then $p(\heartsuit) = p(\spadesuit) = \frac{1}{2}$.

This is as random as it gets. The symmetry of entangled quantum states we have used above is known as entanglement-assisted invariance, or "envariance" for short. It shows that in the quantum world, sources of unpredictability need not be limited to ignorance but can actually involve complete information about the wrong thing.

The origin of such provably perfect $[p(\heartsuit) = p(\spadesuit) = \frac{1}{2}]$ and fundamental randomness is purely quantum. A completely known state of a composite classical system necessarily consists of parts that are also perfectly known. (One says that such a pure classical state is a Cartesian product of the states of its subsystems.) By contrast, quantum entangled states can be completely known, but—as we have just seen—information about a global state can be completely incompatible with information about its parts.

Decoherence is precisely the setting where such transfers of information from local to global happen. (Dieter Zeh pointedly calls this a "dynamical dislocalization of quantum superpositions.") So envariance is based on quantum correlations set up by the same entangling interactions that cause decoherence, but its consequences are apparent without the usual tools—namely, reduced density matrices and trace—that are traditionally used in analyzing decoherence. This is important, as the trace is, in fact, an averaging procedure. It implicitly uses Born's rule. So starting with decoherence as part of the input would make a derivation (actually, *any* derivation) of Born's rule circular.

Indeed, one can arrive at decoherence—that is, deduce the loss of the significance of the local phases and the emergence of the preferred pointer states—directly from envariance. To see this, imagine an initial state that has a relative phase, $|\heartsuit\rangle_S |\diamondsuit\rangle_\mathcal{E} +$ $e^{-i\varphi} |\spadesuit\rangle_S |\clubsuit\rangle_\mathcal{E}$. Clearly, the phase can be altered by local unitaries at either end; in particular, it can be canceled out by either $|\heartsuit\rangle_S\langle\heartsuit| + e^{i\varphi}|\spadesuit\rangle_S\langle\spadesuit|$ or $|\diamondsuit\rangle_\mathcal{E}\langle\diamondsuit| + e^{i\varphi}|\clubsuit\rangle_\mathcal{E}\langle\clubsuit|$. This implies a loss of coherence—decoherence—since phases of the components of an entangled state have no influence on the local state of either S or $\mathcal{E}$.

Decoherence means the impossibility of interference and leads directly to the additivity of probabilities. This is an important conclusion, as otherwise the additivity of probability *amplitudes*—the quantum principle of superposition—would disallow adding probabilities.

The envariant derivation of equiprobability in perfectly entangled states leads to Born's rule, which says that the probability of detecting a quantum state in a measurement is given by the square of the absolute value of its amplitude in the initial superposition, $p_k = |\psi_k|^2$. All one needs to do to get to the case of unequal amplitudes is a bit of simple algebra: one can always find cases that involve more potential outcomes but have equal absolute values of the amplitudes, and that are therefore amenable to the equiprobability proof we have described. Such fine-graining then leads one directly to Born's formula—one simply adds equal probabilities corresponding to one of the original coarse-grained outcomes, and $p_k = |\psi_k|^2$ follows.

Moreover, one can *prove* the additivity of probabilities using this envariant strategy. This is because (as I have noted above) envariance leads directly to decoherence: the validity of the quantum principle of superposition is suspended. This is in contrast to Gleason's proof, which relied on the additivity of probabilities, an assumption that is far from obvious for quantum systems, where it is at odds with the quantum superposition principle—that is, the additivity of complex amplitudes—as evidenced by interference. Fine-graining assumes continuity rather than additivity: it assumes that as states become infinitesimally close to one another, probabilities they predict will also become infinitesimally close.

The envariant derivation of probabilities and Born's rule shows that randomness in a quantum world is inextricably tied to information. Indeed, one can extend the derivation to show that the observer (who is made out of quantum components) will not be able to predict the outcomes of sequences of local measurements. Thus, in spite of the deterministic evolution of the global state vector, local observers will think that measurement outcomes are random.

QUANTUM PROBABILITIES

Quantum probabilities: subjective or objective?

C HANCES ARE, so to speak, that your local radio station offers up some ludicrously accurate-sounding predictions. "Today's chance of rain," the announcer would proclaim with the sure voice essential to his profession, "is sixty-eight percent."

Not sixty, not seventy. Sixty-eight. I often wonder how this information may influence the actions of the average listener. ("I'm only taking an umbrella with me once it gets past seventy-three," or, "My threshold for going to the park is sixty-four. I'm staying home today.") I tend to wonder less, however, about how such numbers come into being. I picture a horde of weather people—not the nice-looking talking heads you see on television, but pasty meteorologists huddled together somewhere in a room full of equipment—feeding a bucketload of readings into a computer. The computer then gurgles and sputters for a while before coughing up the magic percentage, fooling us into thinking that the notoriously unpredictable discipline of forecasting the weather has joined the ranks of high-precision engineering.

This deceptive lure of device-aided numerology reminds me of my experience teaching laboratory classes. My freshman students would take a few crude measurements using wooden yardsticks, old-school balances, and fogged-up analog thermometers. Then they'd turn to their trusty pocket calculators—nothing seems to instill in students a stronger feeling of infallibility than the use of a calculator—and punch in the measurement results and proudly tell me the answer as a number with no less than a dozen digits after the decimal point. They felt they had indeed determined the value of a physical quantity with incredible accuracy.

In everyday parlance, statements of probability are often merely vague estimates. If I told you that the probability of occurrence of such and such event is five percent, then—depending on the situation at hand—you might not actually read much into this exact numerical value. Instead, you'd simply infer that you shouldn't really keep your hopes up for the event to actually take place. And this feeling wouldn't be much different if I had told you that the probability was three percent, or eight percent. "Unlikely to happen," you'd conclude.

There *are*, however, several ways of injecting meaning into the exact value of probability, all of which have been variously promoted and criticized. Equal likelihoods, for example, are often inferred from some form of symmetry argument. Laplace based such an approach on subjective indifference: if I hide a marble in one of my hands, you won't care if I transfer the marble from one hand to the other before you get to pick one hand. It is in this sense that you'll assign—from your personal point of view—equal probabilities to the alternatives "the marble is in the left hand" and "the marble is in the right hand." Besides such egocentric conceptions of probability-from-symmetry, people also often invoke physical symmetries to deduce equal likelihoods, as in the case of tossing a perfectly symmetrical coin or die, or when drawing identical balls from an urn.

Then there's the altogether different idea of associating probabilities with relative frequencies. This approach has been quite popular, perhaps because it helps us bypass the philosophically charged question of whether it is at all possible to meaningfully and objectively speak of the probability of a single event. And while symmetry arguments are limited to the assignment of the same probability value to all alternatives, relative frequencies can take arbitrary values, and so can the corresponding probabilities.

But whichever route we take, we may ask: are probabilities grounded in something in this world? Are they some kind of physical property? Can their value be objectively right or wrong?

Laplace intended his conception of probability-as-indifference to be subjective from the start. But what about the other two approaches we've mentioned—that is, equal likelihoods from symmetries of the physical situation, and probabilities from relative frequencies? Are the probabilities thus defined objective or subjective?

At a first glance, the answer would seem clear: objective, of course! After all, the symmetrical shape of the die is a feature of the physical world. And relative frequencies result from counting events—surely an objective procedure. But the matter is more subtle. The problem is not that the shape of the die or the frequencies of events lack in objectivity. It is rather, as some have argued, that these observations don't necessarily lead to a noncircular derivation of an objective notion and value of probability. (Have a look at Bruno de Finetti's 1931 essay *Probabilism* for a classic version of the argument.) How could that be?

Let's first consider the case of the die. When we state that all sides are equiprobable because of the die's symmetrical shape, then we've already made a judgment about the kinds of factors that we do (and do not) consider pertinent to the outcome of a toss. This judgment requires us to do two things. First, we need to know all the circumstances that might be relevant to the outcome. Second, we'll need to decide whether each of these circumstances actually has a causal influence.

The need for the first task suggests that the resulting probabilities would have to have an inevitably relative character, because our assessment of the equivalence of the sides of the die is contingent on the set of circumstances considered—the known circumstances, we may say. For example, we may be unaware of a slight imperfection in the material of the die, and so the sides of the die could be regarded as equal only relative to a set of circumstances that does not include this flaw.

And what about the second task? Shouldn't we at least be able to arrive at objective statements about whether each of the circumstances we have previously identified can influence which side of the die comes up? It would certainly seem so. On the other hand, there's a whole school of thought devoted to challenging this sentiment. De Finetti, for example, famously retorted with a forceful "No!" His point was that any statement about causal relationships comes down to our subjective judgment: the world throws events at us, and the best we can do is take note of conjunctions between these events and draw up our own internal maps of what we believe amounts to causal connections. But subjective these maps will remain. And therefore, de Finetti contends, probabilities inferred from symmetries—that is, from the equivalence of cases as judged by their differing only in circumstances that are regarded as causally unrelated to the outcome—ought to be regarded as subjective.

It is often said that the objective character of probability manifests itself in the observation of relative frequencies that approach, in the limit of a great many repetitions of the experiment, the (objective) probability value. It is in this sense, it is claimed, that we can prove, a posteriori, the correctness of an a priori evaluation of probability.

But this argument is not without difficulties. Take the classic example of a coin toss. Having assigned a probability of one-half to heads, we'll toss the coin many times. Will we observe the exact same number of heads and tails? Most likely, we won't. The usual response now is: just keep tossing, and it will be increasingly *unlikely* that the relative frequency of getting heads will deviate *much* from a value of one-half. But short of just taking this statement as a definition of probability, what can we possibly mean by the word "unlikely"? How do we quantify the term "much"? And in what sense, then, could we ever claim that the observation of relative frequencies actually confirms the supposedly objective value of a probability assignment, without plainly decreeing from the start that probability is evaluated via frequencies?

This is no place for me to take sides in the unceasing philosophical quarrels fueled by such questions. Instead, the little musings above are simply intended to give you a hint of the fact that the question of whether probabilities can be given an objective meaning is generally full of pitfalls. And I haven't even mentioned quantum theory yet! Indeed, you may now say, isn't the story of probabilities an altogether different one in the quantum setting?

There's certainly one sense in which the story takes a new turn. In classical physics, probabilities didn't really enjoy fundamental status. They seemed ultimately dispensable. They were just casual expressions of our own ignorance, of practical limitations on our making statements of certainty, of a coarse-graining process that ignores—by deliberate choice or practical necessity—the finer details of the situation at hand. By contrast, probabilities take center stage in quantum mechanics. The point of the theory that makes contact with our observations is formulated in terms of probabilistic statements. Does this mean that probabilities have become, in some sense, more objective than in the classical scenario?

There are at least two ways to read this question. We may take the term "objective" to refer to the question of objective chance—that is, to the question of whether the probabilistic character of quantum theory suggests that nature coughs up events in a fundamentally random fashion, and also to the question of whether we can transcend the probabilistic description in favor of a deterministic account. In this case, we're mostly treading on territory already explored in the previous chapter (see Question 5, *Randomness*).

But there's another aspect to the character of quantum probabilities. This is, in essence, the issue discussed above: are the probability values coming out of quantum theory to be understood as something objective and physical, or are they merely expressions of our subjective judgments?

First, let's note that the possible answers we may give to this question are not necessarily dependent on how we choose to reply to the question of randomness. Strictly speaking, we're dealing with two separate issues here: one is concerned with probability assignments, the other with the manner events come into being. For instance, you may well believe in a fundamentally indeterministic universe (that is, in "objective chance") and yet deem all probability-value assignments purely subjective. There's no contradiction here.

Second, at the risk of stating the obvious, it's worth mentioning that attitudes toward the objective-versus-subjective issue of quantum probabilities will likely be influenced by, or be even directly related to, the interpretation of quantum states (see Question 4, *Quantum States*). The reason is simply that there exists a one-to-one correspondence between quantum states and probabilities: not only do quantum states fix probabilities via Born's rule, but probability assignments for a suitably chosen set of measurements also uniquely determine the quantum state. So if you think that quantum states represent facts about the world, then there's a good chance that you'll take quantum probabilities to be objective as well. Conversely, if you believe that, say, any notion of objective probabilities—irrespective of whether we're talking about classical or quantum theories—is ultimately doomed, then you'll probably gravitate toward a subjective, epistemic interpretation of quantum states.

You'll find that the answers below play out in full detail these manifold relationships that inform our interviewees' take on the nature of probabilities. You'll be the judge, but to me at least one thing seems fairly certain: probabilities will continue to mystify us for some time to come, and quantum theory has probably shifted rather than lifted the fog.

GUIDO BACCIAGALUPPI · This depends, as usual, on the interpretational approach to quantum mechanics one adopts. Before making sweeping generalizations, I should probably run over a brief summary of some philosophy of probability, so as to avoid misunderstandings.

Subjectivists about probabilities claim that probabilities are nothing but degrees of belief, and that they are at most subject only to various forms of rationality constraints, to avoid various forms of inconsistency in one's beliefs or in updating one's beliefs in the light of new evidence. In this approach, what might otherwise appear to be an objective probability becomes the intersubjective agreement among different agents' degrees of belief. This approach, and its viability, were originally established by Bruno de Finetti in the 1930s.

Subjective probabilities are recognized as legitimate by everyone, including believers in objective probabilities. The latter, however, take it as rational that one should ground one's subjective probabilities on the objective ones (if known). Moreover, this requirement, known as David Lewis's Principal Principle, acts as a powerful tool for analyzing what objective probabilities might be in the first place.

Opinions differ as to whether objective probabilities in this sense (also called chances) might exist also in a deterministic setting, for example, for classical coin tosses. At least in a pragmatic sense, they do exist also in such settings, as even the most committed subjectivists will have to recognize—if they admit that as a shortcut for determining their priors, they might actually inspect the coin to see if it is evenly weighted!

Returning to quantum probabilities, these have traditionally been regarded as the paradigm of objective probabilities, and there is a sense in which probabilities are, indeed, objective in all three of the classic approaches to foundations: pilot-wave theory, spontaneous-collapse theories, and Everett interpretations.

Probabilities in the de Broglie–Bohm pilot-wave theory have the same status as in classical statistical mechanics. The fundamental theory is deterministic, and probabilities are epistemic, that is, they arise due to ignorance. Still, given some assumptions on initial conditions, one can confidently use the (objective) wave function to guide one's choice of subjective probabilities, and in this sense probabilities are objective chances even in the de Broglie–Bohm theory.

Probabilities in spontaneous-collapse theories are primitives of the theory, and they have the same objective status as the irreducible probabilities of standard collapse formulations of quantum mechanics (say, those of Dirac or von Neumann). Perhaps they are most naturally analyzed in terms of propensities (if they make sense!), but perhaps different options are open.

Probabilities in Everett have long been the subject of debate, and in recent years there have been proposals, in particular by Deutsch and by Wallace, to define and justify them in terms of rational-decision theory for splitting agents. While these probabilities are at first sight subjective, the claim is that rational constraints will determine them uniquely as fixed by the wave function. In this sense, Everettian probabilities turn out to be the best example to date of objective chances as defined in Lewis's Principal Principle!

The subjective view of probabilities, instead, has been traditionally neglected in the philosophy of quantum mechanics, despite the fact that de Finetti's views have been very influential within the philosophy of probability itself. This long-unexplored avenue has been at last taken up in the "Quantum Bayesian" view advanced by Caves, Fuchs, and Schack, which is directly modeled on de Finetti's ideas, so that now quantum states become entirely subjective entities. As Fuchs puts it, when you walk out of the room, the quantum state of the system walks out with you!

The various approaches in the foundations of quantum mechanics thus map neatly onto the various approaches in the philosophy of probability.

ČASLAV BRUKNER · I do not think that drawing a clear-cut line between subjective and objective quantum probabilities is possible and useful. This is because each of the options is justified, but for different reasons.

Quantum probabilities are subjective in the sense that one can embrace a generally personalist, decision-theoretic view of probability, which enables intelligible reasoning using statements of uncertainty. This broadly Bayesian approach is not specific to quantum theory, but is applicable to any probabilistic theory that aims to maintain coherence of one's probabilistic beliefs in the light of newly acquired relevant data.

Quantum probabilities are, I believe, also objective (or irreducible) in the sense that they cannot be understood as stemming from one's subjective ignorance about some preexisting "real state of affairs."

Quantum probabilities are subjective (again) because observers may disagree about which pieces of hardware in the laboratory are given a quantum description. This is the essence of the Wigner's-friend thought experiment, to which I devote my answer to Question 11 (see page 217).

Let me comment on the objective, or irreducible, aspect of quantum probabilities. Once we accept that probabilities are irreducible, the role of the observer is explicitly introduced into the theory. This is for the simple reason that she, by choosing the measuring device, can decide on the basis of her free will which measurement context will be realized in the actual run of the experiment. But due to the randomness of the individual quantum outcome, she cannot influence which particular outcome will occur in the chosen context. Zeilinger put the point this way: "The observer has a qualitative but not a quantitative influence on reality." Therefore, the observer in quantum mechanics has a participatory role in forming reality. By contrast, in a theory describing observation-independent reality, like in classical physics, the observer has only a passive role, as her actions can always be interpreted as revealing the values of physical quantities that all coexist and are independent of which experiment is actually performed.

The reader may object that my explanations are anthropocentric and that I overestimate the role of the observer. Let me be clear: I am not saying that quantum theory makes sense, or is valid, only if observers are there. The "measurement context" can be induced by the prevalent basis of the environment surrounding the quantum system, without invoking any observers. Yet the mere possibility that an observer *can* choose the measurement context, isolating the quantum system from environmental interactions that select a preferred basis, is exactly what gives her a fundamental role in the act of observation. This is a major intellectual step forward over naive classical realism.

JEFFREY BUB · In my answer to Question 4 (see page 94), I characterized a quantum state as a credence function or a bookkeeping device for keeping track of probabilities. In one sense, this is a subjectivist interpretation of quantum probabilities. As I see it, though, the interpretation is objective, or intersubjective, because the credences specified by the quantum state are uniquely determined, via Gleason's theorem, by objective correlational constraints on events in the nonclassical quantum event space defined by the subspace structure of Hilbert space. So, in the sense of Lewis's Principal Principle, Gleason's theorem relates an objective feature of the world, the nonclassical structure of objective chances, to the credence function of a rational agent.

The objective chances here need not be interpreted as irreducible modalities, or propensities, or necessary connections in nature, but can be understood in a metaphysically minimal Humean or Lewisian sense as simply features of the pattern of actual events: numbers satisfying probability rules that are part of the best system of such rules, in the sense of simplicity, strength, and fit, characterizing the "Humean mosaic," the collection of everything that actually happens at all times.

ARTHUR FINE · To continue where Question 5 left off. Quantum theory provides tools for modeling phenomena that are well-defined experimentally. The probabilities built into these models via Born's rule are contextual. That is, they are not probabilities for what Bell called "beables" (for example, for an atom being here rather than there), but probabilities for outcomes of specific, complete measurements. No measurement is complete unless the result is available to a keeper of the record. The record could be a mechanical printout. Nevertheless, availability means available to us. Thus, subjectivity is built into the toolkit of quantum theory, regardless of the interpretive stance that one takes to the theory as a whole. (Is this true for dynamical-collapse theories? Maybe my argument here is too quick?) Heisenberg emphasized this aspect of the theory, referring to the wave function as having both an objective and a subjective aspect. It was this subjective aspect of the quantum probabilities that bothered Einstein most (the "risky game with reality"), and not just the fact of probabilities (indeterminism).

CHRISTOPHER FUCHS · "Subjective" is such a frightening word. All our lives we are taught that science strives for objectivity. Science is not a game of opinions, we are told. That diamond is harder than calcite is no one's *opinion*! Mr. Mohs identified such a fact once, and it has been on the books ever since.

In much the same way, quantum theory has been on the books since 1925, and it doesn't appear that it will be leaving any time soon. That isn't lessened in any way by being honest of quantum theory's *subject matter*: that, on the QBist view, it is purely a calculus for checking the consistency of one's personal probabilities. If by subjective probabilities one means probabilities that find their only source in the agent who has assigned them, then, yes, quantum probabilities are subjective probabilities. They represent an agent's attempt to quantify his beliefs to the extent he can articulate them.

Why should this role for quantum theory—that it is a calculus in the service of improving subjective degrees of belief—be a frightening one? I don't know, but a revulsion or fear does seem to be the reaction of many if not most upon hearing it. It is as if it is a demotion or a slap in the face of this once grand and majestic theory. Of course, QBism thinks just the opposite: for the QBist, the lesson that the structure of quantum theory calls out to be interpreted in *only* this way is that the world is an unimaginably rich one in comparison to the reductionist dream. It says that the world has excitement, risk, and adventure at its very core.

Perhaps the source of the fear is like I was taught of "that marijuana" in my little Texas town: use it once, and it will be the first step in an unstoppable slide to harder drugs. If quantum probabilities are once accepted as subjective, somewhere down the line Mr. Mohs's scale will have to disappear in a great puff of postmodern smoke. There will be no way to enforce a distinction between fact and fiction, and the world will be anything our silly imaginations make up for it!

The first symptom is already there in a much more limited question: if quantum probabilities are subjective, why would an agent not make them up to be anything he wants? Why not pull them from thin air? The defense to this little question is the same as the defense against the "inevitable" postmodern horrors. My colleague Marcus Appleby put his finger on the issue sharply when he once said, "You know, it is *really hard* to believe something you don't actually believe!" Why would one assign arbitrary probabilities—ones that have nothing to do with one's previous thoughts and experiences—if the whole point of the calculus is to make the best decisions one can? The issue is as simple as that.

GianCarlo Ghirardi · Once more, within the standard scheme, the probabilities are basically subjective, as evidenced by the systematic reference to the outcomes of measurement processes and the essential role of the observer. The probabilities disappear in Bohmian mechanics if one assumes a complete specification of the initial situation. But, as already stated, this amounts to assuming that the hidden variables can be controlled, a hypothesis that nobody would take as appropriate, since it implies that quantum mechanics is de facto an incomplete theory that might be easily falsified.

In collapse models, the probabilities are, in a certain sense, always objective. In line with the fact, however, that these theories agree fully with quantum mechanics concerning all microscopic processes, a microsystem can be in a state for which there are nonzero probabilities of obtaining different outcomes if the system is subjected to a measurement, and for which there are nonzero probabilities of inducing perceptually different final situations as a consequence of the interactions between microscopic and macroscopic systems. In particular, one should never forget the fact that measurement processes do not enjoy any particular logical or ontological status within collapse theories. In spite of the fact that a superposition of microscopically different states has precise probabilities of triggering amplification processes leading to macroscopically distinct affairs, the final situation turns out to be, in any case, macroscopically and perceptually definite. The basic quantum nonepistemic probabilities at the microlevel become objective macroscopic facts, which occur in agreement with the probabilities of standard quantum mechanics. It makes no difference whether the micro-to-macro amplification has been purposely produced by a human being, or whether it occurred naturally—for instance, because different trajectories of a microsystem triggered an avalanche process in a Geiger counter, in a supernova, or in the perceptual apparatus of a conscious being.

SHELLY GOLDSTEIN · In versions of quantum mechanics with irreducible randomness, quantum probabilities are, of course, objective. But even in versions of quantum mechanics that are entirely deterministic—for example, even in Bohmian mechanics—the quantum probabilities, while in a sense subjective, can also be regarded as objective. I need to explain.

It might appear inevitable that quantum probabilities be subjective in Bohmian mechanics. After all, Bohmian mechanics is deterministic, and as such it would seem that for Bohmian mechanics, probabilities can arise only because of our ignorance of initial conditions and thus must be subjective. But in a Bohmian universe, the deterministic dynamics typically produces random patterns of events. For example, after sending, one by one, a great many quantum particles into a double-slit arrangement, an interference pattern described by quantum (i.e., $|\psi|^2$) probabilities will typically emerge. Such patterns are entirely objective and have nothing to do with our knowledge or lack of knowledge.

But it does happen to be the case that in a typical Bohmian universe the sort of knowledge of initial conditions that would make it possible for us to know beforehand what would happen in an experiment involving quantum randomness is absolutely unattainable. So it is natural to wonder what probability measure, if any, gives the subjective probabilities—about, say, positions of particles—that correspond to our lack of detailed knowledge.

Now, it's not entirely clear to me what should be meant by such subjective probabilities. Nonetheless, in view of the objective character of the $|\psi|^2$ quantum probabilities (which, in Bohmian mechanics, describe empirical distributions—patterns of relative frequencies over a collection of real-world systems), it is hard for me to imagine doing better than regarding the subjective probabilities as given by the quantum probabilities as well. (Insofar as physics is concerned, however, the question of what are the subjective probabilities in Bohmian mechanics can be pretty much ignored: the empirical distributions suffice for more or less all physical uses of probability.)

DANIEL GREENBERGER · The conventional wisdom of quantum theory is that the wave function represents only what happens to an ensemble of particles and has no meaning for an individual particle. One prepares an ensemble of particles with the same initial conditions, and one does the same experiment over and over, and $|\psi|^2$ represents the probability of what will happen in a given experiment. Only a frequency interpretation is viable. Well, this is one way of looking at the problem, but it takes most of the fun away. For example, in the case of Schrödinger's cat, instead of thinking of the cat as being in a superposition of half-alive and half-dead, one instead merely says that half the time a measurement will reveal the cat to be alive, and half the time it will be dead. One cannot think about it as in a superposition, since one only measures the resolution of the situation. But actually, the wave function has more subtlety than that.

That interpretation tends to give the impression that there is something truly objective about the wave function itself. But one can always make a gauge transformation on the system. The gauge-invariant quantity is $p - (e/c)A$, where A is the vector potential, and even if there is no magnetic field present, one can always choose A to be the gradient of some potential $\varphi(x)$, which will change ψ to $\psi e^{i\varphi}$. Then even the wavelength of the wave will change, so that the wavelength is not an objective quantity. What is objective depends on the phase difference between two paths to the same point—in other words, the interference pattern, which is what one measures. But the actual waves that are interfering are not themselves objective, in the sense that one can make each of them look very different by making a gauge transformation.

If there is a moral here, it is that the wave function is nonlocal in a special sense. It describes the sum-total of all experiments that can be done with a given Hamiltonian, even those that one is not contemplating performing. The outcomes of all these experiments must be consistent, and that tells you that the wave function is sensitive to what is going on everywhere in space and time. For example, in the Aharonov–Bohm effect, one can have a magnetic field present in a region where the wave function is zero, but if one did do an experiment in that region, one would feel the magnetic field, and the wave function is affected by the outcome of such an experiment, even in an experiment that avoids the region. So while probabilities themselves are objective, the wave function is much more subtle than that and contains information relating to other experiments that one hasn't done. Most of the standard interpretations don't worry about this, and tend to be incomplete in this sense.

On the question of whether an individual particle can be represented by a wave function, if so it cannot be in the sense of a probability representation, since one is not performing a series of experiments and counting up the relative frequencies. Rather, one is finding out something about the individual system, so such an interpretation necessarily involves the concept of information. Feynman gives the argument that if you have a pure state, then there must be some measurement that has a hundred percent probability. (He gives the example that in a one-particle diffraction pattern, every single particle knows to avoid a minimum.) Therefore, this tells you certain information is available for each particle and should be contained in the wave function, which should pertain to individual particles.

But also there are many experiments whose interpretation seems more natural when relating to information about individual particles. If you create a singlet state of two electrons, and you measure one to be spin up, then you know for sure that the other will be spin down. You can say that in an ensemble, when you measure the spin of one, you know the spin of the other member of the ensemble, but it is much more natural, after measuring the state of particle A, to say that you have learned something about particle B, its partner, rather than about one member of an abstract pair. So talking about individual particles is sometimes the natural thing to do, and talking about ensembles is sometimes the natural thing to do. Both are viable interpretations in certain circumstances.

LUCIEN HARDY · There are three main interpretations of probability: (1) that it corresponds to subjective degrees of belief (this is sometimes called the Bayesian approach), (2) that it corresponds to an objective property (sometimes called a propensity) possessed by a system, (3) that it should be regarded as the relative frequency in the limit of an infinite number of trials. In my opinion, none of these interpretations are satisfactory.

The problem with the objective and frequency interpretations is basically the same. Imagine that the probability is p. In both of these interpretations we want to assert that in the limit of a large number of trials, the relative frequency will, with high probability, be very close to p. But this makes reference to the notion of probability itself and so is circular.

In the Bayesian (or subjective) approach, an agent starts with a prior belief that he updates when he obtains new data. If the agent has prior beliefs that are too definite, then Bayesian updating will leave them unchanged. For example, if an agent believes that the probability for a coin to come up heads is equal to one, then he will, under Bayesian updating, continue to believe that the probability is one, even if it comes up tails five hundred million or so times out of one billion trials. In reality, when presented with such overwhelming data, an agent is likely to undergo something like a personal crisis and change his belief to something less than probability one (closer to a probability of a half). With the given prior, however, there is no way the mathematics of Bayesian updating can guide him through this crisis.

If such a crisis were only likely to happen to people who held unduly firm beliefs (such as probability one), then this would not be a problem. After all, it is unscientific to have no doubts at all. Unless ones prior has a nonzero probability for every conceivable eventuality, however, it is always possible to conceive of subsequent data that will lead the agent to such a crisis.

Further, it is completely impractical to attempt to list every conceivable possibility (and allocate a probability). One pertinent example is the following. Imagine we have a machine that outputs the numbers zero and one, and suppose agent A believes that the distribution is identically and independently distributed (this would be the case for subsequent tosses of a coin). Zero probability is given to distributions that do not have this property. This is a standard assumption in proving the de Finetti theorem, which states that subjective beliefs will converge on relative frequencies (which is essential if the Bayesian account is to be able to account for the sort of phenomena we see in quantum theory). Unbeknownst to agent A, however, the machine is actually outputting the digits of π in binary notation. After ten million digits, agent B points this out to agent A. But agent A is committed to Bayesian updating from his original prior and cannot take this into account. Agent B can now consistently win bets against agent A. The only hope agent A has is to undergo a personal crisis of the sort mentioned above, but there is no way the mathematics of Bayesian updating can guide him through this. A subjectivist account of probability ought to have mathematics to guide an agent through such a crisis.

Given the problems with interpreting probability, it is something of a disaster for modern physics that probabilities appear to have to play such a fundamental role. Ultimately, I am holding out for a deeper idea with respect to which the notion of probability is effective or emergent. One attempt in this direction is that by David Deutsch, David Wallace, and others to show how probability can emerge in the context of the many-worlds approach. This attempt is interesting, since the many-worlds interpretation is actually completely deterministic (the wave function evolves unitarily). We have the apparent emergence of probabilities in a situation where, fundamentally, there are no probabilities. While I do not think this approach is successful in doing what it sets out to do, it is the only attempt of this nature I know of. My own hope is actually for the opposite. I hope that the concept of probability will emerge in the context of a theory that is radically random along the lines I outlined in my answer to Question 5 (see page 118).

ANTHONY LEGGETT · I tend to side, rather generally, with the "objectivist" (frequentist) interpretation of the concept of probability, and I see no particular reason to make an exception in the case of quantum mechanics. Of course, this means that I have to regard, for example, the idea of a "wave function of the universe" as meaningless. But I believe there are already other good reasons to draw this conclusion.

TIM MAUDLIN · Under one interpretation, this question is the determinism question again: one could say that in a deterministic world all probabilities are based on ignorance of the exact initial conditions. But it is more plausible to argue that some objective probabilities are compatible even with determinism. To say that a perfectly uniform cubical classical die has a one-sixth chance of landing on each of its faces when shaken in a box is to make a claim grounded in the geometry and physical composition of the die and the laws of physics, not in anyone's ignorance. Similarly, the claim that an ice cube in warm water is overwhelmingly likely to melt appears to be objective: after all, it is a plain objective fact that such ice cubes always do melt. So if we agree that the laws of physics are compatible with the ice cube not melting, the high probability of melting is grounded in some physical circumstance that goes beyond the narrow content of the laws. The interesting question is what exactly the relevant physical circumstance is.

At the end of the day, the probabilistic predictions of quantum theory are too precise and robust and accurate to have any important dependence on subjective considerations. Those predictions would correctly describe the world irrespective of anyone's beliefs or desires or ignorance. In this sense, the probabilities are objective.

DAVID MERMIN · In a message in a bottle that I tossed into the sea about fifteen years ago—the "Ithaca Interpretation of Quantum Mechanics" (IIQM)—I firmly declared quantum probabilities to be objective properties of the physical world. The bottle was noticed by Chris Fuchs, who introduced me to subjective probabilities and to his collaborators Carl Caves and Rüdiger Schack. I found their point of view so intriguing that I have left the bottle adrift ever since, but in thinking about it today, I wonder why I was so readily persuaded that their view of probability was incompatible with mine.

In declaring quantum probabilities to be objective, I had in mind two things. First, that the role of probability in quantum mechanics is fundamental and irreducible. Probability is not there just as a way of coping with our ignorance of the underlying details, as in classical statistical mechanics. It is an inherent part of how we can understand and deal with the world. Second, that probabilistic assertions are meaningful for individual systems, and not just, as many physicists would maintain, for ensembles of "identically prepared" systems. I believe Fuchs et al. would agree with both propositions.

I also explicitly rejected Karl Popper's promotion of "propensities" into objective properties of the systems they describe. It was not my intent to reify probability, or if it was—fifteen years later it is hard to be sure—I hereby disassociate myself from the foolish person I might then have been. Admittedly, my IIQM motto that "correlations have physical reality" (though correlata do not) sounds dangerously like a Popperian reification of probability. But it is not. In my two IIQM papers, I used the phrase "has physical reality" to mean "can be accounted for in a physical theory," particularly when I insisted that conscious experience has reality, but not physical reality.

Thinking about this today, I see that to be compatible with the point of view of Fuchs et al., I should also have maintained that correlations have physical reality but not reality. "Physical reality" is not, as I seem to have implicitly maintained fifteen years ago, just a subset of "reality." Neither is contained in the other. Conscious awareness belongs to reality and not to physical reality, but correlation belongs to physical reality and not to reality. Putting it like that, I now see that this goes a way toward reconciling the IIQM not only with Fuchs et al., but also with Adan Cabello's demonstration that whatever the sense in which correlations have physical reality, it cannot be that their values are EPR "elements of reality."

So I would say that quantum probabilities are objective in the sense that they are unavoidable. They are intrinsic features of the quantum formalism—not just an expression of our ignorance. And they apply to individual systems and are not just bookkeeping devices for cataloguing the behavior of ensembles of identically prepared systems.

But because quantum mechanics is our best strategy for organizing our perceptions of the world, quantum probabilities have a strategic aspect. Strategy implies a strategist, and in that sense quantum probabilities are subjective.

Strategic as the use of probability may be, the fact that a free neutron has a slightly less than fifty-fifty chance of decaying within the next ten minutes strikes me as just as objective a property of the neutron as the fact that its mass is a little less than 1,839 times the mass of an electron. Of course, one can, and some of my friends do, conclude from this that dynamics itself (in which mass is a parameter, and out of which emerges the half-life) is as subjective a matter as probability. Wary as I am of reification, I'm not ready to take that step.

LEE SMOLIN · Objective, but not fundamental.

ANTONY VALENTINI · From a de Broglie–Bohm point of view, the situation is more or less the same as in classical statistical mechanics. At the fundamental level, there is no such thing as probability. The universe contains a huge number of degrees of freedom evolving according to deterministic equations of motion. In principle, that's all there is to it. In practice, for large numbers of similar and approximately independent systems, it's useful to work with a distribution of configurations and to consider the theoretical limit of an infinite ensemble. This is only a practical tool. The interpretation of that distribution depends on what approach you take to probability theory. This leads to interesting questions in the foundations of probability theory, but those questions have nothing particularly to do with pilot-wave theory. They arise in a similar way in ordinary classical statistical mechanics.

Because the theory is fundamentally deterministic, it may seem natural to characterize a probabilistic description as "subjective." On the other hand, the statistics we see in the lab are properties of the actual configuration of our universe, so in that sense they are "objective."

Questions about the foundations of probability theory arise not only in statistical mechanics, but in any application of probability theory or statistical inference—for example, to genetic populations on earth or to the distribution of galaxies in deep space. De Broglie–Bohm theory has nothing new to add to such debates, and so I try to avoid them.

We should avoid getting distracted by such questions in a de Broglie–Bohm context, when the focus should be on finding evidence for the details of the underlying dynamics. There's a parallel with atomic physics in the late nineteenth century. Boltzmann's central belief was that everything was made of atoms, and that macroscopic physics could be reduced to atomic physics. In retrospect, it's a pity that he got distracted by controversies relating to the foundations of probability theory, as well as by questions concerning time reversal, and so on, when the priority was to find evidence for atoms. Similarly, while I agree that conceptual questions about the meaning of probability are interesting, I think that in the context of pilot-wave theory they are at best distracting us from more important issues, and at worst obscuring the physics of the theory—which is fundamentally a nonequilibrium physics that violates quantum mechanics.

As an example of the sort of thing I mean, some people in quantum foundations talk as if it is problematic to consider probabilities for the "whole universe." And yet cosmologists not only do so every day, they are also busy testing primordial probabilities experimentally by measuring temperature anisotropies in the cosmic microwave background. By making statistical assumptions about a theoretical "ensemble of universes," cosmologists are able to test probabilities in the early universe, such as those predicted by quantum field theory for vacuum fluctuations during inflation. One can question what the ensemble of universes refers to. Is it a subjective probability distribution? Or, is the universe we see in fact a member of a huge and perhaps infinite ensemble, as is the case in theories of eternal inflation? Those are interesting questions, but only tangentially related to the ongoing experimental tests. This point is related to my critique, in my answer to Question 4 (see page 103), of supposed problems with contingency for the universal wave function. I don't see why people working in quantum foundations should worry about such matters in a cosmological context, when cosmologists do not.

DAVID WALLACE · Objective, definitely. I'm much more confident of that than I am of any particular interpretation.

I think this is another place where the abstractness of quantum information can be a bit misleading. It can seem kind of tempting to suppose that when we talk about the probability of a qubit being measured to be in a certain state, we're just talking about our subjective assessment. But quantum probability doesn't just apply to qubits, it applies to the *half-life of uranium-235*, and I really can't make sense of the idea that the decay rate of uranium isn't some fact about the world. When I say, "You can make nuclear weapons out of plutonium, because it has a really high probability to undergo fission in such-and-such situations, so we shouldn't let terrorists get hold of it," am I *really* not saying anything objective about plutonium? This is one of the places where I find the situation in the field kind of confusing, because some really smart people who I respect a lot seem happy with saying this, and I can't understand that. But then, I think people often say that about supporters of the many-worlds theory.

I suppose I should point out that there are ways and ways for probability to be objective. According to Everettian quantum mechanics, it's identified with mod-squared-amplitudes of the branches, so it's objective, but can't be defined except in situations where decoherence gives us a branching structure. According to (most) hidden-variables theories, it's derived from the probability distribution over the hidden variables (and so *that* has to be objective). According to dynamical-collapse theories, it's written into the equations.

ANTON ZEILINGER · Quantum probabilities are objective in the sense that everyone with the same information agrees to use the same quantum state to calculate probabilities. This does not mean that the quantum state has to be the same for everybody describing or observing the same specific experiment. Quantum states are dependent on the specific information an individual observer has. For example, for spatially separated, entangled systems, quantum states can be dependent on the reference frame in which an observer moves. Or, in the Schrödinger-cat case, the quantum state the cat would use is certainly different from the quantum state used by an external observer.

WOJCIECH ZUREK · Subjective probabilities are (since Laplace) thought to be "the fault" of the observer, of his subjective ignorance about the objective state existing out there. To insist on discussing our (quantum) universe in terms of classical

concepts is dangerous, and this distinction between subjective and objective probabilities is a perfect example of where our language fails us.

My favorite view of the origin of probability is based on the symmetries of entanglement ("entanglement-assisted invariance," or "envariance"; see my answer to
Question 5, page 122). There, the need to use probabilities arises from the observer's
knowledge of the wrong thing—for example, of an entangled state of a bigger composite system. Because the observer knows the entangled state of the whole, one can
show that he cannot predict the outcome of a measurement on a part.

But this is not really subjective ignorance in the sense of Laplace or Bayesians.
Probabilities follow as a consequence of symmetries of a known quantum state, and
these symmetries are an objective property of that state. So in that sense, probabilities are objective. They represent objective ignorance: information that is objectively
inaccessible to the observer, because he is in possession of the complementary information.

This objective ignorance should be contrasted with the subjective ignorance often
invoked in discussions of probability. In quantum theory, one might say that Bohr's
complementarity principle mandates objective ignorance, and envariance, a quantum
symmetry of entangled states, allows one to quantify its extent and its nature.

Still, objective ignorance and the resulting probabilities derive from what the observer knows. Does that make objective ignorance and quantum probabilities subjective?

I think many of us have way too much confidence that our everyday language
can capture everything that we will ever encounter in our quantum universe. Clearly
(and as Bohr insisted!), it should capture whatever crosses into the classical realm.
But mathematics is the language of quantum theory, and trying to translate it into
everyday language is often simply impossible.

QUESTION 7

THE MEASUREMENT PROBLEM

———

The quantum measurement problem: serious
roadblock or dissolvable pseudo-issue?

AH, YES, THE MEASUREMENT PROBLEM! No other subject has served so reliably as a catalyst for foundational debates. Every interpretation of quantum theory owes its existence to somebody making a new assault on this perennial problem. And then there are those who dismiss concerns about the measurement problem with a deflating "*What* measurement problem?" (or go as far as to bluntly suggest that the problem merely "refers to a set of people"). But even they must make their claim in the context of a particular interpretation if they are to legitimately justify their a priori dismissal. And so every foundationalist will have to grapple with the measurement problem one way or the other and will at some point be called to testify.

But what exactly *is* the measurement problem? I have found that everyone seems to have a somewhat different conception of the affair. One way of identifying the root of the problem is to point to the apparent dual nature and description of measurement in quantum mechanics. On the one hand, measurement and its effect enter as a fundamental notion through one of the axioms of the theory. On the other hand, there's nothing explicitly written into these axioms that would prevent us from setting aside the axiomatic notion of measurement and instead proceeding conceptually as we would do in classical physics. That is, we may model measurement as a physical interaction between two systems called "object" and "apparatus"—only that now, in lieu of particles and Newtonian trajectories, we'd be using quantum states and unitary evolution and entanglement-inducing Hamiltonians.

What we would then intuitively expect—and perhaps even demand—is that when it's all said and done, measurement-as-axiom and measurement-as-interaction should turn out to be equivalent, mutually compatible ways of getting to the same final result. But quantum mechanics does not seem to grant us such simple pleasures. Measurement-as-axiom tells us that the post-measurement quantum state of the system will be an eigenstate of the operator corresponding to the measured observable, and that the corresponding eigenvalue represents the outcome of the measurement. Measurement-as-interaction, by contrast, leads to an entangled quantum state for the composite system-plus-apparatus. The system has been sucked into a vortex of entanglement and no longer has its own quantum state. On top of that, the entangled state fails to indicate any particular measurement outcome.

So we're not only presented with two apparently mutually inconsistent ways of describing measurement in quantum mechanics, but each species leaves its own bad taste in our mouth. When confronted with measurement-as-axiom, many people tend to wince and ask: "But ... what counts as a measurement? Why introduce a physical process axiomatically? What makes the quantum state collapse?" And so on. But measurement-as-interaction delivers no ready-made remedy either. As we have seen, the interaction leads to nothing that would resemble the outcome of a measurement in any conventional sense of the word.

What, then, is the correct and more fundamental description? Measurement-as-axiom? Measurement-as-interaction? Both? Neither? Is the measurement problem an issue of physics? Of formalism? Of epistemology? Or is it a beast of an altogether different nature? Is it a dire warning that perhaps something is irrevocably rotten at the very core of quantum mechanics, something that could prompt this theoretical edifice to collapse at any moment, like a house haphazardly erected on swampy grounds? If that's the case, what tools do we have at our disposal for injecting strength and resilience into the theory's foundations? Or is there, in fact, no cause for alarm once we learn how to properly look at the whole affair? Perhaps it is all just a red herring.

If all these queries hadn't already amounted to weighty matter, the central issue brought to light by the measurement problem spreads in fact well beyond the confines of measurement proper. For whenever we apply the quantum-states-plus-unitary-evolution description to a collection of interacting systems, these systems become entangled. The final quantum state, then, represents a monstrous abstract soup from which we no longer seem to be capable of picking out any definite properties for the individual systems. And because the state is governed by the Schrödinger equation, its evolution is deterministic and in principle reversible. So where in this picture do we find the definite physical quantities and actual events of our experience? Is the seemingly objective, irreversible occurrence of those events just an illusion? Or ought events to be regarded as a notion physically prior to quantum states, and would we therefore be putting the cart before the horse in trying to extract events from quantum states?

GUIDO BACCIAGALUPPI · I believe one can be justified in adopting two attitudes to the measurement problem: one can take it to be a burning issue in foundations of quantum mechanics, or one can postpone judgement and focus on other issues instead. In this sense, it is a roadblock only for certain lines of research—although it is, by all standards, a serious one. In no sense, however, is it a pseudo-issue.

For instance, one might safely ignore the measurement problem if one is interested in solving the reconstruction problem (taking measurements or some equivalent notion as primitive), just as one might safely ignore the reconstruction problem if one is interested in solving the measurement problem. The choice is purely pragmatic, and dictated by one's interests or by one's sense of how fruitful a certain line of research might be. The idea that the measurement problem should be dissolvable because some philosophical position takes the notion of measurement—or of classical context, or of subject and object, or what have you—as fundamental, is just closing the door on potentially new developments in physics out of philosophical prejudice.

Of course, one might judge that such new developments are too speculative to pursue seriously (for instance, given the plethora of proposed solutions to the measurement problem), or too difficult to follow up experimentally (despite recent proposals regarding pilot-wave theories and collapse theories). But this will always be a pragmatic motivation for not considering the measurement problem a fruitful starting point of investigation, not a justification for dismissing it in principle.

At a more general level, I think the whole discussion about whether measurements in quantum mechanics are indeed problematic somewhat misses the point. Measurement interactions are only one of many examples of quantum interactions that lead to superpositions of macroscopically distinct states. Nature has been producing macroscopic superpositions for millions of years, well before any quantum physicist cared to artificially engineer such a situation.

The key concept here is decoherence. Environmental interactions tend to produce superpositions of classically distinct states. This raises the issue of how one could describe a classical regime in quantum mechanics, quite irrespective of the existence of measuring apparatuses. For instance, many chemical reactions—especially in organic chemistry—depend on molecules having certain shapes, but the shape of a molecule in general is only a feature that emerges at the level of components of the quantum state, thanks to decoherence. Or, genetic mutations induced by natural radioactivity can magnify quantum phenomena to the macroscopic level, quite analogously to the case of Schrödinger's cat. Or, think of the quantum description of classically chaotic systems: also here, it appears that classically chaotic trajectories can be recovered in quantum mechanics only at the level of decohered components.

In all of these and many other cases, superpositions of classically (and often macroscopically) distinct states arise spontaneously, due to the system of interest becoming entangled with its environment. The minimal interpretation of quantum mechanics has nothing to say about these cases, except that if we were to perform a measurement on these systems, we would observe classical behavior. It is literally a case of "The moon is not there if nobody looks" (at least not Hyperion, a moon of Saturn, whose dynamics is indeed chaotic)! If decoherence and its applications had been developed early in the history of quantum theory, then the idea that measurements play a special role in the theory might not have risen to such prominence, and the foundations of quantum mechanics would have focused instead on the problem of how to derive a classical regime within the theory.

ČASLAV BRUKNER · There are two measurement problems: the "big" measurement problem and the "small" measurement problem. The big measurement problem is the problem of explaining why a particular outcome—as opposed to one of the other possible outcomes—occurs in a given run of an experiment. The small measurement problem is the problem of explaining how outcomes come about in a measurement in the first place. It addresses the question of what makes a measurement a measurement.

The proposed solutions to the big measurement problem purport to solve the problem either by introducing "hidden" causes, such as in Bohm's hidden-variables theory, or by denying that measurements have definite outcomes, as in the Everett interpretation. There is no need for any of that. I would like to claim that the big measurement problem is a pseudo-issue arising from the failure to accept the possibility of having irreducible probabilities for measurement outcomes. This problem is not specific to quantum theory. It is inevitably present in any theory with irreducible probabilities, simply because any rational reason for the outcome would contradict the very idea of irreducible probabilities.

The small measurement problem is more subtle. If the state of an electron is measured by means of a heavy atom, which is itself measured by a cluster of macromolecules, and so on, and the result is finally recorded in a computer and the observer's mind, at what stage of this chain can we say that the measurement takes place? Bell was sarcastic about this: "What exactly qualifies some physical systems to play the role of 'measurer'? Was the wave function of the world waiting to jump for thousands of millions of years until a single-celled living creature appeared? Or did it have to wait a little longer, for some better qualified system ... with a Ph.D.?"

I will now give my account of the small measurement problem and explain why I think Bell's sarcasm is misplaced. In any study, one must concede the need for primitive notions. If quantum mechanics is understood as a fundamental theory of observations and observers' actions upon these observations, then measurement should be introduced as a primitive notion, which cannot be a subject of a complete analysis in principle. Just as an axiom in mathematics, quantum measurement serves as a starting point for deducing and inferring other propositions in the theory. But it can only be motivated informally, through an appeal to intuition and everyday experience. In the words of Peres: "While quantum theory can in principle describe anything, a quantum description cannot include everything. In every physical situation something must remain unanalyzed."

What does, however, remain a subject of scientific analysis in the quantum measurement process has to do with the following question: what makes a photon counter a better device for detecting photons than a beam splitter? The crucial ingredient of any measurement process is the evolution of the initial quantum state of the composite object-plus-apparatus system into a linear superposition of quantum states associated with macroscopically distinct states. The latter are defined as states that can still be differentiated even if the measurement precision is poor and one performs coarse-grained measurements. For example, if only a few spins of a large magnet are flipped, its entire quantum state will change into an orthogonal one, but at our macroscopic level we will still perceive it as the very same magnet. Only if a sufficiently large number of spins are flipped, such that a macroscopically distinct state is reached, do we perceive it as a new state of magnetization—that is, as a new "fact." A photon bouncing off a beam splitter won't result in macroscopically distinct states under standard laboratory conditions—although, as Peres warned us, "You should never underestimate the skill of experimental physicists!"—but the interaction of the photon and electrons in the photon counter will.

There are two important features of macroscopically distinct states. First, under sufficiently coarse-grained measurements, any superposition of them appears as a classical mixture, indicating a loss of coherence. Second, the macroscopic states are robust: they are stable with respect to small perturbations, such as those caused by repeated observations. This gives rise to a level of intersubjectivity among observers. Yet if quantum mechanics is universally valid, it is in principle possible to undo the entire measurement process. An observer capable of fully controlling the degrees of freedom of the measuring apparatus could decorrelate the apparatus from the measured system, erasing the information about the measurement result (having solid research funds and a Ph.D. in experimental physics might be of help in this). From this perspective, "irreversibility" in the quantum measurement process only means that it is extremely difficult to reverse the process.

JEFFREY BUB · Fundamentally, the measurement problem is the problem of connecting *probability* with *truth* in the quantum world, that is to say, it is the problem of how to relate quantum probabilities to the objective occurrence and nonoccurrence of events. The problem arises because there appears to be a difficulty in reconciling the objectivity of a particular measurement outcome with the entangled state at the end of a measurement.

In quantum mechanics, conditionalizing on a measurement outcome requires updating the credence function represented by the quantum state via the von Neumann–Lüders rule, which—as a non-Boolean, or noncommutative, version of the classical Bayesian rule for updating an initial probability distribution on new information—expresses a necessary information loss on measurement. This is Bohr's "irreducible and uncontrollable" measurement disturbance. It is a generic feature of nonclassical, i.e., non-simplex theories that satisfy a no-signaling constraint.

Just as Lorentz contraction is a physically real phenomenon explained relativistically as a kinematic effect of motion in a non-Newtonian space-time structure, so the change arising in quantum conditionalization that involves a real loss of information should be understood as a kinematic effect of *any* process of gaining information of the relevant sort in the non-Boolean probability structure of Hilbert space, considered as a kinematic framework for an indeterministic physics (irrespective of the dynamical processes involved in the measurement process). Given the no-signaling constraint, cloning an arbitrary extremal state is impossible, and since perfect cloning is possible if and only if nondisturbing measurement is possible, there can be no deeper explanation for the information loss on conditionalization than that provided by the structure of Hilbert space as a nonclassical probability theory (or information theory). The definite occurrence of a particular event is constrained by the kinematic probabilistic correlations represented by the subspace structure of Hilbert space, and only by these correlations—it is otherwise free.

From the perspective of the information-theoretic interpretation sketched above, there are two distinct measurement problems in quantum mechanics: what Pitowsky has called a "big" measurement problem and a "small" measurement problem. The "big" measurement problem is the problem of explaining how measurements can have definite outcomes, given the unitary dynamics of the theory: it is the problem of explaining *how individual measurement outcomes come about dynamically*. The "small" measurement problem is the problem of accounting for our familiar experience of a classical, or Boolean, macroworld, given the non-Boolean character of the underlying quantum event space: it is the problem of explaining the *dynamical emergence of an effectively classical probability space of macroscopic measurement outcomes* in a quantum measurement process.

On the information-theoretic interpretation, the "big" measurement problem is a pseudoproblem that arises if we take the quantum pure state as the analogue of the classical pure state, that is, as the "truthmaker" for propositions about the occurrence and nonoccurrence of events, rather than as a credence function associated with the interpretation of Hilbert space as a new kinematic framework for the physics of an indeterministic universe, in the sense that Hilbert space defines objective probabilistic or information-theoretic constraints on correlations between events. The "small" measurement problem is ultimately a consistency problem. In special relativity, one has a consistency proof that a dynamical account of relativistic phenomena in terms of forces is consistent with the kinematic account in terms of the structure of Minkowski space-time. An analogous consistency proof for quantum mechanics would be a dynamical explanation, taking account of decoherence, for the effective emergence of a classical, i.e., Boolean, event space at the macrolevel, because it is with respect to the Boolean algebra of the macroworld that the Born weights of quantum mechanics have empirical cash value.

ARTHUR FINE · One of the most ancient philosophical questions (Heidegger thought is was *the* question) is this: why is there something rather than nothing? In terms of events rather than substances, the question would be: how come anything happens at all? That question is the measurement problem. In some guise or other, the problem challenges any version of quantum theory. In the standard version, entanglement of a measured object with the measuring instrument disallows definite outcomes for either. If we add on the environment, then the entangled superposition of the whole likewise is incompatible with definiteness of any part. Resort here to decoherence does not really help, for in the best case (where convergence is rapid) the trace over environmental degrees of freedom only tells us that something *seems to happen* from a certain perspective. Thus, decoherence can sometimes help understand why it seems to us that things happen, but it does not address the question as to why anything *does* in fact *happen* (if in fact it does).

Some nonstandard versions of the quantum theory claim to have no measurement problem. In de Broglie–Bohm, for example, particles always have definite positions and follow trajectories determined by a state-dependent velocity field. This way of describing things, however, is not fundamental. Fundamentally, we have a universal wave function that need not evolve, and in terms of which all events (i.e., particle configurations), everywhere in the universe, are already fixed. That is, fundamentally, in de Broglie–Bohm nothing need happen at all. The appearance of change comes about in defining the wave function for a subsystem, by fixing its configuration relative to the configuration of all other particles in the universe. Holding those other particles fixed, this conditional wave function appears to evolve, even though, from the perspective of the universe, nothing needs to be happening. Thus, fundamentally, in de Broglie–Bohm, as in standard quantum theory, the resolution of the measurement problem is merely perspectival; we still do not understand why anything happens at all. The same is true in many-worlds (or other no-collapse) versions, where once again everything is already determined by the universal wave function. It is customary—for example, in many worlds—to say that as a result of a measurement, all the possible outcomes become actual (whatever that means). In fact, in many worlds nothing "becomes" at all, since the branches of the universal wave function (with respect to any basis) are there, like the function itself, for all time, without collapse.

This is not the place to survey the alternatives to no-collapse theories (primarily, dynamical collapse as well as relational and informational accounts). But it is appropriate to mention the cost of these alternatives, which is an understanding of the wave function that is radically different from the standard view. Is this a roadblock for these theories, a serious one?

CHRISTOPHER FUCHS · I remember giving a talk devoted to some of the points in this interview at a meeting at the London School of Economics seven or so years ago. In the audience was an Oxford philosophy professor, and I suppose he didn't much like my brash cowboy dismissal of a good bit of his life's work. When the question session came around, he took me to task with the most proper and polite scorn I had ever heard (I guess that's what they do). "Excuse me. You seem to have made an important point in your talk, and I want to make sure that I have not misunderstood anything. Are you saying that *you* have solved the measurement problem? This problem that has plagued quantum mechanics for seventy-five years? The message of your talk is that, using quantum information theory, *you* have finally solved it?" (Funny the way the words could be put together as a question, but have no intended usage but as a statement.) I don't know that I did anything but turn the screw on him a bit further, but I remember my answer. "No, not me; I haven't done anything. What I am saying is that a 'measurement problem' never existed in the first place."

The "measurement problem" is purely an artefact of a wrong-headed view of what quantum states and/or quantum probabilities ought to be—that they ought to be either (better) objective properties themselves or (worse, but still relatively acceptable) subjective ignorance of some deeper, observer-independent, agent-independent, measurement-independent events. The measurement problem—from our view—is a problem fueled by the fear of thinking that quantum theory might be just the kind of *user's manual* theory for individual agents (contemplating the consequences of their individual interactions with quantum systems) that we have described in the previous answers. Take the source of the paradox away, we say, and the paradox itself will go away.

Jim Hartle already put it fairly crisply in a 1968 paper:

A quantum-mechanical state being a summary of the observers' information about an individual physical system changes both by dynamical laws, and whenever the observer acquires new information about the system through the process of measurement. The existence of two laws for the evolution of the state vector becomes problematical only if it is believed that the state vector is an objective property of the system. If, however, the state of a system is defined as a list of [experimental] propositions together with their [probabilities of occurrence], it is not surprising that after a measurement the state must be changed to be in accord with [any] new information. The "reduction of the wave packet" does take place in the consciousness of the observer, not because of any unique physical process which takes place there, but only because the state is a construct of the observer and not an objective property of the physical system.

Quantum Bayesianism's contribution has only been in making the point of view absolutely airtight, making it clear that "information" is (and must be) a subjective notion, choosing a language for expressing this in the most calming terms possible, and showing that the whole thing has some bite for proving theorems and moving physics itself forward.

GIANCARLO GHIRARDI · For me, the quantum measurement problem—or what I would prefer to call the quantum macro-objectification problem, in order to stress that it emerges, and cries for a clarification, even in the absence of conscious observers—represents a serious roadblock of the standard theory. I would like to present some synthetic comments on the most well-known solutions, and I would like to discuss how the large majority of them are characterized by imprecise, purely verbal statements that do not help to identify the heart of the problem. The difficulties are not related, as it has been variously suggested, to adhering to a priori prejudices about reality, or to resorting to oversimplified schemes for accounting for the emergence of different macrosituations (the von Neumann scheme for ideal measurement is often mentioned in this context). Instead, they are rooted in the fact that the theory does not contain any objective element or parameter that, in a formally precise manner, would locate the shifty border between deterministic and stochastic natural processes, between micro- and macroprocesses, and between reversible and irreversible processes. So let me comment on some of the most popular proposals.

Let's first turn to the suggestion that the conscious observer may play a fundamental role. This idea, suggested by von Neumann and supported for a certain time (but ultimately abandoned) by Wigner, is based on the assumption that physical happenings and conscious perceptions are fundamentally different processes. All material objects obey the linear and deterministic quantum mechanical evolution laws, while the act of becoming conscious, typical of human beings, is a radically different process that is not accounted for by the standard dynamics; essentially, it is not a physical process. Apart from the peculiar aspect of a proposal that radically undoes the Copernican revolution by putting conscious beings at the very center of all natural history, one immediate objection derives from the recognition that our present knowledge does not allow us to clearly identify what entities are to be regarded as conscious (as Bell has put it: "a single-celled living creature ... [or] some better qualified system ... with a Ph.D.?"). This is one of the many faces of the "shifty split" characterizing the conceptual foundations of quantum theory.

The consideration of decoherence meets analogous difficulties. It amounts to the claim that logically incomprehensible final situations—namely, superpositions of macroscopically distinguishable states—actually occur. But because of our present (and also future?) limited technical skills, so the claim continues, one can approximately identify a pure state with a statistical mixture. The objections are obvious. In the history of science, it has never happened that a theory leads, from the point of view of our definite perceptions, to a meaningless situation, and that at the same time, we can get a conceptually clear and meaningful description of reality if we accept to account for the implications of the theory by resorting to an approximation.

Another crucial formal aspect to be taken into account is that contrary to the case of classical mechanics, in quantum mechanics the correspondence between statistical ensembles and the statistical operators describing them is infinitely-many-to-one. When one replaces the embarrassing pure state by a statistical operator, what makes it legitimate to interpret this operator as describing a statistical ensemble whose members have precise properties that match our definite perceptions? Just to give one example, let's consider the infamous case of Schrödinger's cat. When one replaces the superposition with a statistical operator, what justifies interpreting this operator as describing an ensemble whose members are either alive or dead, rather than as a statistical ensemble of systems associated with appropriate superpositions of alive-cat states and dead-cat states?

This point is so conceptually important and unescapable that even the most serious supporters of decoherence-based interpretations have plainly admitted that, after all, there is no objective reason to make the above (natural) choice. For instance, Erich Joos and Dieter Zeh, two eminent representatives of the decoherence position, point out that "the use of the local density matrix ... already assumes a local description," that "the locality assumption may perhaps be justified by a fundamental (underivable) assumption about the local nature of the observer," and that "no unitary treatment of the time dependence can explain why only one of [the] dynamically independent components is experienced." Let me say that this statement puts the supporters of the decoherence approach in a position that's reminiscent, to some extent, of the situation of those people who make reference to the act of conscious perception as a means for solving the measurement problem.

The ensemble interpretation is an approach that has been supported by Leslie Ballentine, among others. All the remarks I have made concerning the decoherence approach apply also to this point of view. But here is a further comment: in modern technological applications of quantum theory, such as quantum cryptography, quantum teleportation, and quantum computation, one is always dealing with an individual physical system. The statement that quantum theory is not a theory of individual systems but rather deals exclusively with ensembles seems irreconcilable with our need to systematically deal with individual processes. And, let me add, I believe that there is only one universe, and not a statistical mixture of them.

An interesting, and at first sight promising, solution, which makes some contact with decoherence-based strategies, is the so-called decoherent-histories approach to quantum mechanics. The basic goal of this interpretation is to allow for the consistent and coherent use of classical logic in making statements about quantum systems.

This is an appealing move. Unfortunately, in my opinion, the proposal does not work. In the approach, one considers families of histories that represent the evolution of physical systems, and one then puts precise conditions on the allowed histories. These conditions demand that the histories mutually decohere, which amounts to a precise mathematical requirement. In a series of papers, Angelo Bassi and I have proved that this leaves us with only two alternatives. Either, one makes absolutely precise that only a restricted and limited class of histories ought to be considered, such as those making claims exclusively about the positions of macroscopic objects. But this amounts, in some sense, to (tacitly) assuming that for macroscopic objects, one cannot measure noncommuting observables. Or, one can instead maintain (as Robert Griffiths and others do) that any conceivable decoherent set of histories—even the most illogical from the point of view of our experience—ought to be taken into account. In this case, however, it turns out to be impossible to attach logically consistent truth values to the statements of the theory.

Many-universes and many-minds interpretations deny the occurrence of wave-packet reduction by associating the macroscopically different states in a superposition either to different universes, or to different and coexisting perceptions. One might claim that in many-universes interpretations all potentialities become actual in different and unconnected branches of the universe. Alternatively, in the many-minds interpretation, all potentially possible perceptions are associated with infinitely many perceptual branches of our minds. I do not want to discuss these interpretations: even though they are in a certain sense consistent, to me they seem to be too much removed from what one ought to require from a scientific theory. As far as the branching processes of the universes or minds are concerned, I would like to draw the reader's attention to a recent paper by Hilary Putnam. Putnam points out that the many-universes interpretation is incapable of accounting for the statistics of physical outcomes we find when performing repeated experiments in our universe—which is, after all, supposedly one of the branches described by the theory.

I won't analyze modal interpretations in any detail. While they are interesting in their own right, they consider two kinds of statistical ensembles that are treated in rather different ways. The specific statistical mixture made to correspond to the pure state of a composite system does not have an ontological status that would in any way resemble that of the statistical mixture that we can produce by preparing a bunch of physical systems in different states. In fact, after having claimed that the pure state can be read as a statistical mixture, the supporters of modal interpretations are forced to accept that for studying the system's future evolution, one must use the pure state rather than the statistical mixture. This is an essential aspect of the interpretation, and it's partially responsible for the need to resort to what's known as modal logic.

Bohmian mechanics represents a deterministic completion of quantum theory, and it's predictively equivalent to quantum theory concerning the position distributions of all particles of the universe. It leads to a lucid and fully satisfactory solution of the measurement problem, and personally I attach great relevance to it. (Bell once called it "a great loss" that the de Broglie–Bohm picture is "generally ignored and not taught to students.") The only reason why I prefer the collapse theories over Bohmian mechanics is that collapse theories lead to predictions different from those

SHELLY GOLDSTEIN   151

of standard quantum mechanics, thus making them testable. Let me also mention that if one adopts the mass-density interpretation for collapse theories, there exist strong analogies between these theories and Bohmian mechanics, as has been recently suggested by Shelly Goldstein, Detlef Dürr, Roderich Tumulka, and Nino Zanghì.

Collapse theories and the GRW theory have already been repeatedly discussed, so I won't comment on them again. Let's simply recall that they assume that there exist, beside the standard quantum evolution, spontaneous stochastic localization processes acting on all particles in the universe. Such localizations occur quite rarely for an elementary particle (once every ten millions years for a nucleon) and therefore don't affect microscopic systems. But there is a trigger mechanism built into the theory: if a nearly rigid body is in a superposition of two states peaked in different regions of space, then the localization of any of the body's constituents leads to a localization of the entire object. This implies that superpositions of macroscopically distinct spatial states are spontaneously suppressed by the universal dynamics in about one millionth of a second. The phenomenological parameters that characterize the frequency and the accuracy of the localization events define, in a precise way, the borderline between quantum and classical physical processes. As such, they identify the region where collapse theories depart from the predictions of standard quantum mechanics.

SHELLY GOLDSTEIN · Neither. It is rather a serious problem (namely, the problem that quantum measurements typically don't have results) that arises when one makes what seems to me to be the serious mistake of regarding, contra Einstein, the wave function of a quantum system as the complete description of that system. If the wave function provides only a partial description, there need be no such problem.

And in Bohmian mechanics, there is no measurement problem—as in any version of quantum mechanics involving, in addition to the wave function, local beables describing the configuration of matter in space, and in particular the orientation of pointers and the like. In fact, in Bohmian mechanics, not only do measurements invariably have results (this is trivial), but also in quantum measurement situations the measured system's wave function typically collapses in the usual textbook way (this is not so trivial; it seems paradoxical but in fact is pretty easy to see and understand).

Similarly, there should be no measurement problem either for proponents of strict textbook (Copenhagen) quantum mechanics or for advocates of quantum information theory. Quantum mechanics for them should be about certain macroscopic variables, presumably those that convey "information" or results of observation. For them, the wave function of a system should not be regarded as providing a complete description of that system—if indeed it is regarded as part of the physical description at all.

Of course, one might nonetheless insist on the wave function's being the complete description of a quantum system. This has the virtue, after all, of arguably allowing quantum theory to assume its simplest form: a theory involving a single entity, the wave function, and a single equation, Schrödinger's equation, with all the measurement postulates eliminated from the formulation of the theory. If one insists on this, then the measurement problem looms large. And, of course, the resolution of the measurement problem takes on a dramatic character: Everett's many worlds. Whether this approach genuinely succeeds—and, in particular, whether it really resolves the measurement problem—remains controversial.

The most common response to the measurement problem is to appeal to decoherence. I suppose that those who think that the measurement problem can be resolved in this way, without the invocation of many worlds or anything beyond the wave function, would be inclined to regard the measurement problem as a dissolvable pseudo-issue. I find the pure decoherence approach to the measurement problem utterly incomprehensible. The only way to begin to make sense of it is from a strictly positivist perspective. But from such a perspective, it would make no sense to take the wave function seriously enough for there to be a measurement problem to begin with.

DANIEL GREENBERGER · I don't think that the quantum measurement problem has a real solution. The problem, of course, is that a coherent, pure quantum state cannot turn into a mixture. Most of the solutions that have been proposed are what John Bell called FAPP (for all practical purposes), which in this case means that the system gets imbedded into a larger system, whose internal degrees of freedom are so complicated that one cannot keep track of the coherence, and at some arbitrary point one gives up and calls the system incoherent, and one then says a measurement has been made. This is what happens with the decoherence solutions, where the environment plays the role of the larger system. But even a detector can play this role. This is a good FAPP solution.

But there is a problem with it. When one has an entangled state and one observes only one of the entangled particles, one sees a random distribution of outcomes. It is only when one observes correlations between the two entangled particles that one sees coherence. So one part of an entangled pair looks to be incoherent. Still, there is the possibility that someone thinks he has made a measurement and that the system he has is completely incoherent—for example, he has measured his system, which, say, has two states, 0 and 1, and has found a completely random distribution of 0s and 1s. Then, a week later, someone drops by and says that he too has made the same measurement on his system, which happens to be on Mars, and they find that their results are completely correlated: either they both get 0 or they both get 1 on each measurement. It turns out that each measurement they made that they thought was random was on one of an entangled pair, and that the other fellow had the other member of the pair.

The moral of this is that one never really knows whether his system is truly random or part of an entangled pair. So what one takes for a random measurement is not really random. It is not even approximately random. It is completely correlated with another particle, only you don't know it.

It seems to me that this shows that the concept of something being random is completely subjective. As a subsystem, it is completely random, but as part of a larger system, it is completely correlated, and there is no way to tell which it really is. There doesn't appear to be any solution to this problem. One can never objectively tell whether a system you have is random. And therefore there is no way to tell whether a measurement has been finished. This says that there are only FAPP solutions to the problem, and it is pointless to try to develop schemes to actually solve the measurement problem beyond the FAPP level.

This might seem far-fetched, but in practice, it turns out that we are producing better and better correlations all the time. Entangled states were exotic until a generation ago, and today they are a readily produced resource. It is even possible, by the process known as entanglement swapping, to produce particles that have never met but are entangled. I think this process will go on, and we will soon produce mesoscopic states—semimacroscopic states—that are correlated, and that will push back the idea of measurement even further.

LUCIEN HARDY · The measurement problem has been called "the reality problem" by Philip Pearle. This is a better name for it. We perceive objects in the world as being in definite states. A door is either open or shut, a given ball either is in a given box or it is not. The wave function, however, can have superpositions of these things, suggesting that the door can be simultaneously open and shut at the same time, and that the ball can be both in the box and not in the box at the same time. The reality problem is that there is a discrepancy between the version of reality we perceive, and the version presented to us by the most obvious interpretation of the wave function. This is a serious problem for any attempt to provide an ontological interpretation of quantum theory.

The reality problem stems from asserting that the wave function be given ontological status. It is an open question as to whether we have to do this (see my answer to Question 4, page 99). The work of Alberto Montina tends to suggest that we can avoid this, but only if we allow a kind of nonlocality in time (so that the state at time $t + \delta t$ is not determined by the state at time t). Indeed, the fact that we expect indefinite causal structure in a theory of quantum gravity suggests that we should even give up the notion of a state at time t evolving in time as a fundamental concept. It is, in fact, possible to do quantum theory in a very natural way without using the notion of an evolving state (as has been shown in my causaloid and duotensor papers). It would be interesting to study the reality problem in the context of this more general formulation of quantum theory.

We might argue that we can dissolve the reality problem simply by taking an operational approach to understanding the wave function. This is appealing, but it cannot be quite as simple as this. To address the reality problem, we have to be concerned with reality. What is the underlying reality these probabilities supervene on? In the classical case, we have underlying ontic states. An operational state is simply a list of probabilities for the system being in each of these underlying ontic states. The fact that we have quantum interference means that the situation cannot be so simple in quantum theory. We will see interference regardless of what approach we take to understanding the quantum state.

What is really needed is an interpretation providing an ontology that accounts for why we perceive objects in the world as being in definite states. The pilot-wave

154 QUESTION 7: THE MEASUREMENT PROBLEM

model, the many-worlds approach, and collapse models all do this (modulo the various problems these interpretations face). But none of these interpretations is sufficiently compelling to me. To solve the measurement problem, I think we need to do some serious off-read research to go beyond quantum theory, in the direction of a theory of quantum gravity. That this will solve the reality problem is more or less an article of faith for me at this stage. But there are reasons to think that out of the mixture of conceptual tensions at play in the forming of a theory of quantum gravity will come a solution to the reality problem. There is a delicate interplay between notions of causal structure and reality. For example, the assertion that a particle is either here or there means nothing if there is no definite causal structure with respect to which a particle can be in one place or another. The "no-matter-of-fact-ness" that leads to quantum interference and causes the measurement problem in the first place may help dissolve it when applied to causal structure. This would be somewhat similar to the way Einstein resolved the problem of understanding how the speed of light can be the same in all frames, by removing the notion of absolute simultaneity from our ontological understanding of reality.

Whatever solution we eventually adopt to the measurement problem, it should not be put in by hand, but rather emerge naturally from more fundamental considerations.

ANTHONY LEGGETT · Very definitely a serious roadblock. Here is an argument that I have given in several other places and that I believe adequately demonstrates this contention.

It is essential to distinguish between the *interpretation* of the formalism of quantum mechanics, and the *evidence* that a particular interpretation is correct or, more relevantly, incorrect. At the microlevel, for example, in a Young-type slits experiment performed with electrons or atoms, we have the following state of affairs:

(1) It is an experimental fact that the probability of a result obtained when both paths A and B are open is not the sum of the probabilities obtained when only A or only B are open. This very strongly suggests that whatever positive statement (if any) we choose to make about the situation, there is one negative statement we can certainly make, namely, that *it is not the case that each individual atom of the relevant ensemble chooses either path A or path B.* (I'm well aware, of course, that there exist some interpretations, for example, the neo-Bohmian one, that formally deny this statement. But it seems to me that this is a purely verbal maneuver and in no way resolves the measurement paradox, which merely needs to be restated in slightly different language.)

(2) The formalism of quantum mechanics gives a simple account of this kind of experiment. Nonzero probability amplitudes are simultaneously assigned to path A and to path B, and we then solve the relevant time-dependent Schrödinger equation and invoke the Born probability rule.

Taken together, considerations 1 and 2 very strongly suggest the following conclusion. When, in the description of this kind of experiment, the quantum formalism assigns nonzero amplitudes to two different possibilities (paths), then *it is not the case that each individual atom of the relevant ensemble chooses either one possibility or the other*. (Incidentally, in giving physics-department talks on this subject, I often start by asking for a vote on this point—or, more accurately, on item 1 above. I almost invariably get a large majority in favor.)

It is crucial to distinguish, in the above argument, between the conclusion we have drawn concerning the illegitimacy of a certain interpretation of the formalism, and the *evidence* for that illegitimacy, namely, the experimentally observed phenomenon of interference.

Now, following Schrödinger, let us consider a thought experiment in which the quantum-mechanical description of the final state, as obtained by appropriate solution of the time-dependent Schrödinger equation, contains simultaneously nonzero probability amplitudes for two or more states of the universe that are, by some reasonable criterion, macroscopically distinct (in Schrödinger's example, this would be "cat alive" and "cat dead"). Of course, just about everyone, including me, would accept that because of, inter alia, the effects of decoherence, it is likely to be impossible, at least for the foreseeable future, to experimentally demonstrate the interference of such states. (On the other hand, as the late John Bell was fond of pointing out, the "foreseeable future" is not a very well-defined concept. In fact, as late as 1999, not a few people were confidently arguing that because of the inevitable effects of decoherence, the projected experiments to demonstrate interference at the level of flux qubits would never work. In this case, the "foreseeable" future lasted approximately one year. As Bell used to emphasize, the answers to fundamental interpretive questions should not depend on the accident of what is or is not currently technologically feasible.) But the crucial point is that the formalism of quantum mechanics itself has changed not one whit between the microscopic and macroscopic levels. Are we then entitled to embrace, at the macrolevel, an interpretation that was forbidden at the microlevel, simply because the evidence against it is no longer available?

I would argue very strongly that we are not, and would therefore draw the conclusion: also at the macrolevel, when the quantum-mechanical description assigns simultaneously nonzero amplitudes to two or more macroscopically distinct possibilities, then it is not the case that each system of the relevant ensemble realizes either one possibility or the other.

Needless to say, this conclusion is disturbing. I believe there are only two possible ways out. Either quantum mechanics must break down, at some level between that of the atom and that of our own direct consciousness, in favor of a theory of some quite different nature whose details we cannot at present imagine. Or we must embrace the "extreme statistical" interpretation outlined in my response to Question 3 (see page 79), and thus refuse to make any statements at all about the interpretation of the quantum formalism, even negative ones, at either the micro- or the macrolevel.

TIM MAUDLIN · Neither. The measurement problem is a useful reminder of the inadequacy of the standard textbook accounts of quantum theory as physical theory. The measurement problem arises from the demand that the observable behavior of laboratory equipment be explicable from a description of the microscopic constituents of the equipment and the physical laws that govern those constituents, irrespective of the behavior being categorized as a "measurement." This is a demand that any acceptable physical theory must be able to meet.

The problem is not a pseudo-issue, because the standard account contains the term "measurement" in its fundamental axioms. So there is a real issue of how to avoid that. Neither is it a roadblock, since there are several distinct theories that can solve it and treat "measurements" as the same sort of physical interaction as any other. The virtue of the measurement problem is that it focuses our attention on these successful physical theories.

DAVID MERMIN · It's a pseudo-issue. But I have not dissolved it entirely to my satisfaction. So while I see no roadblock, I do feel the need to drive slowly past some unfinished construction, attending to signals from the people with flags.

Today "the quantum measurement problem" has almost as many meanings as "the Copenhagen interpretation." I mention only two of them. The first is how to account for an objective physical process called the collapse of the wave function, which supersedes the normal unitary time evolution of the quantum state in special physical processes known as measurements. I believe that this version of the problem is based on an inappropriate reification of the quantum state. So are efforts to eliminate the special role of measurement through dynamical modifications in standard quantum mechanics that make an appropriate rate of collapse an ongoing physical process under all conditions.

The quantum state is a calculational device, enabling us to compute the probabilities of our subsequent experience on the basis of earlier experiences. Collapse is nothing more than the updating of that calculational device on the basis of additional experience. This point of view is the key to resolving this form of the quantum measurement problem. I look forward to the day when some clear-headed gifted writer has spelled it out so lucidly that everybody is completely convinced that there is no such problem. (I'm convinced. But I'm not completely convinced.)

A second question going under the name "quantum measurement problem" is whether there can be quantum interference between quantum states that describe macroscopically distinct physical conditions (sometimes called "cat states"). If such interference is not just hard to observe but strictly absent, then quantum mechanics must break down in its answers to questions of sufficient complexity, asked of systems of sufficient size. Size alone is not the issue, since quantum mechanics works brilliantly in accounting for all kinds of classically inexplicable behavior in the gross behavior of bulk materials. Indeed, the appropriate definition of "macroscopic" in this setting is far from obvious.

The fact that the unavoidable entanglement of a macroscopic system with its environment renders manifestations of quantum interference effectively unobservable is a good practical rejoinder to those who seek an answer from a macroscopic breakdown of quantum mechanics. But decoherence does not directly address the question of whether anything actually changes when the superposition is replaced by a mixed state, beyond an abstract representation of our practical ability to acquire knowledge. And it is subject to the same kinds of time-reversal problems that plague statistical-mechanical derivations of the second law.

Seeing quantum interference effects with carbon-60 molecules is an experimental tour de force. But I would have been astonished if interference had been demonstrably absent. My impression is that those who did the experiment did not expect it to reveal a breakdown of quantum mechanics. They did it because it was there, like Mount Everest, challenging somebody to take it on.

LEE SMOLIN · A very serious roadblock to taking the standard quantum formalism as a fundamental theory, and pointing to the formalism's instead being an approximation to a deeper and truer description.

ANTONY VALENTINI · The measurement problem is often stated as the problem of the interpretation of a quantum superposition, such as Schrödinger's cat. That is inaccurate and misleading. Among other things, it allows for the facile response that the wave function refers only to a statistical ensemble. While such a "statistical" or "epistemic" interpretation might turn out to be correct, it does *not* solve the true and deep measurement problem, which is the problem of what happens to macroscopic realism at microscopic scales. In quantum physics, we have definite states of reality at the macroscopic level but not at the microscopic level. There is no precisely defined boundary between these two domains. Therefore, standard quantum theory is fundamentally ill-defined.

An apparatus pointer in the lab, for example, points in a definite direction. Particles, on the other hand, generally have indefinite positions. How many particles are required to make a "macroscopic" pointer? How many are required to cross the line from microscopic fuzziness to macroscopic definiteness? There is no precise dividing line between the microscopic and the macroscopic. And all macroscopic equipment is built out of microscopic systems. How do definite states arise from indefinite ones? There's a temptation here to think in terms of emergence, but there can be no continuous transition from indefinite to definite states of reality. Either something exists or it does not.

Some people say that there are "degrees of reality," that one object can be "more real" or "less real" than another. One sometimes hears physicists ask if a rock somewhere out in deep space is real when no one is looking at it. But such talk misunderstands the nature of the word "real." When we say that "*X* is real," we simply mean that "*X* exists." If I say there is a rock out in deep space with no one looking at it, I have already stated that the rock exists, that is, I've already said the rock is real. To then suggest that perhaps the rock is not real, because no one is looking at it, is to contradict oneself. An analytical philosopher would probably convey this point by saying that "real" is a quantifier, not a property. But the point is simple enough, and is the basis of much elementary reasoning, both in physics and outside of it.

Others try to evade the measurement problem by claiming that the usual notion of ontology depends on a "God's-eye view" of the world. But that is mistaken. For example, if a piece of macroscopic apparatus, with a dial and a pointer, has a particular setting and pointer reading, this is not dependent on anyone's (or God's) "viewpoint." It is simply a fact about the dial and pointer. Facts require neither a human observer nor a deity. They are facts, whether or not we—or God, or whoever—is there to know them.

Other attempts to avoid the issue include recourse to nonclassical logic. One can invent a mathematical structure that violates some rules of formal logic and *call* it a "logic." But everyone still uses so-called classical logic in order to reason and argue. I think there's a misuse of words here. Logic is logic. If there's a contradiction, for example, in a thought experiment such as that of Wigner's friend, then it won't do to dismiss it as a failure of classical logic. The contradiction comes from clear thinking and requires a clear answer.

Finally, some people say that the concept of objective reality must be abandoned even at the macroscopic level. But we each know that we exist, and if we have any sense, we will know that other minds exist as well. There is a world out there, containing other human beings, as well as things like tables and chairs, and pieces of equipment with dials and pointers.

To make quantum mechanics a precise theory, we must posit the existence of something that extends into the microscopic domain. This can and has been done in various ways, involving hidden variables, or many worlds, or collapse theories. It remains to be seen which, if any, of these proposals is correct.

To suggest that the measurement problem is a pseudo-issue is to say that the simple question "What is real?"—or equivalently, "What exists?"—does not require an answer. When people say that, they are being inconsistent, because they themselves talk about "what is real" or "what exists" when it comes to things like the outcomes of experiments in their laboratories, or what car they own. Everyone uses the notion of definite states of objective reality at the macroscopic level, including in the laboratory—when it is asserted, for example, that we really did find a certain wavy pattern of dots on a photographic film in an interference experiment. It's only at the microscopic, or quantum, level that there is controversy over what is real. To say that we don't need a notion of microscopic reality at all, while at the same time using a notion of macroscopic reality whenever one describes an experiment, is to ignore the self-evident ambiguity in the dividing line between microscopic and macroscopic, and to ignore the resulting self-evident ambiguity in what one is saying.

DAVID WALLACE · In a sense, I don't think those are incompatible possibilities. Working out that something is a dissolvable pseudo-issue can be really hard work—just look at the difficulties that Einstein had thinking about general covariance, or that people thinking about black holes had thinking about the coordinate singularity on the event horizon. Something can be a serious roadblock *until* the conceptual insight that lets us dissolve it.

I actually think that's basically what the situation is in quantum mechanics, in that the measurement problem arises because it looks like you can't take the wave function literally as a description of reality without getting a flat contradiction with observations, because of Schrödinger cats and the like: the theory predicts that we ought to see the world in a superposition of macroscopic states, and it seems that we don't. And the relevant conceptual insight was Everett's: what would it really look like if the world *was* in a superposition of macroscopic states. Once you start thinking that way, you start to see that it's not *obvious* that we don't see the world looking like that, because, of course, if we looked at a cat in a superposition, we'd end up entangled with it and becoming part of the superposition. That doesn't dissolve the *problem*, but you might say that it dissolves the *paradox*. It changes it from, "The world can't possibly be like quantum mechanics says it is, what do we do??!!!" to, "Okay, what *exactly* does quantum mechanics say the world is like, and is it like that?" And that gets us into decoherence theory and the like.

Maybe the thing I should say is that it's not an *easily* dissolvable pseudo-issue! The measurement problem maybe ought to be called the macroreality problem—how can quantum mechanics be reconciled with observed macroscopic reality? It's not at all *obvious* that it can. I think if you think hard about it, along Everettian lines, and play around with decoherence theory and the quantum theory of big open systems, you can basically establish that it can. But that took a lot of hard work by a lot of people. And if it turns out all to fall apart for some reason, then I'd go right back to thinking of the measurement problem as a roadblock.

ANTON ZEILINGER · The quantum measurement problem is neither a serious roadblock nor a dissolvable pseudo-issue. Rather, it is just a *Scheinproblem* (pseudoproblem) that arises if one does not realize that quantum states represent information. From that point of view, what would be more natural than changing the representation of information (i.e., the quantum state) when one obtains new information (i.e., a measurement result)?

WOJCIECH ZUREK · Neither. It is a very good question. It should be celebrated. Measurement—perception—is the place where physics gets personal, where our role and our capabilities as observers and agents of change in the universe (and our limitations as entities subject to the laws of physics) are tested—or, rather, where we get put in our place.

I believe that quick solutions, and I include both the Copenhagen interpretation and many worlds here, have a tendency to gloss over the real mystery, which is how do we—that is to say, how does life—fit within the quantum universe. I think we have managed to constrain the possible answers (for example, through research on decoherence), but I believe there is more to come.

The virtue of the focus on quantum measurement is that it puts issues connected with information and existence at the very center. This is where they should be. The disadvantage of having quantum measurement as the focus is that it has turned into a cliché, and that many people have preconceived ideas on what would be an acceptable solution.

So there are dangers here. This is why it was important for decoherence to look at the bigger problem: the problem of the emergence of the classical from the quantum substrate *in general*, rather than just in the case of measurements. And to a degree, one could even claim that the measurement problem has dissolved as a result of the progress in our understanding of the emergence of the classical realm. In that sense, one can regard decoherence as the solution to the "impersonal" part of the measurement problem.

But in spite of this, I think there is still a core of the problem—the one that touches on the relation between information and existence—that remains a conundrum. This is, as I said earlier, where quantum physics gets personal, where we want to understand how what we find out turns out to actually be there, how—as Wheeler put it—"bit" relates to "it."

QUESTION 8

BELL'S INEQUALITIES

What do the experimentally observed violations
of Bell's inequalities tell us about nature?

J OHN BELL'S WISDOM holds appeal for foundationalists of all stripes. Just count
how many times he's mentioned in this book! What makes Bell such a darling of
the quantum-foundations community? Well, first of all, there's Bell's trademark
wit, famously on display in *Speakables and Unspeakables in Quantum Mechanics*, a
collection of classic essays that has achieved pop-culture status. Mention Bertlmann's
socks at a conference, and no one will be puzzled.

But Bell is known for more than his prose. His arguably most lasting legacy is
couched not in words but in steely mathematics: his famous inequalities, first published in 1964, rang in a new era in quantum foundations. Up to that point, foundational considerations amounted to wordy debates seemingly immune to resolution by
mathematical or experimental means. Take Bohr's writings on the interpretation of
quantum theory as an example: lambasted by some—and cherished by others—not
only for a style that's notoriously difficult to penetrate, but also for their virtual absence of explicit models and calculations. Bell's analysis, by contrast, casts a deep
foundational question into a form amenable to experimental decision. The question,
in broad outline, is this. To what extent can we cook up a richer description of physical reality than provided by textbook quantum mechanics, and what properties would
such a description have?

For many people, the nothing-but-the-quantum-state picture just doesn't seem to
quite satisfy their longing for the kind of description familiar from previous physical
theories. For example, suppose we prepare every member in an ensemble of systems
in the exact same quantum state. Does this mean that a measurement of some observable on each of these systems will give the same result? Of course not. Far from
it, in general a whole range of values will be thrown at us, and we are powerless to
predict which result we'll get in each instance. The quantum state is of no help in
this regard. On the other hand, what the quantum state *does* do so exceedingly well
is tell us the statistical distribution of the results.

But can't we say more about what's going on deep down in the bowels of nature?
Why shouldn't we try to give a more detailed account than that given by quantum states? An account that would allow a physical system to possess, independent

of measurement, definite values of many quantum-mechanically incompatible observables. Measurements would then simply reveal these preexisting values—even though nature hisses "Hands off!" whenever we attempt to find out these values simultaneously.

The legendary Einstein–Podolsky–Rosen paper of 1935 played an important role in bolstering such sentiments. EPR considered two spatially separated systems described by a particular entangled quantum state. They looked at measurements performed on the individual systems, leading, in the way prescribed by the quantum formalism, to the reduction of the wave function. Crucially, EPR also assumed a principle of locality: that a measurement performed on one of the systems must not instantaneously alter the physical reality of the other, distant system. (As Arthur Fine has emphasized, there's an additional assumption implicit in this principle: that it is at all possible to assign some independent reality to the second, unmeasured system, despite its entanglement with the first system.)

From these premises EPR deduced that each system must be regarded as simultaneously possessing definite values for more physical quantities than permitted by quantum mechanics. So the description afforded by quantum states must be necessarily incomplete, the reasoning went, and quantum mechanics could not possibly be the whole story.

The EPR argument stirred a debate that rages to this day. It helped motivate the search for a more fundamental, "complete" description, where additional hidden variables would charter a well-defined course on the vast ocean of the potentialities represented by the quantum states. Such hidden-variables theories would then explain the behavior of individual systems in a classically causal (and usually deterministic) manner. This, in turn, would render the probabilistic nature of the quantum-state description a good deal less mysterious, because the statistical element would now reflect merely the failure of quantum states to fully describe the richer going-ons beneath. Hidden-variables theories would also do away with the quantum–classical split that assumed such a principal role in Bohr's and Heisenberg's interpretations of quantum mechanics. As Bell once put it, "It is this possibility, of a homogeneous account of the world, which is for me the chief motivation of the study of the so-called 'hidden variable' possibility."

The conceptual relationship between hidden-variables interpretations and quantum theory is often compared to the relationship between classical mechanics and classical statistical mechanics. In classical statistical mechanics, probabilistic elements arise simply as a consequence of deliberate coarse-graining over microscopically well-defined states of affairs. But there's one important difference between the classical scenario and the quantum scenario. In the classical scenario, specific microscopic configurations can be experimentally prepared at the experimenter's choosing (at least in principle). By contrast, if we could do the analogous thing in the quantum setting—namely, if we were able to prepare systems with prescribed values of the hidden variables—then we could also prepare ensembles of such systems with any statistical distribution of values of observables we like. But this would mean, in general, that the particular distribution given by the quantum state would be empirically inadequate. It is for this reason that the additional variables must remain hidden, that is, fundamentally inaccessible to us. So while from God's perspective no room may be left for chance and he may no longer need to play dice, at the level of the user interface the statistical element of quantum theory would continue to retain (some of) its fundamental character.

Now, what shape could a hidden-variables theory possibly take? One might think that there would be few constraints and therefore a whole zoo of possibilities to choose from. But, as it turns out, it's no piece of cake for such a theory to match all the statistical predictions of quantum mechanics.

In 1932 John von Neumann laid out a set of requirements that hidden-variables theories ought to meet and then showed that no such theory could be made to work. His argument became quickly accepted as a definitive proof of the impossibility of any hidden-variables theory. Von Neumann was widely believed to have settled the matter in a mathematically sharp and unassailable way, and over the next two decades his proof never received much scrutiny. When Bohm presented his hidden-variables theory in the early 1950s, several (albeit inconclusive) attempts were made to clarify the relevance and scope of von Neumann's proof. If Bohm's theory indeed worked as claimed, how could the proof uphold its reputation as a sweeping no-hidden-variables-at-any-price demonstration? Something had to give.

In an article published in 1966—but in fact written before the 1964 piece—Bell pointed out that one of the properties von Neumann had chosen for his hidden-variables scheme were overly restrictive. Later, he would call the assumptions underlying von Neumann's argument "nonsense" and dismiss the proof as "silly" and "not merely false but *foolish*." But as recent reappraisals of von Neumann's argument (for example, by Jeff Bub) have shown, Bell's harsh words are not entirely fair, because von Neumann's argument does in fact successfully exclude a certain class of hidden-variables theories. The point is that it does not exclude *all* hidden-variables theories, Bohm's theory among them.

In one important aspect, Bohm's theory didn't quite match the spirit of a completion of quantum mechanics as envisioned by those who saw themselves as following in the footsteps of EPR. Recall that a linchpin of the EPR argument was a notion of physical locality: that whatever I do to a system *here* should not be able to instantaneously influence the physical situation of another system *over there*. To hold fast to this intuition was a crucial principle for Einstein. In Bohm's theory, however, each particle's velocity depends on the instantaneous positions of *all* other particles. And this means that the outcome of a measurement performed on a particle will in general depend on what a myriad of other, arbitrarily distant particles are doing at that very moment.

Struck by this radical nonlocality of Bohm's theory, Bell wanted to find out "whether *any* hidden variable account of quantum mechanics *must* have this extraordinary character." This was the issue Bell confronted head-on in his 1964 paper. He chose a model similar to that used by Bohm in his version of EPR, namely, an entangled pair of spin-½ particles. Bell considered pairs of spin measurements (one per particle, along some direction in space) that would determine statistical correlations between the particles—say, the joint probability of finding the first particle "spin up" along one axis and the second particle "spin down" along another axis.

Bell then defined a fairly general conceptual framework for a hidden-variables scheme. This amounted to a set of assumptions about how the probabilities would depend on the presumed complete state and on the kinds and outcomes of measurements performed on the two systems. What was so useful about this scheme was that it didn't require specifying the explicit form of the complete state: the hidden variables were just abstract placeholders.

Of course, Bell's main curiosity was aimed at the locality business, and so a central assumption of his scheme was a locality condition on the probability functions. Chiefly, Bell assumed that the probability of a particular outcome of a measurement on the first particle should depend only on the complete state and the choice of that measurement. In particular, the probability was taken to be independent of the choice and outcome of any measurement performed on the second particle.

Bell then worked out clever mathematical combinations of expectation values for different joint measurements. When these expectation values are calculated using the assumptions of Bell's local hidden-variables framework, then the values of the combinations will always be bounded from above; in other words, the combinations become inequalities. But if we use standard quantum mechanics instead, these bounds can be exceeded ("violated"). So the predictions of Bell's local hidden-variables framework become statistically distinguishable from those of standard quantum mechanics. In other words, a hidden-variables theory based on Bell's assumptions cannot fully reproduce the statistical predictions of quantum mechanics. We are thus faced with two incompatible theoretical schemes. Which one correctly models nature?

The Bell inequalities made it possible to apply the experimenter's toolbox to a foundational issue—although the world had to wait until 1972 for the first experiment to demonstrate a violation of the inequalities. Bell's result had both psychological and practical impact. It eventually helped promote quantum foundations to an area of investigation that was at least partially concerned with questions that could be settled by experiment. In principle, self-consciously down-to-earth physicists no longer had to flinch at the mention of a foundational problem (though, arguably, most continued to do so). Instead, at least some of them could legitimately devote intellectual and financial resources to what Abner Shimony has christened "experimental metaphysics." The past and present work of Anton Zeilinger's group in Vienna, to name just one example, beautifully demonstrates the possibilities and power of such fundamental quantum experiments.

From a practical point of view, the experiments motivated by Bell's work soon took on lives of their own. They inspired a new generation of experimental techniques for tinkering with quantum phenomena—phenomena that had previously been thought of as destined to remain the stuff of mere thought experiments (see also Question 11, *The Experiment of My Dreams*). And thus Bell-type experiments sowed the seeds for what has now grown into a full-blown "second quantum revolution": the delicate ability to manipulate and control physical systems at the quantum level, something that is also a prerequisite for the realization of fashionable future technologies like quantum computers.

To date, the statistics gathered by numerous experiments have consistently shown a violation of the Bell inequalities, and loopholes are now largely (but not completely) closed. What does this mean for our picture of how nature may or may not operate? Appearances notwithstanding, this is a nontrivial question. After all, any experimentally confirmed violation of a Bell inequality merely implies, in the first instance, that Bell's particular hidden-variables scheme—the set of assumptions used in deriving the inequality—fails to account for the observed statistics. But can we say more? Which of Bell's assumptions will have to go? If Bell's notion of locality is the sticky issue, can we conclude that nature herself must be nonlocal? Or is the upshot something else altogether? Can there be interpretation-neutral answers to such questions?

Guido Bacciagaluppi · There may yet be more to be discovered about what the violations of the Bell inequalities tell us, but what can be safely said is that they show that the distant correlations present in nature cannot be understood in terms of what seem to be quite general forms of local models. (That is, provided the current loopholes in the experiments do not turn out to be significant after all.) I am happy to call that nonlocality. Others prefer to restrict the use of that term to denote some form of action-at-a-distance, as exemplified in particular by pilot-wave theories (which are the only fully worked-out example of hidden-variables theories we have). This is purely an issue of terminology, and the fact remains that the distant correlations lack any straightforward explanation in local terms. And the explanations that are in fact provided by the different approaches to interpreting quantum mechanics are of the most different kinds.

Pilot-wave theories embrace action-at-a-distance at the fundamental level, so that the distant correlations are explained by means of full-blown nonlocal mechanisms.

Spontaneous collapse theories vary in their explanatory power with regard to the correlations. Indeed, it appears, roughly speaking, that the more relativistically invariant a collapse theory is, the less explanatory it becomes in this regard. Using EPR electron pairs as an example, standard nonrelativistic collapse is an explicit mechanism whereby a measurement on Alice's side of an entangled pair will cause the whole entangled state to collapse. Thus, in particular, it will cause the state of Bob's particle to collapse from a mixed state to an eigenstate of spin (one that can be known, even if initially only by Alice). This is a nonlocal mechanism that explains why a subsequent measurement on Bob's side yields the results it does. At the opposite end of the spectrum, one can take a (putative) collapse theory with collapse defined along the forward light cone, as sketched back in 1969 by Hellwig and Kraus. In such a theory, the collapses on Alice's and Bob's sides appear to conspire to produce the correct results, so that when the two forward light cones meet, the resulting state is the same. The theory thus has the same mysterious kind of distant correlations as standard quantum mechanics.

Everett theories also have a way of explaining the correlations, which invokes the global structure of wave-function branching, or rather—if one takes the definition of "branches" to be partly conventional—the global features of the causal structure arising from wave-function branching, in which certain local components of the wave function will be dynamically insensitive to certain distant components of the wave function on the intersection of the relevant forward light cones.

A lot of research is currently devoted to trying to understand the nature and significance of the specific violation of the Bell inequalities arising in quantum mechanics (why $2\sqrt{2}$, rather than anything else between 2 and 4?), and this may uncover further deep lessons about nature from the experimentally observed violations of the Bell inequalities.

ČASLAV BRUKNER · Bell's theorem is a no-go theorem that states that no "local causal" or "local realistic" theories can ever reproduce all of the predictions of quantum mechanics. The desire for a local causal theory is based on the following three assumptions:

(1) There exist "causes" that determine measurement outcomes, or probabilities of outcomes, for all possible experiments that could be performed on an individual system, no matter whether any experiment—and which experiment—is actually performed (and so, in this sense, would be "real").

(2) The actually measured outcome (or the probability for the outcome), and equally those outcomes that could be potentially measured, can only be influenced by local causes (that is, other events in the backward light cone) and not by any event in spacelike separated regions ("locality").

(3) The experimenter's choice of the measurement setting is independent of the causes that determine the actually measured outcome ("freedom of choice").

From these three assumptions one can derive Bell's inequality, the violation of which has been confirmed in experiments with entangled particles. There is thus no way of getting around Bell's theorem. At least one of the assumptions 1–3 must be wrong. Needless to say, any of these choices requires a radical revision of the ruling philosophical view among most scientists and is in sharp contrast to our everyday experience. Let me discuss each of these possibilities individually.

If one abandons assumption 3, what one actually believes is that there is a mechanism that determines the observer's choices of measurement settings in advance and correlates them to the actual measurement outcomes in such a special way that Bell's inequality is violated (and by exactly the amount predicted by quantum mechanics—no less and no more!). A Laplace's demon–like ("superdeterministic") theory may be invented in which such conspiracies inevitably occur, but it becomes grotesque when one thinks that the choice of measurement setting does not need to be made by the experimenter's free will but could instead be made by an automaton calculating the parity of the number of cars passing the laboratory within n seconds, where n is given by the fourth decimal of the cube of the actual temperature in degrees Fahrenheit, multiplied by two to the power of the number of kids my next-door neighbor has, and so on. I will not comment further on this view.

The second possibility is to preserve the core of classical realism but to deny locality, by assuming that "causes" for the actual and potentially measured outcomes (or their probabilities) depend on actions in spacelike separated regions, and in particular on the choices of observables measured alongside in these regions. Such nonlocal causal explanations always add something to the structure of quantum theory, such as particle trajectories supplemented with a nonlocal potential in the case of de Broglie–Bohm's theory. There is a sense in which such an addition is like a Ptolemaic attempt to cling to the picture of heavenly bodies moving on trajectories centered at the earth by adding epicycles upon epicycles. The de Broglie–Bohm particle trajectories are rarely computed; attempts at doing physics (for example, ionization by escape) with them fail miserably; they require a preferred reference frame; they are not needed to explain any observational phenomenon but are themselves unobservable according to the theory; and so forth. Irrespective of all of this, it appears to me that the most convincing sign of failure of the hidden-variables approach so far is that it could not elicit any new phenomenology that might lead to a progressive research program toward answering Wheeler's famous question, "Why the quantum?"

The main misunderstanding surrounding Bell's theorem arises from a failure to recognize the significance of irreducibility of quantum probabilities *irrespectively* of the relative space-time arrangements of the individual observations on the constituents of the composite system. Some researchers are willing to accept that an outcome measured on a single (local) system is irreducibly probabilistic. But then they are surprised to realize that one runs into a conflict with quantum predictions when the same is not recognized for *correlations* between outcomes measured on several such systems. Bell's "local causality" condition is exactly this: one respects irreducible probabilities locally but does not allow for "intrinsic" probabilities for correlations; they are assumed to be always reducible to—to be the product of—probabilities for outcomes on individual systems. Confronted with the experimental violation of this condition, some researchers are driven to strange solutions: they accept an acausal explanation for an individual quantum system but hold fast to a causal (albeit nonlocal) one for correlations.

There is an alternative view, which treats probabilities for outcomes on individual systems and probabilities for correlations on an equal footing. The guiding idea is that *all systems of the same information-carrying capacity are equivalent*. For example, every elementary system—whether a single qubit or a subspace of two qubits describing correlations—has an equivalent set of states, transformations, and measurements. This seems to be the most natural assumption if one imposes no prior restrictions on a probability theory and preserves the full symmetry between all possible elementary systems.

In logical terms, this means the following. We can think of the two basis states $|0\rangle$ and $|1\rangle$ in the qubit state $|\psi\rangle = \alpha|0\rangle + \beta|1\rangle$ as two binary propositions about an individual system, such as (1) "The outcome of measurement A is 0," and (2) "The outcome of measurement A is 1." An alternative choice could be a pair of propositions about joint properties of two qubits, such as (1') "The outcome of measurement A on the first system is 0, and the outcome of measurement B on the second system is 0," and (2') "The outcome of measurement A on the first system is 1, and the outcome of measurement B on the second system is 1." This pair of propositions corresponds to two basis states, $|00\rangle$ and $|11\rangle$ respectively, which can also be used to span the full state space of an abstract elementary system. But this system now consists of two qubits, and the possible states include entanglement: $|\psi\rangle = \alpha |00\rangle + \beta |11\rangle$. Most importantly, the equivalence of systems with the same information-carrying capacity, as well as entanglement resulting from this equivalence, are independent of the relative space-time arrangements of measurements on the individual constituents of a composite system. Thus, Bell's theorem arises.

JEFFREY BUB · Modulo certain loopholes, which most people expect will eventually be closed as experimental techniques become more sophisticated, the experimentally observed violations of Bell's inequalities tell us that we live in a world in which there are nonlocal correlations that are inconsistent with any explanation in terms of common causes.

Consider the problem of simulating a PR box: how close can Alice and Bob come to simulating the correlations of a PR box for random inputs if they are limited to local resources? Imagine a simulation game in which they are allowed to confer on a strategy in advance, but that once the game starts they are separated and allowed to communicate only with a moderator who gives them inputs, 0 or 1 randomly, to which they are supposed to respond in such a way as to simulate the PR-box correlations. Suppose also that they are allowed to keep any data prepared during the strategy phase, such as lists of shared random numbers they might generate, or other shared instructions. If the outputs of a PR box are expressed in units $a = \pm 1, b = \pm 1$, then

$$\langle 00 \rangle = p\,(\text{same output} \mid 00) - p\,(\text{different output} \mid 00),$$

where $\langle 00 \rangle$ is the expectation value of the outputs if the x and y inputs are both 0, and $p\,(\text{same output} \mid 00)$, etc., denotes the conditional probability. So:

$$p\,(\text{same output} \mid 00) = \frac{1 + \langle 00 \rangle}{2},$$

$$p\,(\text{different output} \mid 00) = \frac{1 - \langle 00 \rangle}{2},$$

and similarly for input pairs 01, 10, 11. It follows that the probability of a successful simulation is given by:

$$p\,(\text{successful sim}) = \frac{1}{4}\Big[p\,(\text{same output} \mid 00) + p\,(\text{same output} \mid 01)$$

$$+ p\,(\text{same output} \mid 10) + p\,(\text{different output} \mid 11)\Big]$$

$$= \frac{K}{8} + \frac{1}{2},$$

where $K = \langle 00 \rangle + \langle 01 \rangle + \langle 10 \rangle - \langle 11 \rangle$ is the Clauser–Horne–Shimony–Holt (CHSH) correlation.

Bell's locality argument in the CHSH version shows that if Alice and Bob are limited to classical resources—that is to say, if they are required to reproduce the correlations on the basis of shared randomness or common causes established before they separate (after which no communication is allowed)—then $|K_C| \leq 2$, so the optimal probability of success is ¾. Evidently, if Alice and Bob agree in advance to both respond 0 for each round of inputs, they will be correct for ¾ of the rounds, on average (because the outputs are required to be the same for ¾ of the rounds). To recover random outputs that are independent of the inputs, they could generate a sequence of random 0s and 1s in the strategy phase and each keep a copy. If they respond to each input according to the random number in the sequence, they will be correct for ¾ of the rounds, on average. Bell's argument simply proves that this strategy is optimal with local resources. The shared random numbers are the common causes of the correlations. If Alice and Bob are allowed to base their strategy on shared entangled states prepared before they separate, then the Tsirelson inequality requires that $|K_Q| \leq 2\sqrt{2}$, so the optimal probability of success limited by quantum resources is approximately .85. For the PR box, $K = 4$, so the probability of success is, of course, equal to one.

The experimentally observed value of .85, rather than .75, tells us that nature allows nonlocal correlations that are more like PR-box correlations than classical correlations.

ARTHUR FINE · Generically speaking, the Bell inequalities result from a standard way of modeling the observables and probabilities of a quantum system in a given state. That way represents the observables as random variables over a common space, whose single distributions agree with the quantum single distributions in that state, and whose joint distributions coincide with the quantum joint distributions for commuting pairs of observables in that state (as defined by the usual trace formalism). The assumptions of a local, noncontextual hidden-variables theory yield such a random-variables representation, as do other sets of assumptions on which various derivations of Bell inequalities are based. (Indeed, in the simple two-by-two and three-by-three cases, satisfaction of the CHSH inequalities is equivalent to the existence of such a representation.) Thus, the observed violations of Bell's inequalities tell us that no such random-variables representation is empirically adequate. Since there are probability models that duplicate the quantum probabilities if we give up the requirement on joint distributions (using single distributions of suitable product observables instead to capture the joints), the experimental evidence points to the quantum joint probabilities as problematic. If one takes a realist attitude toward probabilities, it would be reasonable to conclude from the experimentally observed violations of Bell's inequalities that where quantum theory does not mandate joint distributions (namely, where observables do not commute), those probabilities simply do not exist in nature and should not be introduced in our probabilistic models.

In any case, this much is clear: satisfaction of the Bell inequalities makes certain joint distributions for observables that do not commute well-defined. General theorems indicate that this contradicts the linear geometry of the state space. That is one way to read, for example, the Bell–Kochen–Specker theorem. Thus, any set of assumptions that leads to the satisfaction of the Bell inequalities is inconsistent with the mathematical framework of quantum theory.

These are not the lessons drawn by many, who focus just on the locality assumption featured in many accounts of the Bell inequalities, and who conclude that nature is nonlocal. Certainly drawing this conclusion is premature in the absence of having good reason to eliminate alternative ways of blocking the inequalities. Apart from the preceding reflections on how to treat joint probability, at a minimum one would have to pay serious attention to sources of contextualism (for instance, perhaps the distribution of "hidden variables" is subject to some moderate global constraints) and not simply play it down it rhetorically, using persuasive terms like "free will" or "conspiracy." Likewise, it would be good to refrain from the dismissive language of "loopholes" to minimize the failure, up to now, to perform correlation experiments where the measurement events are spacelike and where, *also*, both the coincidence errors *and* the failures of detection are sufficiently low. The repeated failure of tight experimental refutation of the Bell inequalities along these three dimensions simultaneously may well point to new quantum restrictions, similar to the uncertainty relations, that limit observability in certain contexts. Perhaps, in time, that is what we may learn from the Bell experiments.

CHRISTOPHER FUCHS · Oh, something wonderful. One of my favorite movies of all time is Frank Capra's *It's a Wonderful Life* staring Jimmy Stewart and Donna Reed. If you ask me, the message of quantum theory's necessary violations of the Bell inequalities is the same as the message of this movie—that our actions matter indelibly for the rest of the universe (pluriverse).

In the movie, the protagonist George Bailey proclaims in a moment of anguish, "I suppose it'd have been better if I'd never been born at all." It was the idea George's guardian angel Clarence needed for saving him from suicide. "You got your wish. You've never been born." The story then develops with George seeing how disturbingly different the world would have been without his presence, so much so that by the end of it he wants to live again. As Clarence told it, "You've been given a great gift, George—a chance to see what the world would be like without you." George came to realize how integral his life and his actions were to the rest of the world around him. "Strange, isn't it," Clarence says, "Each man's life touches so many other lives. And when he isn't around, he leaves an awful hole, doesn't he?"

The received wisdom on the Bell-inequality violations for the vast majority of the quantum-foundations community is that it signals nature to be "nonlocal"—that Einstein's spooky action-at-a-distance is alive and well and, to use a word used in your question, "observed." But action-at-a-distance has always been only one of two possible explanations for the violation. The other is that quantum measurement results do not preexist in any logically determined way before the act of measurement. Asher Peres would say, "Unperformed experiments have no results," and we've already heard William James—"Each detail must come and be actually given, before, in any special sense, it can be said to be determined at all." It is this option that fits most naturally within the framework of Quantum Bayesianism, with its two levels of personalism: personal probabilities, whose concern is the agent's expectations for the personal consequences of his actions on an external quantum system. On this view, the place where a quantum measurement outcome "happens" is exactly at the agent who took the action on the quantum system in the first place.

There is a coterie within the quantum-foundations wars (which included John Bell himself and has modern spokesmen in David Albert, Nicolas Gisin, and Travis Norsen) that claim that the *only* implication of the Bell-inequality violations is non-locality—in other words, that it is *not* the dichotomous choice between nonlocality and "unperformed experiments have no results" (or both) that we have been claiming. But their arguments hold no water for the Quantum Bayesian. This is because they all inevitably accept the EPR criterion of reality (or a moral equivalent to it) out of hand—key to this particularly is that they all elide the difference between "probability-one" and "truth." Quantum Bayesians are so stubborn about probabilities being personal degrees of belief that they hold fast to the point even for probability-one statements. "If ... we can predict ... with probability equal to unity ... the value of a physical quantity, then there exists an element of reality corresponding to that quantity." That is the sort of thing I am talking about. It's buried in a hundred different forms in a hundred different treatments of Bell's great result—sometimes it's hard to spot, but it's always there.

But if there is indeed a choice, why does QBism hold so desperately to locality while eschewing the idea of predetermined measurement values? The biggest reward, of course, is that it gives the option to explore "it's a wonderful life," but one can give more strictly academic arguments. Einstein, for one, did it very well:

> If one asks what is characteristic of the realm of physical ideas independently of the quantum-theory, then above all the following attracts our attention: the concepts of physics refer to a real external world, i.e., ideas are posited of things that claim a "real existence" independent of the perceiving subject (bodies, fields, etc.), and these ideas are, on the one hand, brought into as secure a relationship as possible with sense impressions. Moreover, it is characteristic of these physical things that they are conceived of as being arranged in [space-time]. Further, it appears to be essential for this arrangement of the things introduced in physics that, at a specific time, these things claim an existence independent of one another, insofar as these things "lie in different parts of space." Without such an assumption of the mutually independent existence (the "being-thus") of spatially distant things, an assumption which originates in everyday thought, physical thought in the sense familiar to us would not be possible. Nor does one see how physical laws could be formulated and tested without such a clean separation. ...
>
> For the relative independence of spatially distant things (A and B), this idea is characteristic: an external influence on A has no *immediate* effect on B; this is known as the "principle of local action" .... The complete suspension of this basic principle would make impossible the idea of (quasi-) closed systems and, thereby, the establishment of empirically testable laws in the sense familiar to us.

The argument has nothing to do with an unthinking wish to retain Lorentz invariance (as it is often presented): it is much deeper than that. It is about the autonomy of physical systems and about doing science.

In the ellipses I chose for the Einstein quote, one part I hid was Einstein's claim for field theory: "Field theory has carried out this principle to the extreme, in that it localizes within infinitely small ... space-elements the elementary things existing independently of the one another that it takes as basic." I did this because I would say field theory is only a half-hearted expression of the principle. Take a solution to the Maxwell equations in some extended region of space-time, and focus on a compact subregion of it. Can one conceptually delete the solution within it, reconstructing it with some new set of values? It can't be done. The fields outside the subregion (including the boundary) uniquely determine the fields inside it. The interior of the subregion has no identity but that dictated by the larger outside world—it has no real autonomy.

Quantum theory on the other hand, we Quantum Bayesians believe, carries the principle of independent existence to a much more satisfactory level. Wigner and his friend really do have separate worlds, modulo their acts of communication—and so of all physical systems one to another. That, we think, is the ultimate lesson of the Bell-inequality violations. It signals the world's plasticity; it signals a "wonderful life." With every quantum measurement set by an experimenter's free will, the world is shaped just a little as it participates in a kind of moment of birth.

The historian of philosophy Will Durant said it perhaps better than anyone before or since:

> The value of a [pluriverse], as compared with a universe, lies in this, that where there are cross-currents and warring forces our own strength and will may count and help decide the issue; it is a world where nothing is irrevocably settled, and all action matters. A monistic world is for us a dead world; in such a universe we carry out, willy-nilly, the parts assigned to us by an omnipotent deity or a primeval nebula; and not all our tears can wipe out one word of the eternal script. In a finished universe individuality is a delusion; "in reality," the monist assures us, we are all bits of one mosaic substance. But in an unfinished world we can write some lines of the parts we play, and our choices mould in some measure the future in which we have to live. In such a world we can be free; it is a world of chance, and not of fate; everything is "not quite"; and what we are or do may alter everything.

GIANCARLO GHIRARDI · The many experiments that have proved beyond any doubt that Bell's inequalities are violated point to an absolutely fundamental, revolutionary, and unexpected aspect of natural processes, namely, that nature is not locally causal.

One can find various statements in the recent literature that try to make some other assumption—for example, realism, hidden variables, or counterfactual definiteness—responsible for the violation of Bell's inequality. I would like to stress that such statements completely miss the target. I have known John Bell in person, and I have seen him, on various occasions, become terribly upset at suggestions that the derivation of his inequality may require assumptions other than locality. As it can be easily deduced from all of his writings, Bell always emphasized that whenever he was using the hypothesis of determinism or realism, he was actually assuming locality, and he then derived these further assumptions from the logical conjunction of locality and the validity of the perfect quantum correlations.

SHELLY GOLDSTEIN · The experimentally observed violations of Bell's inequalities tell us that nature is nonlocal. More precisely, they tell us that for any theory to predict those violations, it must be a nonlocal theory. They tell us that the nonlocality involved in the global collapse of the wave function in textbook quantum theory can't be eliminated by a suitable reformulation of quantum theory—provided only that for that reformulation measurements have results in the usual normal sense of having results.

Whether Everett's many-worlds version of quantum mechanics is nonlocal—and the jury is out on this—is not addressed by Bell's analysis showing that the violations of Bell's inequality imply nonlocality, because for Everett's many worlds, measurements don't have results in the usual normal sense of having results. In Bell's analysis, violation of Bell's inequalities refers to the way certain experiments produce a pattern of definite results, whereas in many worlds the occurrence of definite results is an illusion.

The violations of Bell's inequalities also tell us, of course, that any version of quantum theory involving hidden variables must be nonlocal. But that's not a terribly informative way to describe the implications, since, with the exception of Everett, they tell us that *any* version of quantum theory must be nonlocal. As Bell repeatedly stressed, hidden variables (that determine beforehand the results of the measurements involved in his argument) are not an assumption of his overall analysis, but an inference along the way—in fact the conclusion, assuming locality, of the EPR part of his two-part argument. The second part of that argument is the derivation of Bell's inequality, demonstrating the incompatibility with the experimental predictions of quantum mechanics of those very hidden variables derived in the EPR argument by assuming locality.

DANIEL GREENBERGER · Bell's theorem is based on the idea that there exist sets of instructions that determine future events as consequences of past events. The violations of Bell's theorem tells us that there exist situations that do not follow from such sets of instructions. This relates directly to our ideas of classical causality, which are based on the future being determined by specific conditions in the past. Instead, in quantum theory we have entangled states, where neither state is determined until it is measured. What is determined is the correlation between the states.

For example, in a singlet state, we have two particles, one spin up and the other spin down. Neither is in a definite state, but we know that if we measure one to be spin up in a certain direction, the other will be spin down. Such states violate the Bell inequalities, in the sense that if you make a series of measurements in arbitrary directions on these correlated states, on the average the results will be more correlated than any set of instructions could possibly explain. For three particles, we also have GHZ states, where, if you measure two particles, you know for certain the state of the third, although none is in a definite state beforehand. But the states are one hundred percent correlated. And yet you can construct states from these that are so correlated that in a single measurement, quantum mechanics says the result must be "yes," while any classical set of instructions says the result must be "no." The lesson seems to be that the correlations are the important thing here, not the individual states. I believe that, but I can't say that I understand the implications of that statement.

LUCIEN HARDY · The first experiment violating Bell's inequalities was performed by Stuart Freeman and John Clauser in 1972. They used an atomic-cascade source, where an atom decays via two levels and emits two photons along the way. They saw a statistically significant violation of the inequalities, in agreement with quantum theory. But the experiment suffered from two loopholes. First, it required making the fair-sampling assumption—that the sample of photons detected at the two ends are fairly sampled from the whole ensemble. It is very easy to construct local hidden-variables models exploiting this loophole (as was first pointed out by Philip Pearle) that can violate the Bell inequalities. The second problem with this experiment was that the measurements at the two ends of the experiment were chosen in the backward light cone of the photons being emitted from the source. Consequently, there is the possibility of this choice affecting the properties of the photons when they are emitted from the source. This is called the causality loophole.

There have been various attempts to close the causality loophole, starting with the experiment of Alain Aspect, Jean Dalibard, and Gérard Roger in 1982. They used acousto-optical switches to vary the measurement being made at each end. These switches are periodic and, as pointed out by Anton Zeilinger, the period was unfortunately chosen just such that the measurements had been switched back to the original setting by the time the photons had arrived at the measurement devices.

Since then, better experiments have been performed in the groups of Anton Zeilinger and Nicolas Gisin. There, the measurement choice was switched randomly. Even in this case, there remains the possibility that the switching is not truly random, and that a sufficiently conspiratorial hidden-variables model could exploit this.

One possibility of closing this loophole is to use humans to make the choices at each end. Such an experiment would be just about feasible with current technology. It seems rather unlikely that Bell's inequalities would cease to be violated when humans did the switching. If this did happen, however, then the consequences for our understanding of the world would have ramifications far beyond quantum foundations (could a computer that passed the Turing test for consciousness also pass this test?). For this reason alone, it seems to be worth doing such an experiment.

It has been much harder to close the first loophole (and remove the need for the fair-sampling assumption), because this requires having sources of entangled particles that are virtually guaranteed to arrive at the appropriate detectors. Detectors of very high efficiency are also needed. The first experiment to close this loophole was an ion-trap experiment in the group of David Wineland in 2001. The two ends of this experiment were, however, only three micrometers apart. In 2003 Christoph Simon and William Irvine had an ingenious idea for doing loophole-free tests of the Bell inequalities. Entanglement is postselected between two ions (or atoms) at the two ends. This is done by a measurement on the photons emitted from the atoms when they reach a central location. In the case of an appropriate outcome for this measurement at the center, the two atoms have the required entanglement. It is easy to do high-efficiency measurements on atoms, and in this way the fair-sampling loophole can be closed. This experiment was performed in the group of Christopher Monroe in 2008 over a distance of one meter.

What is required now is an experiment closing both loopholes, ideally with humans doing the switching. I anticipate that this will happen in the next five years (though maybe not with humans choosing the settings). I fully expect that quantum theory will survive this test—though, secretly, I hope it will fail, because then we would start to have a lot of fun.

There would be two major consequences of a violation of Bell's inequalities that closed the above two loopholes. The first would be practical. We could expect to effectively exploit quantum entanglement in information-processing tasks, such as in reduced communication complexity or quantum cryptography. The second is fundamental. We would have to admit that nature is nonlocal in the sense outlined by Bell. Fotini Markopoulou has pointed out that from the point of view of quantum gravity, this need not be such a big problem. If space-time is emergent, then locality need not be a fundamental property of the world.

Markopoulou and Lee Smolin have constructed a hidden-variables model for quantum theory in which it is suggested that space-time emerges after a flattening-out of a much more connected graphical structure. The small number of nonlocal connections that remain correspond to the nonlocality seen in quantum theory. While we can debate the merits of this particular model, the idea is a good one. Perhaps the most important lesson one can take from the fact that quantum theory is nonlocal is that space-time is not fundamental.

ANTHONY LEGGETT · The more careful discussions of the Bell inequalities (for example, the 1979 *Scientific American* article by d'Espagnat) define the class of theories that can be shown to satisfy the inequalities—that is, the class of "objective local" theories—by the conjunction of the following three postulates:

(1) *Microscopic realism.* Each individual photon (or whatever physical entity is considered) carries with it the information necessary to determine, possibly statistically, its response to any measurement that may be carried out on it.

(2) *Einstein locality.* No supraluminal propagation of causal influences.

(3) *Induction.* No "retrospective" causation. (Note that this postulate is a consequence of postulate 2 only within a theory constrained by the postulate of Lorentz invariance and, arguably, by some other implicit assumptions. In a more general theory, postulate 2 might hold in some special reference frame—for instance, the cosmic rest frame—and the inequalities would still follow for experiments conducted with sources and detectors stationary in that frame.)

It is possible (and I personally prefer) to replace the first postulate with:

(1') *Macroscopic counterfactual definiteness.* The hypothetical "outcome" of measurements that were not in fact performed can be treated as a fixed fact about the world.

In view of the experimental violation of the Bell inequalities, we are forced to con-clude that at least one of the premises 1'–3 (or 1–3) is false. Interestingly, most popular writers on the subject seem to plump for postulate 2, while most professional physi-cists prefer to deny postulate 1', that is, to concur with the late Asher Peres that "unperformed experiments have no results." I would personally take very seriously the possibility that it is postulate 3 that has to go.

TIM MAUDLIN · Assuming we can accept what we seem to see, namely, that every experiment has a unique outcome (contrary to the many-worlds view) and that the correlations between experiments performed at spacelike separation violate Bell's inequality, then we can conclude that nature is nonlocal. That is, in some way certain events at spacelike separation are physically connected to each other. Einstein's dream of a perfectly local physics, in which the occurrences in any small region of the space-time depend only on what happens in that region, cannot be fulfilled. It is an open question what the implications of this fact are for the relativistic account of space-time.

DAVID MERMIN · They tell us something strange about correlations in the out-comes of certain sets of local tests, independently chosen to be performed on far-apart noninteracting physical systems, which may have interacted in the past but no longer do. Prior to Bell's analysis of such quantum-theoretic correlations (and the experimental confirmation of those theoretic predictions), it seemed reasonable to assume that correlations in the outcomes of such tests could find an explanation in correlations in the conditions prevailing at the sites of the tests. Such local conditions can include individual features of the locally tested system, acquired at the time of its past interaction with the other systems; the conditions can also include the weather at the place of the test, the time of each local test, and so on.

Such local explanations can indeed be constructed for any single choice of which local test to perform on each system. But if there is more than one choice of test for each system, then there can be circumstances (revealed by a violation of an appro-priate Bell inequality) in which no single explanation, based on correlation in the locally prevailing conditions, works for all possible choices of local test, even if the choices of local test are made randomly and independently in each local region. This is strange, because the local conditions prevailing at the site of any particular test cannot depend on a random choice of what test to perform far away from that site.

Failure of a Bell inequality fatally undermines the view that all the correlations in all the possible tests can find a single explanation in terms of correlations in condi-tions at the sites of the tests. The conclusions people draw from this vary widely. Those who conclude that the choice of what test to perform in one region does affect the prevailing conditions in the other regions (as it does explicitly in the de Broglie–Bohm pilot-wave interpretation) have embraced nonlocality.

A more conservative conclusion is that it is unreasonable to demand a single ex-planation that works not only for the choices of test that were actually made in each region, but also for the choices of test that might have been made but were not. This is the conclusion of that subset of the quantum-information community with which I sympathize. It is also the conclusion of consistent historians (see my answer to Question 16, page 279), but their apparent conservatism hides their ontologically radical insistence that all the explanations give correct accounts of the tests to which they apply, subject to the proviso that you cannot combine ingredients of one explanation with those of any other, since their validity is in general relative to different "frameworks."

I like Asher Peres's conclusion that unperformed tests have no outcomes: it is wrong to try to account for the outcomes of all the tests you might have performed but didn't. This too is more radical than it appears, since recent versions of Bell's theorem (inspired by Danny Greenberger, Mike Horne, and Anton Zeilinger) show that the outcome of the test you actually performed is incompatible with each and every possible set of outcomes for all the tests you might have performed but didn't. This adds a word to Asher's famous title: "Unperformed experiments have no *conceivable* results."

That addition makes his point just a little harder to swallow. But swallowing becomes easier again if I expand Asher's title further to "Many different sets of unperformed experiments have no conceivable sets of results, *if* the result for each local test has to be exactly the same in every set of results in which that particular local test appears." (The expanded title itself, however, is harder to swallow.) What can it mean to impose such consistency on sets of conceivable data associated with different choices of sets of local tests, when only one set of tests was actually performed?

So for me, nonlocality is too unsubtle a conclusion to draw from the violation of Bell inequalities. My preference is for conclusions that focus on the impropriety of seeking explanations for what might have happened but didn't. Evolution has hard-wired us to demand such explanations, since it was crucial for our ancestors to anticipate the consequences of all possible contingencies in their (classical) struggles.

See also the second of my answers to Question 2, page 53.

LEE SMOLIN · The observed violations of the Bell inequalities imply that there are real physical nonlocal correlations in nature. It seems simplest to suppose these are evidence for nonlocal interactions.

ANTONY VALENTINI · The observed violations of Bell's inequality tell us that locality is violated—if we assume that there is no backward causation and that there are not many worlds.

There is a widespread misunderstanding that Bell's theorem assumes determinism or the existence of hidden variables. In fact, Bell's original 1964 argument had two parts. The first part uses the EPR argument to show that if locality is assumed, then quantum outcomes must be determined in advance. The second part takes this deduction as a starting point and goes on to prove the famous inequality. As Bell himself emphasized, determinism is not assumed; it is deduced from locality in the first part of the argument. So the contradiction is not only between hidden variables and locality. There is a contradiction between quantum theory itself and locality (in the absence of backward causation and many worlds).

If we allow for backward causation, so that future apparatus settings can affect systems in the past, then it seems that nonlocality is not required. It's unfortunate that very little work has been done developing such models to cover a broad range of physics, so we don't know if plausible and attractive theories along these lines exist. As for many worlds, it's a possibility—in my view unlikely, but possible—though whether that theory is well-defined remains controversial.

DAVID WALLACE · Well, for them to tell us anything about nature, we have to accept that it's legit to ask questions about nature (and not just about our experimental apparatus) in the first place. I think it's *obviously* legit; that's what science is for. But I guess a really hardline operationalist about quantum mechanics wouldn't care one way or another about the Bell inequalities.

Having got that out of the way, what the violations of Bell's inequalities *seem* to tell us is that the dynamics of the microworld allows interactions that are faster than light (or slower than light but backward in time, I guess, if that really means anything). If the only interactions in the world are subluminal, Bell's inequalities would be satisfied; they're not, so systems can interact superluminally. End of story. Sometimes people talk about Bell inequalities as if what they rule out is just local hidden-variables theories—maybe even just deterministic local hidden-variables theories—but I think Bell's later work makes it clear just how general they are. (I'm thinking of his paper "Bertlmann's Socks and the Nature of Reality" in particular.)

But I said that's what they *seem* to tell us. I don't think they *actually* tell us that, because there's a tacit premise in Bell's argument: that the results of measurements actually have definite outcomes. That looks pretty innocuous, because if measurement outcomes are macroscopic results, *of course* they're definite. But, of course, that's exactly what the Everett interpretation of quantum mechanics denies. Or, more accurately, measurement outcomes are relative to a branch. And branching (because it's just a dynamical process, namely, the process of decoherence) is a local effect and spreads out at lightspeed (actually, it spreads out at the speed of the fastest interaction that entangles regions with their neighbors, but in practice that's always lightspeed). So if I'm at one end of a Bell-type experiment and you're at the other, I won't be able to ascribe any definite measurement outcome to your measurement until the branching caused by that measurement has reached me, and that happens at lightspeed.

So a better way of putting it is: *if the Everett interpretation is wrong*, violation of Bell's inequalities tells us that there are faster-than-light interactions. And this isn't particularly controversial among people that try to build realist (usually dynamical-collapse or hidden-variables) alternatives to quantum theory. What's slightly more controversial is whether that faster-than-light interaction requires a violation of Lorentz covariance. At first sight it looks like it has to—if we have superluminal interactions and we have Lorentz covariance, it looks as if we can construct closed causal loops. But actually it's a bit subtler, and people—notably Wayne Myrvold and Roderich Tumulka—have played around with so-called hyperplane-dependent collapse theories that try to get a relativistically covariant version of wave function collapse that's compatible with Bell's result. (Tumulka actually has a concrete version, albeit for noninteracting particles). Of course, if you buy the Everett interpretation, then it's of rather theoretical interest if this works. But I'm told not everyone does buy the Everett interpretation.

ANTON ZEILINGER · Bell's inequalities are generally seen to disprove local realism, that is, the joint assumption of locality and realism. Strictly speaking, the experiments still have loopholes open, which I trust will be closed within the next few years. Going back to the reasoning leading to Bell's inequalities, there are a number of other assumptions that come in. These include counterfactual definiteness, or the validity of standard logic. Other assumptions include, for example, that there are no actions back into the past, or that the universe is not completely deterministic. I personally feel that, in combination with the violation of Leggett's inequality and the Kochen–Specker and GHZ theorems, it is most likely the idea of realism that we have to give up—at least in those situations where the respective reasoning applies. To carefully analyze the implications of the various assumptions for the different experiments will be a very important challenge of the future. That way, I expect we will be able to define further experiments, refining our view of the foundations of quantum mechanics. This, in the end, will again lead to novel applications—in the same way as the fundamental experiments that started in the 1970s gave rise to the field of quantum information.

WOJCIECH ZUREK · The basic message is that our universe is quantum. Not classical. In the beginning of the movie *The Wizard of Oz*, the heroine—transported from her home in Kansas to a magical place inhabited by wizards and witches and the like—is told, "You are not in Kansas anymore." I think the import of the observed violations of Bell's inequalities is very much like this. They show that we are not in a classical place anymore.

A corollary to this is a better understanding of the origins of randomness. In the classical realm, we take for granted that systems are entitled to have individual states. They can be correlated, but one can always imagine an underlying pure classical state of a composite system, where each of the subsystems has a state of its own. In other words, a pure classical state in a composite system is always a Cartesian product of pure states of the subsystems. This is not the case in the quantum domain; a pure quantum state of a composite system can be entangled—that is, it is a tensor (rather than a Cartesian) product of the states of the subsystems.

This is the reason why probabilities in the quantum domain can be objective: knowledge of the entangled state of the whole implies objective ignorance of the states of subsystems (see my answer to Question 5, page 122). Environment-assisted invariance is based on this tensor structure of states of quantum composite systems.

Their symmetries allow one to find out outcomes that are equiprobable, which then leads to the derivation of Born's rule that relates probabilities to the amplitudes of quantum state vectors.

QUANTUM INFORMATION

*What contributions to the foundations of quantum mechanics
have, or will, come from quantum information theory?
What notion of information could serve as a rigorous basis
for progress in foundations?*

Q UANTUM INFORMATION is quantum foundations' adolescent nephew, brash
and suntanned, who crashes the party with a teasing, "Whatcha frettin'
about, people?" But the foundations family can't kick him out so easily because he's brought beer for everyone. And while some in the family brush off their
relative's skits as fleeting entertainment and as no match for their adult conversation,
a few eagerly join the youngster upstairs for an intergenerational sit-in.

For many of my age group, research in quantum mechanics has largely meant
research dominated by quantum information theory. It's as if, sometime in the 1990s,
quantum mechanics underwent a complete facelift that transformed the way people
now look at it. Without a doubt, quantum information theory has taught an old
dog some mean new tricks. And it's no longer just the physicists, philosophers, and
mystics who rendezvous with quantum mechanics: gaggles of computer scientists,
engineers, and mathematicians have also jumped on the bandwagon. Schooled in
the new idiom and high on a what-can-information-do-for-you-today optimism,
some avantgarde types have even started planting yard signs proclaiming QUANTUM
MECHANICS = QUANTUM INFORMATION THEORY.

While everyone would rush to confirm intellectual enlightenment as the decisive criterion for judging the relevance of a discipline, some cynics suggest that hard
cash has also played a role in wooing the crowd. Be that as it may, the market- and
security-relevant promises of the quantum-information age—scarce as implemen-
tations may remain for some time to come—have led to a sudden windfall that has
ushered in a quantum renaissance on a scale no one would have imagined thirty
years ago. Now for many the rule of thumb is: if you want to get your research in
quantum mechanics funded, make sure to spice it up with a good dose of quantum
information.

Research in quantum information treats us to a daily barrage of new terminology,
theorems, algorithms, protocols, and experiments, at a pace that's all but impossible
to keep up with. But once we rise above this din of incessant chatter, what's in it
for the foundations of quantum mechanics, and for our understanding of nature at
large? Does the information-theoretical viewpoint have the power to relieve some of
the headaches that have made us queasy since quantum theory's inception? Is quan-

tum information a game-changer for foundations? Or is it a distraction from the real issues, with acute symptoms being obscured by fancy new language? What's the foundational worth of the concept of information? Which particular notion of information might be sufficiently sturdy and noncontroversial to make for a structure capable of elevating us above the quicksand of foundational paradoxes? Can the operational mindset embraced by many quantum informationalists help air out a stuffy room? Or is it an ineffective antidote, perhaps even one that's detrimental to our quest for foundational clarity?

GUIDO BACCIAGALUPPI · Quantum information theory has sparked off a huge interest in foundational issues, especially in the nature of entanglement and nonlocality and in the axiomatic foundations of quantum mechanics. Indeed, some of these topics had been fairly dormant until then (for instance, apart from Solèr's theorem in quantum logic, the reconstruction problem had not seen much progress in recent years). This has contributed, and is likely to continue contributing, to foundations in a lasting way. Prominent examples of such contributions are the work of Popescu and Rohrlich on nonlocality, Rob Spekkens's toy theory that qualitatively reproduces an amazing array of features of quantum mechanics from purely epistemic restrictions on states, and several new results on reconstruction (by Hardy, Goyal, and Chiribella–D'Ariano–Perinotti, to quote a few). We should expect in the future to gain many more insights on quantum foundations coming from quantum information.

That said, I believe it remains to be seen whether quantum information might prove to be an alternative foundation for quantum mechanics, and I shall not venture an answer to the second part of the interview question.

ČASLAV BRUKNER · All our descriptions of objects are represented by propositions. The use of propositions is not a matter of choice, but rather a necessity that is behind each of our attempts to learn something new about nature and communicate this knowledge to others. It is a necessity that we follow constantly and unintentionally. This was repeatedly emphasized by Bohr:

> [H]owever far the phenomena transcend the scope of classical physical explanation, the account of all evidence must be expressed in classical terms. The argument is simply that by the word "experiment" we refer to a situation where we can tell others what we have done and what we have learned and that, therefore, the account of the experimental arrangement and of the results of observations must be expressed in unambiguous language with suitable application of the terminology of classical physics.

Rigorously speaking, the state of a system is nothing else than a construct based on a complete list of (potentially probabilistic) propositions. The propositions from this list could be "The momentum of the system is p," or "The position of the system is x." They could be associated with either a classical or a quantum system. Yet there is an important difference between the two cases. From the theorems of Bell and of Kochen and Specker we know that for a quantum system, one cannot assign definite (noncontextual) truth values simultaneously to all conceivable propositions. It then follows that in attempting to describe quantum systems, we are unavoidably put in the following situation. On the one hand, the epistemological structure to be applied has to be inherited from classical physics: the description of a quantum system has to be given in terms of propositions that use the terminology of classical physics. On the other hand, these propositions cannot be assigned to a quantum system simultaneously. How to join these two seemingly competing requirements?

I would like to suggest using the notion of "knowledge" or "information." Then, although we cannot assign definite truth values simultaneously to all conceivable propositions, we can nonetheless assign measures of information to them—that is, we can quantify how certain or uncertain we are about the truth values of the propositions. The structure of the theory, including the dynamics, can then be expressed in terms of measures of information. To me, this seems to be a change in the epistemological structure of classical physics at the lowest possible cost. And since some costs are unavoidable anyway, I think that the information-theoretical interpretation of quantum physics leads to the smoothest understanding of the theory.

JEFFREY BUB · Before the advent of quantum information, and certainly before Bell's work on nonlocality, research in the foundations of quantum mechanics was largely the province of philosophers or a few philosophically-minded physicists and hardly a respectable field in physics. Now articles on quantum foundations are quite common in journals like *Physical Review Letters*, which accepts less than thirty-five percent of submitted papers. So thinking about quantum mechanics from an information-theoretic standpoint has radically transformed the field of quantum foundations.

To cite just a couple of examples: Toner and Bacon proved that two parties, restricted to classical resources, can perfectly simulate the correlations between the outcomes of projective measurements on an entangled Bell state of two qubits if they are allowed just one bit of classical communication for each round of measurements. In a related result, Barrett and Gisin showed recently that if one party's choice of measurement setting is constrained by one bit—in the sense that the mutual information between local variables and the measurement setting is one bit—then the correlations can be reproduced by local resources. Results such as these have considerably advanced our understanding of the nature of quantum correlations, *the* characteristic trait of quantum mechanics that enforces its entire departure from classical lines of thought, to paraphrase Schrödinger.

I think Shannon's notion of information, suitably generalized for quantum information or information in a nonclassical setting, is the appropriate notion. In a 2010 *New Journal of Physics* paper, Anthony Short and Stephanie Wehner have shown how to define a general measure of information for a broad class of theories, including PR-box theories, that reduces to von Neumann entropy for quantum theories and to Shannon entropy for classical theories.

The question considered by Shannon was how to quantify the minimal physical resources required to store messages produced by an information source, so that they could be communicated via a channel without loss and reconstructed by a receiver. The essential notion underlying Shannon's measure of information is compressibility: information as a physical resource is something that can be compressed, and the amount of information produced by an information source is measured by its optimal compressibility.

Shannon's source coding theorem shows that there is an optimal, or most efficient, way of compressing messages produced by a source (assuming a certain idealization) in such a way that they can be reliably reconstructed by a receiver. Since a message is abstracted as a sequence of distinguishable symbols produced by a stochastic source, the only relevant feature of a message with respect to reliable compression and decompression is the sequence of probabilities associated with the individual symbols. The nature of the physical systems embodying the representation of the message through their states is irrelevant, provided only that the states are reliably distinguishable, as is the content or meaning of the message. Shannon entropy is a measure of the minimal physical resources that are necessary and sufficient to reliably store the output of a stochastic source of messages. In this sense, it is a measure of the amount of information per symbol produced by an information source.

So information in Shannon's sense is a quantifiable resource associated with the output of a (suitably idealized) stochastic source of symbolic states, where the physical nature of the systems embodying these states is irrelevant to the amount of classical information associated with the source. The fact that some feature of the output of a stochastic source can be optimally compressed is, ultimately, what justifies the attribution of a quantifiable resource to the source.

The essential difference between classical and quantum information arises because of the different distinguishability properties of classical and quantum states: only sets of orthogonal quantum states are reliably distinguishable with zero probability of error. The theory of quantum information extends Shannon's notion of compressibility to a stochastic source of quantum states, which may or may not be distinguishable. Classical information is that sort of information represented in a set of distinguishable states—states of classical systems, or orthogonal quantum states—and so can be regarded as a subcategory of quantum information, where the states may or may not be distinguishable. It turns out that a suitable measure of information for probability distributions of quantum states—that is, for mixed states—is the von Neumann entropy, which can be interpreted in terms of compressibility via Schumacher's source coding theorem for quantum information.

ARTHUR FINE · I have to duck answering this question in detail, because my own research has not been focused on the approach to foundations by way of information. All I can say is that, so far, to approach quantum mechanics as a theory of information management has not led to any major shift in resolving outstanding issues in foundations, nor in dissolving them. Rather, while leaving standard issues intact, this approach seems to open up new issues of its own. For example, information management involves updating and, since the usual Bayes rule is not available where observables do not commute, a new "quantum" updating rule must apply (some version of the Lüder's rule). Whereas classically one can give a Dutch-book argument for Bayesian updating, in the quantum case it appears that the updating rule has to be put in by hand (despite some misleading analogies with classical conditional probability). This appears to be the revenge of the measurement problem on information. Here I duck the obvious questions, which most adherents of the informational approach also duck; namely, what is "quantum information," and what is it "information" about?

CHRISTOPHER FUCHS · Here's a variant on your question that I posed to myself nearly ten years ago:

> The task is not to make sense of the quantum axioms by heaping more structure, more definitions, more science-fiction imagery on top of them, but to throw them away wholesale and start afresh. We should be relentless in asking ourselves: From what deep physical principles might we derive this exquisite mathematical structure? Those principles should be crisp; they should be compelling. They should stir the soul. ... Until we can explain the essence of the theory to a ... high-school student ... and have them walk away with a deep, lasting memory, I well believe we will have not understood a thing about quantum foundations. ...
>
> So, throw the existing axioms of quantum mechanics away and start afresh! But how to proceed? I myself see no alternative but to contemplate deep and hard the tasks, the techniques, and the implications of quantum information theory. The reason is simple, and I think inescapable. Quantum mechanics has always been about information. It is just that the physics community has somehow forgotten this.

Well, we've come a long way since then, but I fear that despite all the mixing and mingling of quantum information and foundations that has come about in the meantime, the core message is still being forgotten.

Don't get me wrong; great work has certainly been done. For instance, Rob Spekkens's work already mentioned in Question 3 (see page 72) is a really outstanding example of how to examine the fruits of quantum information for their foundational insights. What quantum information gave us was a vast range of phenomena that nominally looked quite novel when they were first found—people would point out all the great *distinctions* between quantum information and classical information: for instance, "that classical information can be cloned, but quantum information cannot." But what Rob's toy model showed was that so much of this vast range wasn't really novel at all, so long as one understood these to be phenomena of epistemic states, not ontic ones. It is not classical *information* that can be cloned, but classical *ontic states* that can be; classical epistemic states (general probability distributions) are every bit as unclonable as their quantum cousins.

So the great contribution of quantum information for quantum foundations, I would say, is in the mass of phenomena it provides to the epistemic playground. By playing with these protocols, we get a much better feel for the exact nature of quantum states as states of mind (and for QBism, states of belief particularly). The reason I said I feared that the core message is still being forgotten is that despite this, it is amazing how many people talk about information as if it is simply some new kind of objective quantity in physics, like energy, but measured in bits instead of ergs. In fact, you'll often hear information spoken of as if it's a new fluid that physics has only recently taken note of. I'm not sure what the psychology of this is—why so many want to throw away the hard-earned distinction the concept of information affords between what's actually out in the world and what an agent expects of it—but the tendency to ontologize information is definitely there in the physics community and is even more pervasive in the philosophy of science community. I sometimes wonder if it is an expression of a deep-seated longing for an old-style aether. But maybe in the end, the cause will turn out to be no more sophisticated than what happens in a first-year calculus service course, where the majority of students learn how to take derivatives of the standard functions but have no clue what the concept actually means.

GianCarlo Ghirardi · On the one hand, I think that the development of quantum information theory represents a step that's extremely important for our understanding of the laws of nature and for taking full advantage of these laws. The fact that by using quantum superpositions and entanglement, one can develop new algorithms, and that in this new perspective, most of the theorems of classical information theory (Shannon's theorems) have been rederived and appreciably improved, represents remarkable progress in the potential exploitation of the resources that nature has put at our disposal.

On the other hand, I believe that the promising developments of quantum information have played a negative role in terms of our need to face the crucial foundational problems of the theory. In my opinion, the idea that a fundamental theory of the universe is concerned not with what actually *is out there* but only with the information we can have about it, represents a step back—of about eighty years—in the crucial debates on foundational issues.

Let me illustrate this point by referring to specific positions of some brilliant scientists, and to statements that have appeared in the literature. A paradigmatic case is David Mermin, who was among the supporters of Bell's position about quantum theory up to few years ago. But then something changed, as Mermin recounts:

> Until quite recently I was entirely on Bell's side on the matter of knowledge-information. But then I fell into bad company. I started hanging out with the quantum computation crowd, for many of whom quantum mechanics is self-evidently and unproblematically all about information.

To clarify Mermin's remark, let me mention that Bell, in his debates with the advocates of the decoherent-histories approach, made it clear that he did not intend to pay serious attention to people who invoke information, unless they would in advance answer two fundamental questions: "*Whose* information?" and "Information about *what?*" Mermin's reply is clear-cut: "'Information about *what?*' is a fundamentally metaphysical question that ought not to distract though-minded physicists."

Let me summarize what I perceive as the shared position of the quantum-computation crowd. One has at his disposal various black boxes, which correspond to the different gates necessary to build the quantum circuit that implements the quantum algorithm. Most of these gates are represented by standard quantum operations on the state vector and account for dynamically unitary processes. To get the final output from the quantum computer, however, one needs to "read" the outcome from appropriate apparatuses. For this purpose, one has to resort to a fundamental gate, which Mermin calls the "measurement gate," and which differs from the other gates in that it works nonlinearly: it returns different outcomes with certain probabilities, and it leads to wave-packet reduction. Using such a gate in combination with the other standard gates, one can implement all necessary quantum circuits. This argument is precise and correct. But it completely ignores the fact that while our basic theory of natural processes—quantum mechanics—accounts, in exact terms, for the functioning of the unitary gates, the theory is absolutely silent on the working of the measurement gate, whose dynamics actually *contradicts* the dynamics of the fundamental theory one is invoking in implementing quantum-computation procedures.

I believe this is an extremely serious issue. It must be stressed again and again that the nonlinear and stochastic process of wave-packet reduction cries out for a reasonable inclusion within the theoretical scheme—or, otherwise, for making precise the sense in which this process, in spite of its essential role, does not fit into the general picture. I consider such a critical reconsideration crucial, and it saddens me to realize that the new exciting branch of quantum information is obfuscating rather than stressing this pivotal point.

SHELLY GOLDSTEIN · I don't think that quantum information theory has been terribly illuminating for the foundations of quantum mechanics. I'm aware of no precise version of quantum mechanics that has been genuinely suggested by quantum information theory—though some would no doubt point to many worlds. I don't see how it helps clarify what is going on in the two-slit experiment. I don't see any evidence, or any serious suggestion, that it can eliminate quantum nonlocality.

But quantum information theory, and quantum computing in particular, could have great practical value. And it does seem to generate a lot of interesting mathematics and physics. Some of these things, such as the exponential speedup afforded by quantum computation, strongly suggest that the wave function of a quantum system—the object in quantum mechanics that, through its tensor-product structure, is most directly connected with the exponential speedup—must be taken seriously as an objective element of physical reality and not be regarded as merely a reflection of our knowledge.

But I don't anticipate any deep insights about the foundations of quantum mechanics to emerge from quantum information theory. It hasn't happened yet, and I'm skeptical that it ever will. One reason for my skepticism is that the vague notion of information seems entirely inappropriate for the problem at hand. It is too closely connected with human needs and desires. The notion of information seems completely out of place on the atomic and subatomic levels. References to information and the like should not appear in the formulation of fundamental physical theories. I simply find it implausible that we are that important.

If, however, one removes the human element from the notion of information and considers instead mathematical structure and the structural aspects of physical theories, well, then principles concerning this could be illuminating for the foundations of quantum mechanics.

DANIEL GREENBERGER · Some of my thoughts on this are in my answer to Question 6 (see page 133). But here, I would just like to give my feeling that we are asking the wrong questions in information theory. The problem is that we assume that all the possibilities are known, and we are choosing one of them. That does provide some information. An example: what would be the next letter in a word? There are twenty-six possibilities with varying frequencies, and we have to choose one. That does provide information. But there is another type of problem that is more important and that seems to have been ignored.

When a baby has to learn something, he doesn't know what the choices are. He has to do something, observe the feedback, and determine whether that was a desirable action. And that is how he learns. That's really a different form of information. The same kind of situation pertains to evolution. An evolving system isn't aware of its choices, and there is nobody to say what is right and wrong. In a given circumstance, one choice might be best, while in a different circumstance, a different choice would be better. But the system must be able to evaluate the response, so it is clearly receiving information from it. Then it must be able to act on that and remember. So the system must be fairly advanced to be able to profit enough to evolve. But as part of the process, the very choices and ways of choosing them have to be evolving. It seems to me that we will have to generalize the idea of information if we want to learn its creative aspects—rather than merely passing on what is already known.

LUCIEN HARDY · In 2001 I gave a set of postulates from which quantum theory could be derived (see my answer to Question 10, page 206). One of the axioms is something I now call the information axiom: *systems having, or constrained to have, the same information-carrying capacity have the same properties.* The information-carrying capacity of a system is given by the maximum number of reliably distinguishable states that can be prepared (this corresponds to the Hilbert-space dimension in quantum theory). It is a remarkable property of quantum theory that this is true. For example, the spin degree of freedom of an electron has the same properties as the polarization degree of freedom of a photon, and each can be used to carry a qubit. Further, if we take a quantum system with, for example, a five-dimensional Hilbert space but constrain all states to have support only on some given three-dimensional subspace, then we effectively have a system having the same properties as a system having a three-dimensional Hilbert space. The information axiom is also true in classical probability theory.

Quantum information theory forces two useful attitudes. First, there is an emphasis on finite-dimensional systems. This is good, because we are more likely to gain deep structural insights in this mathematically simpler situation than in the case of continuous-dimensional Hilbert spaces. Further, there are good reasons coming from quantum gravity to believe that, fundamentally, finite-dimensional Hilbert spaces describe reality better than continuous Hilbert spaces. Second, compositionality plays a deep role. Much of quantum information is concerned with multipartite systems. This corresponds to the deep idea that we should analyze the world into its components and then see how these components fit back together.

Aside from these specifics, there is a more fundamental contribution that the informational approach makes to physics. This is that it forces what might be called a user-centric approach. Information theory is concerned with what we, the users, input into an experimental situation by choosing experimental arrangements of apparatuses and knob settings on these apparatuses, and what we, the users, read off by looking at the results of measurements. This is an operational approach to physics. The strength of this approach is that it makes physics relevant to our experience. A mathematical formalism cannot be regarded as constituting a piece of physics unless we can say which elements in it correspond to what we see and do in the world. The danger is that this approach mitigates against trying to find a deeper ontological picture of what is happening at the fundamental level. This tension between the operational and realist approaches to physics has existed since the early days of quantum theory (and even prior to this, with the heliocentric theory of the cosmos, thermodynamic versus statistical physics, and relativity theory). If we are concerned with constructing new physical theories, then both approaches have a role to play. The operational approach is particularly good if we want to move forward in a way that is not unduly hindered by our ontological preconceptions. By thinking operationally, Einstein was able to disencumber physics of the deeply ingrained notion of absolute simultaneity. I think that progress in constructing a theory of quantum gravity is most likely to come from operational thinking (which is basically the same as the informational approach). In particular, by thinking operationally, we can construct a more general framework for physical theories that are probabilistic and may have indefinite causal structure. We can then attempt to situate quantum gravity in this more general framework.

Ultimately, we will want to have an ontological understanding of the world and need to move beyond the operational framework to some sort of realist picture. Properly used, however, these two approaches can positively reinforce each other.

ANTHONY LEGGETT · At the risk of perhaps sounding a little parochial, I would say that the most important contribution has been a "political" one, in the sense that it is now rather widely accepted (as it was not in 1967 when I took up my first tenured position) that an active interest in the foundations of quantum mechanics does not disqualify one from being a "proper" physicist. At an even more nuts-and-bolts level, it is improbable that the spectacular recent experiments on flux qubits, which I would regard as one of the more significant advances in foundational studies of the last few decades, would have attracted the funding they have were it not for the prospect of using these systems as elements in a quantum computer.

At a deeper level, I would regard the role of quantum information relative to quantum foundations as analogous to that played by engineering problems in the development of classical thermodynamics in the early and mid-nineteenth century. That is, in both cases the requirement of being able to build a device that actually works forces one to make much more precise concepts that were previously at best qualitatively formulated. In the present case, for example, the demands of practical quantum cryptography have forced a much more precise quantification of the ideas of bipartite and multipartite entanglement. I don't see, however, that quantum information has invented new physics from scratch in the way that the early nineteenth-century engineers did. And I certainly don't believe that a better concept of "information," even if it is possible, is likely to do anything to solve fundamental issues such as the measurement problem.

Tɪᴍ Mᴀᴜᴅʟɪɴ · The notion that quantum information theory or quantum computational theory could contribute to the foundational questions has always puzzled me. Each of these is a perfectly legitimate field of inquiry, but each already presupposes some sort of understanding of quantum theory. Any physical theory will have both information-theoretic and computational consequences. One can ask, for example, how to best use electromagnetic fields to transmit information, or even how to build a computer using the physics of John Conway's Game of Life. That is, information-transmission mechanisms and computing mechanisms are all physical objects, and what they are capable of is limited by the physics that governs them. But one has to settle the physics before the information-theoretic analysis can begin. I have no concept of how one could turn the usual project on its head and derive or explain physics from information theory.

It is sometimes claimed that Einstein's account of special relativity can serve as a model of how physical structure can be derived from some general principles, such as the constancy of the speed of light. But Einstein himself insisted that such a derivation was only a temporary halfway house. If one explicates "the speed of light" by reference to clocks and measuring rods, then ultimately one needs a constructive account of these devices: just as an exact physics cannot mention "measurement" in its foundational postulates, so it cannot mention "clock." Or, to take another example, one can derive many interesting physical consequences from the second law of thermodynamics, but it is not appropriate as a foundational principle. Boltzmann showed how to understand the second law as a probabilistic consequence of the underlying atomic dynamics. In any particular case, it is this underlying dynamics that ultimately accounts for the behavior of a system. The reliability of the second law may derive from very generic features of that dynamics—and one learns a lot from seeing how generic those features are—but the dynamics cannot be replaced by the thermodynamic generalization. Similarly, generic features of quantum dynamics may have interesting information-theoretic implications (such as unitarity implying a "no-cloning" theorem), but it would get things backward to suggest that the no-cloning property somehow *explains* the unitarity. The command "Thou shalt not clone" is not a credible possibility for a fundamental physical law, because it does not specify any particular physical behavior. The Schrödinger equation, in contrast, does.

DAVID MERMIN · I agree with Heisenberg and Peierls that the quantum formalism is a tool we have discovered to express the information we have acquired and the consequences of that information for the content of our subsequent acquisition of information. To the extent that it sharpens and systematizes this point of view, I believe that quantum information theory is the most promising and fruitful foundational approach.

Beyond this, applying the quantum formalism directly to the processing of information itself may get us closer to the heart of what quantum mechanics is all about, than can the informationally less subtle problems addressed in more traditional physical applications of quantum mechanics. At the very least, it provides a refreshingly different set of examples of quantum phenomena.

I am not expert enough in quantum (or classical) information theory to have an opinion on the definition of information most likely to shed light on foundational questions. Slogans like "It from bit" are fun, but don't tell me much without considerable (yet to be provided) expansion. It seems to me that any foundationally illuminating concept of information must be explicit about both the possessors of the information and the content of that information. As John Bell put it, "Whose information?" and "Information about what?"

See also my answer to Question 16, page 279.

LEE SMOLIN · Quantum information theory is a very helpful tool for applying quantum mechanics to problems far from the classical limit. It has been useful in quantum gravity, as shown by works of Fotini Markopoulou and collaborators, where it helps to get at the question of what the emergent local degrees of freedom are.

But so far, to my mind, it has not solved the foundational problems that concern me. It has, however, illuminated them. For example, I'm impressed by the work of people like Rob Spekkens and Chris Fuchs, who use quantum information theory to argue that the quantum state represents information an observer has about a quantum system. This supports the view that the quantum-state description does not correspond to the physical reality of individual systems.

ANTONY VALENTINI · In my view, with the rise of quantum information theory in the 1990s, the subject of quantum foundations was set back by at least twenty years. There are, however, some issues that need to be clearly distinguished.

First of all, quantum information theory is just quantum mechanics applied to certain practical problems. Nothing new is said about ontology, and the usual ambiguities remain. No attempt is made even to address the measurement problem.

On the other hand, quantum information theory has emphasized some aspects of quantum theory that had been unduly neglected. In particular: entanglement, peculiarities of the tensor-product structure of Hilbert space, and general properties of unitary evolution such as the no-cloning theorem. What these features have in common is that they don't depend on details of the system or on what its Hamiltonian happens to be. Some people find this exciting. But, in fact, systems do consist of particles and fields, and these propagate in space-time, and there are various symmetries associated with conservation laws, and so on, and all this remains the basic stuff of physics.

It should also be remembered that entanglement as the fundamental new feature of quantum physics was discussed by Schrödinger as long ago as 1935, quantum cryptography was anticipated by Wiesner around 1969, and explicit statements of the no-cloning theorem date from 1982. Much of the basic and truly important material is not as novel as is often claimed. What has really happened is that these features have turned out to be of technological interest, and the resulting outpouring of funding has generated a huge bandwagon.

As far as fundamental physics is concerned, I see a useful parallel with what happened in general relativity in the 1960s, when people discovered that some important deductions could be made purely on the basis of geometrical arguments, without invoking the details of Einstein's field equations. I mean results like the singularity theorems of Penrose and Hawking. Modern textbooks on general relativity include a chapter on such geometrical methods—containing, in particular, a few key results, such as the singularity theorems, and a few useful theorems about global hyperbolicity and causal structure. But still, most of what we know and understand about Einsteinian gravity comes from analysis of the field equations. Now, the parallel with quantum information theory is clear. It was realized that some important deductions could be made purely from geometrical or kinematical properties of unitary evolution in Hilbert space. The details of the Schrödinger equation or Hamiltonian didn't matter. It will soon be standard for introductory textbooks on quantum mechanics to contain a chapter giving a few key results such as the no-cloning theorem and one or two useful theorems about entanglement. But still, most of what we know about quantum physics comes from analysis of the theory applied to concrete systems of electrons, photons, atoms, and so on, and the detailed structure of the Hamiltonian is of central importance.

But the real damage that has been done is in reviving the misguided idea that physics is only about macroscopic operations and observations. A sort of "neo-Copenhagen" attitude has arisen, with the word "information" playing a role similar to the older word "observation." The usual ambiguities remain. Macroscopic equipment with its definite ontological states plays a fundamental role, while no ontology is provided at the microscopic level, and with no heed paid to the lack of a clear dividing line between those two levels. The measurement problem is simply not addressed.

It is sometimes claimed that "information" is a new fundamental concept. But "information" is synonymous with "knowledge about something." What is the knowledge about? If it is only about macroscopic instrument readings, then it is not knowledge of anything fundamental.

I see quantum information theory as also analogous to thermodynamics. In the late nineteenth century, some people thought that they had found a new approach to physics that focused on the production, transmission, and use of energy, based on general principles that didn't depend on details of the system. In retrospect, of course, gases and liquids are made out of atoms and molecules, and their macroscopic behavior is not fundamental but emergent. Nowadays, some people claim that physics is about the production, transmission, and use of information, based on general principles that don't depend on details of the system. But again, the systems we see are built out of microscopic entities, and the behavior of macroscopic instruments is not fundamental but emergent.

DAVID WALLACE · Quantum information theory brought something completely new to foundations of physics, in that it was the first time people had combined foundationally careful attention to the specifically quantum-mechanical aspects of quantum mechanics with detailed, quantitative exploration of the theory's implications in particular situations.

People had done one or the other before. In particle physics, say, people were absolutely asking foundational questions, but they were mostly using quantum mechanics as a calculational tool—come up with a classical field theory, plug it into the machinery of Feynman diagrams and renormalization-group flows, and see what comes out. And what came out was wonderful, of course, but the quantum mechanics was largely functioning as a black box. Conversely, people in foundations of physics and philosophy of physics were asking foundational questions about quantum mechanics itself, but they were either not doing mathematics at all, or they were proving rather general theorems. They weren't playing with toy models, they weren't calculating much, they weren't exploring quantitatively just what the theory was capable of in various specific situations.

Then quantum information came along, and suddenly we discovered a huge range of things that *could* have been discovered in the 1950s, but weren't—teleportation, dense coding, the no-cloning theorem, entanglement swapping, Shor's algorithm, and so on. And those things haven't just been *practically* relevant—they've really deepened our understanding of what quantum mechanics is as a theory. And that's ongoing, and I'm sure other people answering this question are much better placed than I to go into details.

So, quantum information theory is an amazing tool to explore quantum mechanics. But there's a more ambitious project, which is to say that quantum information theory *is* quantum mechanics—or rather, that quantum mechanics just is a theory about information. Slogans like "physics is information" start getting mentioned at this stage.

I'm much more skeptical about this project. Partly that comes from worrying about whether it even makes sense—we don't think that the world could coherently be made of *opinion* or *belief* or *rumor*, and I'm not at all sure *information* is any better as a building block. (That's not to say that it's a scientifically useless concept—no more is belief a scientifically useless concept—it's just not obviously the sort of concept that can do as a fundamental-level description of reality.)

But more seriously, I don't really see how PHYSICS = INFORMATION squares with what we use quantum mechanics for ninety-nine percent of the time, which is to calculate physical properties of rather specific systems—crystals, metals, plasmas, atomic excitations, mass spectra of hadrons, and so forth. Quantum information hides that away from us, because we study it in an incredibly abstract way that conceals the ultimately physical, dynamical nature of whatever the Hamiltonian is of any given system.

That's exactly what quantum information should do, of course. The brilliant thing about it is that precisely because it *does* abstract away all those aspects of the system, it lets us see general features we'd never have spotted if we'd kept all the messy details in play. But it may be a mistake to treat that as an insight into the nature of reality itself, rather than into the nature of information flow in that reality.

ANTON ZEILINGER · To broaden the question a little bit, I'd say that the most important development in the last years is that it has become generally accepted that information plays a basic role in quantum mechanics. For example, in the two-slit experiment, it is whether information about the path taken is present anywhere in the universe that determines whether the interference pattern shows up.

The notion of information, which is a most fruitful one for the foundations of quantum mechanics, must be a notion that does not assume the preexistence of observed values. Therefore, Shannon's information is not fruitful, because it tacitly implies a realistic interpretation of information. The most fruitful measure currently appears to be the one based on the square of probabilities.

WOJCIECH ZUREK · The great virtue of quantum information is that it places questions about the relationship between existence and information in a practical setting. So a lot of the issues that were regarded as paradoxes, including Schrödinger cats and entangled states, acquire a very different status. Quantum error correction is an example of a (relatively) simple process where such formerly paradoxical phenomena acquire practical significance.

I think a lot of issues (starting with EPR) mix quantum physics with the physics of information. I share the suspicion of many that there is more to be understood there. The fact that even nonrelativistic quantum mechanics "knows" about special relativity (for example, via the prohibition on cloning) is such a hint.

The connection between information and thermodynamics forces one to use something like von Neumann's entropy to express *ignorance*. To the extent to which negentropy is information, we have our answer—we have a sensible notion of information. But perhaps we are missing a better definition of what it means "to know," and, hence, what the nature of information is; entropy is a fairly high-level concept. I would expect a fundamental notion of information to be more primitive.

The other virtue of quantum information is that it provides a different, more direct way of introducing students to quantum physics. I still think that a good dose of the "shut up and calculate" approach—that is, of solving problems, starting with the hydrogen atom and then moving on to the usual applications of quantum theory to physics—is essential. Without this, there is a tendency to focus too much on the measurement problem and to forget that a lot of the universe, a lot of essential physics—atomic, nuclear, condensed matter, and so forth—runs perfectly well without the need for deep explanations of the role of the observer.

On the other hand, I'm skeptical of approaches that aim to derive quantum physics from information as we know it. Information as we know it is a fairly high-level concept. So it is difficult to imagine that fundamental quantum laws—which deal largely with probabilities—will emerge from entities that are less fundamental (since they depend on these probabilities), such as the von Neumann or Shannon information. Nevertheless, I'm firmly convinced that information is a part of the mystery. But I expect that something more primitive and immediate—like symmetries tied to information, such as envariance described in my answer to Question 5 (see page 122)—are more likely to hold the key.

I'm thinking of the interplay between two very fundamental notions, ignorance and information, in their most primitive guise. Ignorance can be objective when it relies on physical symmetries to represent a complete lack of information (again, see the discussion of envariance in my response to Question 5, page 122). The counterpoint is perfect knowledge, which represents certainty about the outcome of a future measurement. Let me now give an example of certainty. In addition to the purely quantum principle of superposition and to the Schrödinger-like unitary evolutions, there is only one other assumption we shall make: we assume (as do textbooks) that the same measurement repeated immediately yields the same outcome. That is, we assume there are states $|\heartsuit\rangle$ and $|\spadesuit\rangle$ of the system that are left untouched by the interaction with the apparatus:

$$|\heartsuit\rangle|A_0\rangle \implies |\heartsuit\rangle|A_\heartsuit\rangle, \qquad |\spadesuit\rangle|A_0\rangle \implies |\spadesuit\rangle|A_\spadesuit\rangle.$$

This is certainty: a remeasurement of the system will yield the same outcome.

Linearity of quantum evolutions means that any superposition $\alpha|\heartsuit\rangle + \beta|\spadesuit\rangle$ of such two predictable states must yield a superposition:

$$(\alpha|\heartsuit\rangle + \beta|\spadesuit\rangle))|A_0\rangle \implies \alpha|\heartsuit\rangle|A_\heartsuit\rangle + \beta|\spadesuit\rangle|A_\spadesuit\rangle.$$

Moreover, quantum evolutions preserve the norm. Consequently, the norm of the state of the whole after the measurement must be the same as before. The only difference between these two scalar products, expressed in terms of $|\spadesuit\rangle$ and $|\heartsuit\rangle$ and the corresponding states of the apparatus, $|A_\spadesuit\rangle$ and $|A_\heartsuit\rangle$, is in the cross term. It has to be the same for every α and β. Therefore, the equation

$$\langle\heartsuit|\spadesuit\rangle = \langle\heartsuit|\spadesuit\rangle\langle A_\heartsuit|A_\spadesuit\rangle \qquad (*)$$

must be satisfied by any two states that can be found out without getting perturbed in the process. (Above, we have recognized that $\langle A_0|A_0\rangle = 1$.)

A natural temptation is to simplify the expression $(*)$ and divide both sides by $\langle\heartsuit|\spadesuit\rangle$. This yields $\langle A_\heartsuit|A_\spadesuit\rangle = 1$. As far as measurements go, this is a disaster: we have just proved that the measurement must have failed, as $\langle A_\heartsuit|A_\spadesuit\rangle = 1$ implies $|A_\heartsuit\rangle = |A_\spadesuit\rangle$, that is, the two states of the apparatus are *identical*, and so the apparatus gained absolutely no information about the states it was supposed to distinguish.

Only when $\langle\heartsuit|\spadesuit\rangle = 0$, one cannot simplify the equation $(*)$. We have just proved that only orthogonal states of the system can be found out without getting perturbed in the process (now $\langle A_\heartsuit | A_\spadesuit \rangle$ can take on any value, including zero, which corresponds to a perfect measurement). So our simple equation has profound consequences: it shows that measurement can transfer data that lead to the prediction of the future outcome with certainty only when the states corresponding to the two possibilities for the original are orthogonal, that is, when $\langle \heartsuit | \spadesuit \rangle = 0$.

We could have also arrived at this conclusion by recognizing that quantum evolutions are unitary: they preserve scalar products. So the scalar product of the states of system-plus-apparatus in $| \heartsuit \rangle \, | A_0 \rangle \Longrightarrow | \heartsuit \rangle \, | A_\heartsuit \rangle$, $| \spadesuit \rangle \, | A_0 \rangle \Longrightarrow | \spadesuit \rangle \, | A_\spadesuit \rangle$ before and after the information transfer must be the same, which immediately leads to our simple yet profound equation above. This is no surprise, because unitarity is a consequence of linearity *and* preservation of the norm.

We have here an example of a derivation that is obviously based on information transfer (from the system to the apparatus) but does not appeal to any of the higher-level concepts, such as Shannon or von Neumann entropies and the like. It does not even rely on probabilities. Indeed, the only two values of the scalar product we have used are zero and one. Both reflect certainty. So one can use this derivation to show how two uncontroversial quantum postulates—namely, (1) states that live in a Hilbert space, and (2) unitarity—result, in the presence of repeatability, in a discrete set of possible orthogonal outcomes. This immediately implies that an arbitrary quantum state cannot be found out, a conclusion usually justified by the controversial collapse axiom.

There are two separate but related conclusions we can now reach. The orthogonality of possible outcomes, as derived above, immediately explains why observables are Hermitian. This is usually postulated in the textbook lists of quantum axioms (although Hermiticity is often a part of the collapse postulate).

The above derivation of the orthogonality of the outcomes shows why it is impossible to find out an unknown preexisting quantum state: the choice of the apparatus predetermines which states of the system can be found out. If the preexisting state does not match that choice, the observer will never know what it was, as the measurement will reprepare the system in one of the states that survive it. So, in the absence of that information, one cannot choose the apparatus that will simply find out the preexisting state. This conclusion carries all the symptoms of collapse and, in particular, of randomness. When we choose an apparatus that yields repeatable measurements, we simultaneously pick out a menu of possible outcomes.

It should be noted, however, that repeatability is usually enforced not at the level of states of the measured quantum system (which is often destroyed by a measurement, precluding repeatability) but rather at the level of the apparatus (which keeps records that can be repeatedly consulted). But either way, repeatability is there on the quantum–classical border. Quantum Darwinism relies on such repeatability: only states that can survive copying can proliferate. And something like repeatability is essential to justify the role of states as tools in predicting outcomes of future measurements. After all, what can be more fundamental than confirming, by measurement, that a state is what it is?

It is significant that the states that can be found out must be distinguishable. In quantum theory, this means that they must be orthogonal, but there is a sense in which, once again, one touches on some very basic notions related to information. Indeed, one could see why it had to be so—why, in any theory, states that can be repeatably measured must be distinguishable—without the help of anything quantum, that is, without even the uncontroversial quantum axioms. By definition, two states that repeatably give the same outcome must be distinguishable, because otherwise remeasurement could yield the "other" outcome, so repeatability would not be guaranteed. Thus, when we insist that repeating a measurement yields the same outcome, we rule out any overlap. The basic idea of certainty is prequantum and pre-information-theoretic, although primitive notions of information and predictability play a key role.

So here we have notions of information that are utterly fundamental and that lead to nontrivial consequences from an extremely natural and very primitive (in the good sense of the word) starting point. Another example of such a primitive starting point is the notion of equiprobability, based not on subjective ignorance but understood as an objective consequence of the symmetry of a quantum correlation. Indeed, perhaps one could use envariance—discussed in my answer to Question 5 (see page 122)—as a primitive, prequantum starting point, as a symmetry requirement that has implications for information.

RECONSTRUCTIONS

*How can the foundations of quantum mechanics benefit
from approaches that reconstruct quantum mechanics
from fundamental principles? Can reconstruction
reduce the need for interpretation?*

F OR THOSE WHO SPEND THEIR DAYS tinkering with quantum mechanics in the way other people make a living repairing cars or closing business
deals, the formalism and axioms of quantum theory quickly become second
nature. Hilbert spaces, normalized vectors, Hermitian operators, complex-valued
partial differential equations: it's as if God had carved the quantum axioms into a
slab of stone—complete with a large golden heading reading THE COMMANDMENTS
OF Ψ—and then lowered the tablet down from the heavens and presented it to us
with his trademark enigmatic smile. "This is *it*. These are the symbols and the rules
that represent how your world works. Don't ask why."

So we heeded his call and eagerly gobbled up the tablet's cryptic messages, taking
the symbols and rules as the source from which all understanding had to flow, and
reading all kinds of strange things into them. And soon the symbols started to feel
more real and authoritative than the rock they were written on.

This, of course, is a rather crude metaphor for the programmatic theory-to-reality
approach that drives the various interpretations of quantum mechanics: the approach
of taking the formalism of quantum theory as the starting point, and then cooking up
some interpretation to make contact with the world around us. This need to connect
formalism and reality has traditionally engendered a whole range of different strategies. One person may lean toward an axiomatic, operational notion of measurement,
another may prefer to promote some of the symbols in the formalism to physical
reality, and yet another may choose to throw some new ingredients into the mix, or
try to tweak the rules. (See the answers to Question 3, *My Favorite Interpretation*,
for a representative cross-section of options.)

But there's also the inverse route, a reality-to-theory approach if you will, that
has become quite the rage lately: reconstructions of quantum mechanics. Why and
whence quantum theory? What is it about this world that forces us to navigate it
with the help of such an abstract monstrosity as quantum theory? These are the
kind of questions that motivate reconstructions. Rather than taking the quantum
formalism as God-given and tacking some interpretation to it, reconstructions urge
a fresh start. They want us to pretend naiveté for a moment and to forget about all

the fancy quantum machinery we've grown so accustomed to. They wipe the slate clean and then try to build up—*reconstruct*—quantum theory's axioms and formal structure from a few fundamental, crisp, and physically intuitive principles.

Many people have embraced reconstructions as an invigorating antidote to the perennial interpretive battles and as a promising strategy for making tangible headway in our understanding of the relationship between quantum theory and the world. (In fact, several of our interviewees—Časlav Brukner, Jeffrey Bub, Christopher Fuchs, Lucien Hardy, and Anton Zeilinger—have put forward their own proposals for reconstructing quantum theory.) Most existing reconstructions are decisively infused with an information-theoretic spirit, indicating the stimulating influence the quantum-information boom of the past years has had—and continues to have—on this field of investigation.

The promise of reconstructions, then, is twofold. First, if we can find the raison d'être of quantum theory's axioms and mathematical structure, we can make the theory look and feel less ad hoc. Second, if we can identify the foundational physical principles that lead us to adopt quantum theory in our dealings with the world, we can also learn something deep about how this world is wired.

And there's another, intimately related aspect: quantum theory can simply be seen as one member in the class of generalized probabilistic theories. Curiously, it turns out that many of these other probabilistic theories exhibit features we had always cherished as uniquely, genuinely quantum. For example, there are theories that give rise to interference effects, violate Bell-type inequalities, obey no-signaling and no-cloning constraints, and share one of the hallmarks of quantum mechanics, namely, that a measurement will, in general, alter the system's state (or, to use the flashy terminology of quantum information theory, that there's a fundamental trade-off between between information gain and state disturbance).

Whether this observation alone has enough power to render quantum theory a little less mysterious, and whether it may even convey an admonition against a hasty reification of the quantum formalism, is a matter of opinion. But it certainly raises some important questions: Why do we use quantum mechanics for describing our world, rather than another member of the family of generalized probabilistic theories? What are nature's characteristic traits and the deeper principles that single out quantum mechanics among its quantum-like competitors? And, by extension, what would the world look like if it was represented not by quantum theory proper but by a variant? Could novel nonquantum predictions of a more general probabilistic theory represent what's happening in some hitherto unobserved domain of nature?

What we'd like to do, therefore, is to find a set of criteria that uniquely picks out quantum theory from the rest of the field. But needless to say, not just any set will make us happy. Generally, we'd want to look for a small number of simple, general, and comprehensible physical principles. These principles should be reasonably natural. They shouldn't feel like we've merely turned all the features of quantum mechanics into "principles of nature" and then rederived the quantum formalism from there. Also, we'd obviously like to minimize the use of any peculiar mathematical assumptions and axioms, because otherwise the resulting theory may feel just as ad hoc as quantum mechanics in its usual textbook presentation.

People have quickly realized, however, that it's far from easy to come up with principles that both meet these desiderata *and* single out quantum theory. The current proposals arguably still fall short of either of these two conditions. But this is nascent research, and much progress has already been made in understanding which principles plainly won't work and which may stay in the race.

I'm sure that you'll concur by now that reconstructions are a field worth plowing. But could reconstructions change the course of quantum foundations? How much can they contribute to making quantum theory less puzzling? To what extent may they be able to tell us more about the anatomy of nature than the current axiomatic form of quantum theory could?

Another kind of question looms on the horizon. Could we find a set of fundamental principles that not only does the job, but also becomes universally accepted as *the* definitive set? Or will we end up with many rivaling sets, and will the choice become simply a matter of personal taste? Will we forever be arguing about the "best" set of principles, just as we've been quarreling for close to a century about the "best" interpretation of quantum mechanics?

Finally, once we've successfully recovered the structure of quantum theory from first principles, does this mean that we no longer need to furnish this structure with an interpretation? Or are we, with respect to the need for interpretation, back to square one?

GUIDO BACCIAGALUPPI · I think the situation is analogous to that in other branches of the foundations of physics.

In the foundations of thermodynamics and statistical mechanics, the situation is rather uncontroversial from this point of view (although there are plenty of other open questions!). Namely, the principles of thermodynamics provide fundamental insights into thermal phenomena (in the broadest sense), while at the same time the reduction of thermodynamics to statistical mechanics is seen as a key component in understanding these same phenomena. The two approaches are both essential and complementary.

In the foundations of space-time theories, and specifically in special relativity, the issue is slightly more controversial but, I believe, equally clear-cut. No one will deny that the principle of relativity and the light postulate provide key insights into the nature of space-time. At the same time, one needs to close the circle—I am consciously borrowing Shimony's phrase—and be able to tell a dynamical story about, say, length contraction and time dilation, as Bell does in his lovely "How to Teach Special Relativity." Indeed, for instance, if there were no stable matter and no rigid rods, then the principles could not even be stated in the first place.

In quantum mechanics, there is a long tradition of reconstruction efforts, traditionally along the lines of quantum logic, convex sets, and operator algebras, but now much more diverse, and they have all contributed, and continue to contribute, fundamental insights into the nature of the world and what makes it a quantum world. Insofar as these approaches have a phenomenological starting point in notions akin to "measurement," however, they need to be supplemented by detailed interpretational stories about how processes such as measurements can take place. Call it top-down versus bottom-up, principle theory versus constructive theory, phenomenological versus fundamental: the two approaches are both essential and complementary.

ČASLAV BRUKNER · Ever since quantum theory was born, physicists and philosophers have tried to interpret it. That the theory is not self-evident and requires an interpretation is rooted in its puzzling aspects that are primarily related to the measurement problem and the violation of Bell's inequalities. A plethora of interpretations has been proposed, without reaching consensus on what the meaning of the theory is. Perhaps the most important lesson to be learned from the frustration of not finding a univocal interpretation is that there might be something intrinsically deficient in the idea of looking for the meaning of a physical theory exclusively on the basis of its formalism ("from the inside")—instead of extracting meaning, along with extracting the formalism, in the course of a derivation of the theory from some deeper physical principles ("from the outside").

Acting on this realization, reconstructions of quantum theory have become a major trend in the foundations of quantum physics over the last decade. The explanatory power of a reconstruction is judged by the reconstruction's ability to account for the origin of the basic principles from which the structure of the theory can be derived. The more these principles stand up to the requirements of simplicity and physical plausibility, the less space is left for arbitrariness in the theory's interpretation.

Much progress has recently been made in reconstructing quantum theory in the context of operationalism, where primitive laboratory procedures, like preparations and measurements, are basic ingredients. It is often said that these reconstructions are devoid of ontological commitments, and that nothing can generally be concluded from them about the ontological content arising from the first principles, or about the status of the notion of physical reality. As a supporting argument, one usually notes that within a realistic worldview, one would anyway expect quantum theory at the operational level to be deducible from some underlying theory of a "deeper reality."

It seems to me that these opinions largely underestimate the important fact that in the great majority—if not in all cases—of known reconstructions, the structure of quantum theory is derived in the context of operationalism. I suggest that this fact in itself contains an important message. The point is that the very idea of quantum states as representatives of information—information that is subject to certain information-theoretical constraints and is sufficient for computing probabilities of outcomes following specified preparations—has the power to explain why the theory has the very mathematical structure it does. Fuchs has made this point forcefully:

"By contrast, who could take the many-worlds idea and derive any of the structure of quantum theory out of it? This would be a bit like trying to regrow a lizard from the tip of its chopped-off tail: The Everettian conception never purported to be more than a reaction to the formalism in the first place."

JEFFREY BUB · The program of interpreting quantum mechanics takes quantum mechanics as true, or very nearly true, and attempts to explain away features of the theory that are puzzling if you accept certain prior assumptions about measurement, or physics, or the nature of reality. So, for example, Bohm's theory shows how to interpret quantum mechanics as a deterministic theory. (An equally important aim for Bohm was to provide a new theoretical framework in which to probe the limits of quantum mechanics, but this aspect of Bohm's work was largely ignored.) As I recall, Bohm often referred to the example of a radioactive atom, where the decay time is random. As he saw it, there must be something different about two atoms in the same quantum state that decay at different times, and since this difference is not reflected in quantum mechanics, the theory must be incomplete. While there is a lot more to Bohm's views on quantum mechanics and physics in general, the 1952 theory was primarily a demonstration, in the face of von Neumann's "no hidden variables" proof, that the phenomena of interference and entanglement don't force us to abandon determinism.

One might compare the situation with special relativity. Lorentz's interpretation of special relativity was an attempt to explain relativistic length contraction and the associated time dilation, taking the underlying kinematics as given by the Euclidean structure of Newtonian space and time and invoking the aether as an additional structure for the propagation of electromagnetic effects. In the conclusion of the 1916 edition of *The Theory of Electrons*, Lorentz writes:

> [Einstein's] results concerning electromagnetic and optical phenomena … agree in the main with those which we have obtained in the preceding pages, the chief difference being that Einstein simply postulates what we have deduced, with some difficulty and not altogether satisfactorily, from the fundamental equations of the electromagnetic field. By doing so, he may certainly take credit for making us see in the negative result of experiments like those of Michelson, Rayleigh and Brace, not a fortuitous compensation of opposing effects, but the manifestation of a general and fundamental principle.
>
> Yet, I think, something may also be claimed in favour of the form in which I have presented the theory. I cannot but regard the aether, which can be the seat of an electromagnetic field with its energy and its vibrations, as endowed with a certain degree of substantiality, however different it may be from all ordinary matter. In this line of thought, it seems natural not to assume at starting that it can never make any difference whether a body moves through the aether or not, and to measure distances and lengths of time by means of rods and clocks having a fixed position relative to the aether.

Given Lorentz's prior assumptions about spatiotemporal structure, relativistic length contraction required a *dynamical* explanation in terms of electromagnetic forces associated with the motion of a body through the aether, the medium of propagation of electromagnetic forces. On Einstein's view, the special theory of relativity provides a *kinematic* explanation of length contraction, and no special dynamical explanation is required, beyond the demonstration that a relativistic dynamics for length contraction, consistent with the kinematic structure of Minkowski space-time, is possible. In this sense, interpretations of quantum mechanics are "Lorentzian" in spirit, insofar as the explanatory problems they resolve are motivated by prequantum principles.

In my view, the interpretation program has led to genuine insights about the foundations of quantum mechanics, but it has run its course and should be regarded as superseded by the reconstruction program.

In 2003 I coauthored a paper with Rob Clifton and Hans Halvorson, "Characterizing Quantum Theory in Terms of Information-Theoretic Constraints." The characterization theorem we proved assumed a C^*-algebraic framework for physical theories, which I would now regard as not sufficiently general in the relevant sense, even though it includes a broad class of classical and quantum theories, including field theories, and hybrid theories with superselection rules. The task here is to answer the question "Why quantum mechanics?" with respect to a class of "foil" theories. The relevant class of theories to consider would seem to be the class of no-signaling theories, which includes theories that violate the Tsirelson bound and transcend the C^*-algebraic framework, because part of the question is: why the Tsirelson bound? One of the most interesting new results in this framework is by Pawłowski, Paterek, Kaszlikowski, Scarani, Winter, and Żukowski, published in a 2009 *Nature* article. Exploiting the power of PR boxes in an ingenious way, the authors showed that stronger-than-quantum correlations violating the Tsirelson bound also violate a principle of "information causality" (that is, the principle that if Alice communicates m bits to Bob, Bob cannot extract more than m bits of information using only local resources; for $m = 0$, this is the no-signaling principle).

ARTHUR FINE · This question mentions reconstructions. That is a term of art much used by John Dewey. One of my favorite quotes from Dewey occurs in his discussion of Darwin, where Dewey says:

> [I]ntellectual progress usually occurs through sheer abandonment of questions together with the alternatives they assume .... Old questions are resolved by disappearing, evaporating, while new questions corresponding to the changed attitudes of endeavor and preference take their place.

Fresh fundamental approaches can introduce new questions. Generally, that would not reduce the need for interpretations (see my response to Question 9, page 184), but it can redirect it. In the best case, pursuing the new questions turns out to be more fruitful than pursuit was of the old. Scientific revolutions have this character, which is why Dewey focused on Darwin. So far, we have not seen comparable dividends from recent axiomatics in the quantum domain. Thus, we do not yet know whether Hardy's axioms, or information-theoretic constraints as in the Clifton–Bub–Halvorson theorem, will fall under the "best case" scenario.

CHRISTOPHER FUCHS · I'm fairly sure I've already lingered on this topic long enough in my answers to earlier questions, but let me reiterate this much. From my point of view, the very best quantum-foundational effort will be the one that can write a story—very literally a story, all in English (or Danish, or Japanese, or what have you)—so compelling and so masterful in its imagery that the mathematics of quantum mechanics in all its exact technical detail will fall out as a matter of course.

By this standard, none of the reconstructive efforts we have seen in the last ten years—even the ones proclaiming quantum information as their forefather—have made much headway. On the other hand, there is no doubt that we have learned quite a lot from some of the reconstructions of the operationalist genre. I feel they contain bits and pieces that will surely be used in the final story, and for this reason, it is work well worth pursuing. For instance, I am struck by the sheer number of things that flow from the "purification" axiom of the operationalist framework of Giulio Chiribella, Mauro D'Ariano, and Paolo Perinotti. It issues a deep challenge to understand its nature from a personalist Bayesian perspective.

Another example is Lucien Hardy's "Quantum Theory from Five Reasonable Axioms." That paper had a profound effect on me—for it convinced me more than anything else to pursue the idea that a quantum state is not just *like* a set of probability distributions, but very literally a probability distribution itself. When I saw the power he got from the point of view that probabilities come first, it hit me over the head like a hammer and has shaped my thinking ever since. (Beware: Hardy would likely not take this to be one of the implications of his paper, but it certainly is what I took from it.) Where, however, Hardy emphasized that *any* informationally-complete set of measurements would do for translating a quantum state into a set of probability distributions, I have wanted to find the most aesthetic measurement possible for the translation. My thinking is that beauty once found has a way of leading us to insights that we would not attain otherwise. Particularly, I am goaded by the possibility that so simple an expression as the one given in my answer to Question 2 (see page 46) might carry the content of the Born rule, that I toy with the idea of it being the most significant "axiom" of all for quantum theory. Indeed, through recent work with Marcus Appleby, Åsa Ericsson, and Rüdiger Schack, we have quite some indication that a significant amount of the structure of quantum-state space arises from it alone.

But! the thing to keep in mind is that no matter how pretty I think this equation is, it cannot live up to my standards for a proper starting point to quantum mechanics. It is after all an equation, and thus has to be part of the endpoint. What is needed is the story *first*!

GIANCARLO GHIRARDI · I believe that quantum mechanics requires neither a reconstruction nor an interpretation. I take the position that it requires a reformulation that makes it internally and logically consistent—and, even more importantly, that allows it to account for our definite perceptions concerning macroscopic events.

SHELLY GOLDSTEIN · The program of reconstructing quantum mechanics from fundamental principles can be illuminating, but not so much for the foundations of quantum mechanics, where the difficulty lies not with why quantum mechanics *is*, but with what quantum mechanics *says*.

Of course, it is very nice to see how a fundamental physical theory emerges from, or is strongly connected with, some compelling fundamental physical principles. But at the end of the day, we must consider where those principles have led us. If they have led us to a clear physical theory, then good for them, but insofar as the foundations of quantum mechanics is concerned, a clear statement of the theory would have been sufficient.

If we had had a clear formulation of quantum mechanics to begin with, then there would have been no need for the subject called the foundations of quantum mechanics, regardless of whether we could derive quantum mechanics from some fundamental physical principles. And if what we extract from the fundamental principles is just plain old standard quantum mechanics, formulated in the usual textbook way, then insofar as the foundations of quantum mechanics is concerned, we will have accomplished precious little, since we still would not know precisely what it is that quantum mechanics says about physical reality.

The goal of deriving quantum mechanics from fundamental principles is a worthy one. But it is not the problem with which the foundations of quantum mechanics is concerned.

DANIEL GREENBERGER · Reconstructing quantum theory from a set of basic principles seems like an idea with the odds greatly against it. But who knows. It has worked before, in the most unlikely of circumstances. Who would have thought that you could base thermodynamics on the Carathéodory principle? It is so different in style from thermodynamics, which itself is so different from the rest of physics. And yet it produces, from a purely mathematical perspective, an alternative that provides an enormous insight into the meaning of the second law.

I might say that statistical mechanics itself provides a kind of reconstruction of thermodynamics, whose insights are deeper than the original, and which is experimentally extremely fruitful. So maybe quantum theory—whose ideas, like those of thermodynamics, are relatively abstract—is rife for just this kind of reconstruction. It's a worthy enterprise. But it takes a special kind of mind, not better or worse, but special. It's not where my talents lie, but I would be curious to see the results.

LUCIEN HARDY · In 2001 I gave a set of axioms for quantum theory. In modern form, these axioms are:

Information.   Systems having, or constrained to have, a given information-carrying capacity have the same properties.
Composites.   Information-carrying capacity is additive, and local tomography is possible for composite systems.
Continuity.   There exists a continuous reversible transformation between any pair of pure states.

Simplicity.   Systems are described by the smallest number of probabilities consistent with the other postulates.

From these axioms we can reconstruct the quantum formalism for finite-dimensional Hilbert spaces. There has recently been some progress in replacing the simplicity axiom by other (possibly more reasonable) axioms by Časlav Brukner, Borivoje Dalić, Lluís Masanes, and Markus Müller.

These axioms are couched in the context of an operational framework consisting of preparations, transformations, and measurements. This raises a difficult question. Why is the operational approach so successful? I find this question quite disturbing and have no good answer. Surely we would expect much of the structure of quantum theory to come from ontological considerations. And yet here we see that it is fixed by a set of operational principles. It is, of course, likely to be the case that we can reason backward from operational properties and fix some aspects of the structure of physical theories. But it is surprising we can get so much.

I think the most important contribution axiomatic approaches like this one can make is that they can help us make progress toward a theory of quantum gravity. To pursue this, we need to be aware that there are always background assumptions in any axiomatic framework for physics. In constructing special relativity from two axioms, Einstein implicitly assumed flat space-time. He had to both identify this and relax it to construct his theory of general relativity. Nevertheless, he was able to take his axioms for special relativity over to general relativity and apply them locally.

In the operational framework used for constructing quantum theory from the above axioms, the preparation–transformation–measurement structure sneaks in via the assumption of definite causal structure. This almost certainly has to be dropped in going to a theory of quantum gravity. With this in mind, I constructed the causaloid framework, which is for probabilistic theories that admit indefinite causal structure. It is possible to formulate quantum theory in this framework, though I am still looking for a compelling set of axioms, like those above, that would do this job more naturally.

The next step is to attempt to formulate general relativity in the causaloid framework—hopefully with a set of simple axioms. Actually, the natural theory to put in the framework is *probabilistic general relativity*. This is the theory we will obtain when we have arbitrary probabilistic ignorance as to the values of measurable quantities in general relativity. Such a theory has not been satisfactorily formulated yet. I am currently working on enhancing the causaloid framework to equip it for this task. Hopefully, it will be possible to formulate probabilistic general relativity within a suitably enhanced causaloid framework with a set of simple axioms.

Once we have quantum theory and probabilistic general relativity formulated within the causaloid framework (or something similar), my hope is that we can construct a theory of quantum gravity by taking some axioms from each of these two less fundamental theories.

In the case that we find an operational theory of quantum gravity—one that is verified in experiments—the most important foundational question to ask is whether it admits a natural realist interpretation. Does a solution to the reality problem (the measurement problem) naturally suggest itself? It is at this stage, at the end of a very long road, that we will see the ultimate payoff from attempts to reconstruct quantum theory.

ANTHONY LEGGETT · I assume this refers to the sort of approach taken, for instance, by Bub–Clifton–Halvorson or by Hardy. I find these attempts very interesting in their own right, but I don't see how they are going to reduce the need for interpretation.

TIM MAUDLIN · "Fundamental principles" in physics ought to refer to the specification of an exact physical ontology (what exists) and a dynamics (how what exists behaves in space and time). Without these "principles," one does not have a clear physical theory at all. And everything else, such as the analysis of interactions in the laboratory ("measurements"), physical capacities for transmitting information, the computational power of physical systems, and so on, is understood in terms of the physical constitution of things and the laws that govern the basic physical items. It is rather misleading to call this "interpretation," or even "foundations": it is rather a description of physics as a discipline.

One of the central contentions of Bohr was that physics had somehow reached a critical point at which this traditional quest for an exact account of the constitution of matter could no longer be continued. The arguments provided for this astonishing conclusion—think, for example, of the claims about "measurements" requiring some sort of interaction with the measuring device that could not be controlled or predicted—simply do not establish the claim. And Bohr's rhetoric about the *impossibility* of a precise account of subatomic matter in space and time is belied by the existence of such accounts: the de Broglie–Bohm theory, for example. Bohr's philosophy was predicated on an untenable division of the physical world into a "classical" macroscopic reality and an *unanschaulich* formal mathematical scheme for representing the microscopic world. But since macroscopic objects are just collections of microscopic constituents, there must be a single unified ontology that encompasses both. "The interpretation of quantum theory" should be primarily a search for such a unified ontology.

Principles that are neither specifications of ontology nor of dynamical laws governing the ontology—such as "The speed of light is constant," "All inertial frames are equivalent," "Entropy cannot decrease," "One cannot clone a physical system," "One cannot predict the exact position and momentum of a particle at the same time"—can serve as useful maxims in two ways. They can lead quickly to predictions in some circumstances, without having to worry about the exact physical description of a system. This method for solving problems relies on taking the principles for granted. In the other direction, if one feels confident that the principle holds, one can use that fact as a guide when seeking ontology and dynamics. But these maxims cannot be reasonably used to *dictate* either ontology or dynamics. That is because their ultimate justification is parasitic of the ontology and dynamics: the justificatory relation here is asymmetric.

Let's illustrate this with an example. The Mother of All Useful Maxims is the principle of the conservation of energy. That principle has almost boundless utility and scope of application. Physicists are (rightly) extremely leery of any proposed physical theory that violates the maxim. Some physicists would even (wrongly) reject any such theory out of hand. But what, exactly, is the source of one's confidence in the principle?

It is not that the principle is literally a foundational axiom of any physical theory. It is not, for example, one of Newton's laws, or a postulate of quantum theory. The confidence that one has in the principle arises from the fact that it *works* across a very wide field of application. Sometimes, it can even be *derived* (via, say, Noether's theorem) from some fundamental dynamical laws. But ultimately, the fact that the maxim works is an *explanandum*, not an *explanans*: it is something to be accounted for by the fundamental dynamics.

The Ghirardi–Rimini–Weber spontaneous-collapse dynamics of the wave function famously violates the principle of the conservation of energy. If that theory is correct, the universe as a whole has been heating up over the past fourteen billion years due to the collapse mechanism. It is incumbent upon that theory to show that the degree of heating postulated is consistent with all known observations—which it is. The violation of the principle is so small that predictions made using the principle will always have been accurate to within experimental accuracy. So the widespread utility of the maxim is explained by the theory, despite the fact that the theory entails that the maxim is false. I cannot see, therefore, that the violation of the principle can be fashioned into a valid objection of the GRW theory. If all of our observations are consistent with the GRW collapses, and if the dynamics explains the utility of the principle as a maxim in the setting in which it is used, then there are no grounds to "fetishize" the principle into an inviolable constraint.

This example illustrates the proper methodological role of nonfundamental maxims. If experience testifies that the maxim holds in certain circumstances and to certain experimental tolerances, then the fundamental theory must account for that, as it must account for all observational data. But the fundamental theory does not need to elevate the maxim above that station, and can imply that the maxim may be violated. In such a case, experimental verification of the predicted violations should be sought. But whether the fundamental theory ultimately entails the exact truth of the maxim or only its approximate truth, the maxim remains nonfoundational: it is only a consequence of ontology and dynamics.

DAVID MERMIN · It is wonderful that all of special relativity follows from the principle that no physical behavior can distinguish among frames of reference in different states of uniform motion, combined with the realization that the simultaneity of events in different places is a convention that can differ from one frame of reference to another. Can the rest of physics—in particular quantum mechanics—be reduced to so economical a set of assumptions?

I doubt it. Even the foundations of special relativity are not captured as compactly as I just claimed. I failed, for example, to mention the assumptions of spatial and temporal homogeneity, and of spatial isotropy. And the fundamental notion of an "event"—a phenomenon whose spatial and temporal extent we can ignore for purposes of the topic currently under discussion—might strike some as irritatingly vague, bringing "us" into the story in a way physics traditionally (and, I increasingly believe, wrongly) tries to avoid. And just what are these human artifacts called "clocks" that play so fundamental a role in the story? In short, it's not as simple as advertised.

Yet quantum mechanics does seem to be floating in the air, in a way that makes relativity seem quite anchored. At least the basic conceptual ingredients of relativity have at first glance a direct intuitive correspondence with familiar phenomena in our immediate experience. The complicating issues for relativity emerge only when one insists on sharpening up these intuitions. In contrast, the basic ingredients of quantum mechanics—states, superpositions, and their linear evolution in time—bear not even a vague relation to anything in our direct experience, while measurement—the only thing that ties the subject to the ground—seems to introduce what John Bell derided as "piddling laboratory operations" at too fundamental a level.

I'm glad people are attempting to reconstruct quantum mechanics from (a few) fundamental principles, but I'm skeptical that they'll succeed without slipping into at least one of their principles something just as much in need of interpretation. The reason I'm nevertheless glad is that having a new and strikingly different formulation of the really puzzling stuff can sometimes be a useful step toward untangling the puzzle.

LEE SMOLIN · This kind of approach is a helpful navigational tool to isolate those aspects of quantum mechanics that could be fundamental, and to separate these aspects from those that could emerge from a more fundamental description. I don't know this area well, but I've found Phillip Goyal's approach illuminating.

ANTONY VALENTINI · I don't think quantum mechanics is a fundamental theory. It's ambiguous. And it's ambiguous because it lacks a microscopic ontology. Any reconstruction that does not provide such an ontology will remain ambiguous and therefore not fundamental. We see this in work over the past decade or so on reconstructing quantum theory from various operational axioms. Those axioms refer only to outcomes of experiments performed with macroscopic equipment. They provide constraints on the statistical properties of those outcomes. This may be of some interest, but only up to a point. Nothing is said about fundamental ontology. Pieces of macroscopic equipment are treated as if they were fundamental or elementary objects, when in reality they are emergent objects built out of atoms, particles, and fields. The pieces of equipment are assigned definite ontological states—the pointers point in definite directions, the knobs and dials on the apparatus have definite readings, and so on—while microscopic systems are not. Nothing is said about the dividing line between the definite macroscopic world and the indefinite microscopic world. Therefore, these operational approaches remain fundamentally vague. They do not attempt to address the measurement problem; therefore, they are of limited interest.

Pilot-wave dynamics, in contrast, does provide a reconstruction of quantum mechanics in terms of a fundamental ontology that is equally valid at the macroscopic and microscopic levels. There are two simple equations of motion, de Broglie's guidance equation and Schrödinger's wave equation. As with other fundamental equations of physics—such as Maxwell's equations or Einstein's field equations—one can try to motivate these equations on the basis of simple physical principles. In the early 1920s de Broglie motivated his guidance equation as a way to unify the principles of Maupertuis and Fermat. The Schrödinger equation is the simplest wave equation that respects the nonrelativistic dispersion relations. Thus, simple physical principles suggest two general equations of motion, which—if an initial Born-rule distribution is assumed—provide a complete and unambiguous reconstruction of quantum mechanics as an emergent equilibrium phenomenology. Though I wouldn't put too much emphasis on the motivating principles: at the end of the day, the basis of the theory is the equations themselves.

As for reducing the need for interpretation, that happens only if we provide an ontology. The question being asked probably refers to reconstruction along operationalist lines, which has become fashionable in recent years. As I've explained, that work does not even attempt to address the measurement problem. People often draw an analogy with special relativity. In 1905 Einstein gave an operational treatment based on macroscopic rods, clocks, and light beams, and he derived the Lorentz transformation from a small number of simple principles. Current work in operational quantum theory seeks to emulate that. In my view, Einstein's famous 1905 paper is the historical source of a serious mistake, whereby macroscopic equipment is given a fundamental role—a mistake that was repeated by Bohr, Heisenberg, and others in the 1920s, with catastrophic consequences. Like any other piece of macroscopic equipment, rods and clocks are not elementary systems; they are emergent objects built out of particles and fields. Our modern understanding of Lorentz invariance, commonly described in textbooks on high-energy physics and quantum field theory, boils down to having a Lagrangian density that is a Lorentz scalar. It's a symmetry of the basic equations. There is no mention of rods and clocks, or of any principle about the speed of light—the photon could, after all, turn out to have a small mass and move at slightly subluminal speeds. I think Einstein's 1905 paper was deeply damaging, and continues to be so. Nor was it necessary. The structure of special relativity was independently derived by Poincaré in 1905, by generalizing the Lorentz invariance of Maxwell's equations to all the laws of nature—precisely the approach that a modern particle physicist would have taken. I see little to emulate in Einstein's first paper on special relativity, and much to deplore. Einstein himself deeply regretted the operational fashion he started in that paper.

I see different formulations of operational quantum theory as analogous to different formulations of thermodynamics. People can argue over whether Kelvin's formulation of the second law is better than that of Clausius, or whether Carathéodory's geometrical approach is to be preferred. But in the end, they are merely talking about different axiomatizations of the same phenomenological theory, none of which bears on the burning issue of fundamental ontology.

DAVID WALLACE · I should say first that I'm not up to speed with recent work on reconstruction. It's a field where there's been lots of very exciting progress in recent years, and I'm not well-positioned to comment on the details.

But in general, I think reconstructions can tell us something interesting about the structure of quantum mechanics, but maybe not as much as their proponents sometimes hope. They certainly do a lot to help us understand the logical structure of quantum mechanics, and what happens to that structure if, say, we use reals or quaternions instead of complex numbers, or swap tensor products for Cartesian products, or whatever. And it's very often the case that something that's fairly opaque from one perspective on quantum mechanics is much more transparent from another perspective. The equivalence principle in general relativity is like that—once you understand that principle, various results that would have been calculationally horrific become really obvious.

But beyond that, I'm not sure how much we gain by rederiving the theory from "fundamental" principles, or even what it means for those principles to be "fundamental" in the first place. Take the analogy with special relativity, which often gets used in these discussions. Yes, we can understand why the Poincaré symmetry group applies by deriving it from the relativity principle and the light postulate. But we can equally well understand the relativity principle as a consequence of the dynamical fact that the symmetries of fundamental physics include the Poincaré group. Which route is more fundamental? I'm not sure that's a very fruitful question.

(It's tempting to say that the fundamental principles are in some sense "natural" or "intuitively reasonable." But our intuitions about what's reasonable and natural don't have such a great track record at predicting how fundamental physics turns out.)

I'd also say that I don't see how reconstruction could reduce the need for interpretation. Ultimately, however we reconstruct quantum mechanics, we're either going to end up saying (1) that the mathematical structure thus reconstructed represents physical reality faithfully (in which case we end up with the Everett interpretation, or something like it), or (2) that it represents physical reality incompletely or inaccurately (in which case we need to fix it, which leads us to hidden-variables or dynamical-collapse theories), or (3) that it's not in the business of representing physical reality at all (which leads us to operationalist or neo-Copenhagen or physics-is-information approaches). I say a bit more about this in my answer to Question 3, page 83.

ANTON ZEILINGER · I expect that the ultimate reconstruction has to start from very simple fundamental principles that are intuitively clear—very much in the same way as, for example, in the general theory of relativity, where we have the equivalence principle.

Like with any theory, there are two levels of interpretation. The first level is an operational one, connecting the symbols of the theory to operations in the laboratory. On this level, in quantum mechanics we have the Born interpretation, which associates probabilities with the squares of the amplitudes. On this level of interpretation, there is no disagreement.

On the second level of interpretation, we ask questions about the meaning of the theory for our understanding of the world and of our role in the universe. On this second level, there is strong disagreement in the case of quantum mechanics. Once we have arrived at a full understanding, we will also have broad agreement at the second level of the interpretation of quantum mechanics.

WOJCIECH ZUREK · These sorts of questions can be answered with any degree of confidence only after the fact—after a really successful, simple, and compelling derivation of quantum physics from something deeper has been accomplished. I have not yet witnessed such an "aha" moment in this respect. In particular, I think the goal of providing a derivation of quantum theory that matches in clarity Einstein's derivation of special relativity is a noble one, but, again, we are not there yet.

By contrast, I think there are fairly convincing derivations of classical perceptions from purely quantum axioms. I think this is the most promising direction of research, and I believe that very significant progress in this program is underway.

Indeed, I feel that the purely quantum postulates of quantum theory—namely, (1) the superposition principle (i.e., states that live in a Hilbert space), and (2) the linearity of evolutions (which explains, with very little extra input, unitarity)—are so simple and so natural that they are good candidates for fundamental principles. They have the quality of Einstein's postulate that the speed of light is the same in every frame, a postulate that yields special relativity.

Moreover, when supplemented with just one more small piece—specifically, (3) repeatability, the requirement that a repeated measurement yields the same outcome—the postulates 1 and 2 allow one to derive the essence of the "unnatural" textbook axioms that deal with measurements (for example, the essence of wave-packet collapse, as well as Born's rule yielding probabilities). In particular, symptoms of collapse follow as a consequence of the prohibition of cloning: in a quantum universe, repeatability (postulate 3 above) means that there are states that can alter the state of the apparatus. (See my answer to Question 9, page 194, for more on this.) These developments go beyond decoherence, as decoherence takes for granted Born's rule, which is used to justify the physical significance of reduced density matrices, a crucial tool in the practice of decoherence.

In any case, I have a feeling that when (if ever) we will see the light, it will shine on more than just quantum theory.

THE EXPERIMENT OF MY DREAMS

If you could choose one experiment, regardless of its current technical feasibility, to help answer a foundational question, which one would it be?

KURT VONNEGUT ONCE REMARKED that "novels that leave out technology misrepresent life as badly as Victorians misrepresented life by leaving out sex." Similarly, one could say that quantum foundations that leave out experiments run the danger of missing a good deal of the fun.

When Schrödinger fantasized abouts cats penned up in boxes, he saw this scenario merely as a case of reductio ad absurdum. And what's happening in today's laboratories? Numerous experimentalists are busy breeding Schrödinger "kittens," taking Schrödinger's original thought experiment as an inspirational blueprint for hands-on tabletop demonstrations. Nimble scientists have created superpositions of two radiation fields with opposite phases, containing several dozens of photons. They have observed superpositions of microampere currents running in opposite directions around a superconducting ring. And most recently, scientists have turned their attention to a particularly intriguing variety. Truly quantum-*mechanical* kittens would, once realized, consist of a tiny beam or lever in a superposition of two different positions.

Other species formerly of the "gedanken" category have also made the leap into experimental reality. Take the double-slit experiment with particles as an example. Lauded by Feynman as the demonstration of a phenomenon embodying the "heart of quantum mechanics," it has enjoyed a venerable career as a textbook illustration of the quirks of quantum theory. Yet its actual experimental realization happened surprisingly late. In 1961 the experiment was first carried out with electrons. But it took almost three more decades to ensure that only one electron was crossing the apparatus at any given time—an important requirement if one wants to convincingly show that the observed interference pattern cannot be attributed to interactions between different particles. After such a slow start, in recent years the double-slit experiment has been elevated to soaring new heights. Markus Arndt and his colleagues at the University of Vienna keep sending pretty much anything they can get their hands on through their diffraction gratings, including some biomolecules. And every time, a handsome interference pattern appears. Next target: a virus.

Undoubtedly, all such experiments play a crucial role in confirming the predictions of quantum theory for systems of ever-increasing size. This is no trivial matter. We humans are quick to trust our extrapolations, and many of us tend to take for granted that, "in principle," we should be able to apply the quantum formalism to anything we like, and that, "of course," quantum mechanics will give us the right answer. But despite all the pep talk about how quantum mechanics has been so spectacularly confirmed by experiment, thus far this confirmation has happened over only a tiny range of the scales found in our universe. So doing experiments in new regimes helps bolster our confidence in quantum theory. And along the way, it helps us come up with new ways of tricking nature into showing us its quantum face.

At the end of the day, how exactly may experimental evidence act as an arbiter of foundational disputes? The tests of the Bell inequalities are usually presented as classic success stories. But we shouldn't forget that even here—as the answers to Question 8, *Bell's Inequalities*, demonstrate—there's no single accepted view of the conclusiveness and meaning of the experimental results. What about Schrödinger kittens and macroscopic interference phenomena? If experiments keep validating quantum theory when applied to larger and larger systems, collapse models might get squeezed out, but the rest of the field—all the many interpretations that make predictions identical to those of standard quantum mechanics—will arguably come through unscathed. Some people conjecture that such experimental developments will have the power to skew the field of contenders in favor of certain interpretations, such as the Everett view, but this might well turn out to be wishful thinking.

And who knows, we might wake up someday to news about an experiment that has produced data that can only be interpreted as the result of a breakdown of quantum mechanics. Of course, it wouldn't necessarily be obvious that we've indeed hit upon something truly fundamental—rather than just maxed out the capabilities of our experimental setup, or overlooked (or underestimated) one more source of noise and decoherence. These are practical difficulties, however, and presumably ones that could eventually be surmounted. Needless to say, if any experiment was to ever demonstrate, beyond doubt, a clash with the predictions of quantum theory, it would be a lightening bolt not only for quantum foundations, but for the whole of physics. In the face of such revolutionary turmoil, chances are that we'd be swamped by new questions without getting answers to all the old ones, some of which may well continue to be pertinent.

In exploring how experiments may conclusively resolve foundational issues, it's helpful to return full-circle to the subject that opened this chapter: thought experiments. Let's imagine we're omnipotent experimenters, free to rise above the petty constraints of current technology and able to carry out any experiment we deem worth doing in the name of foundational enlightenment. The hope is that in this way, we will see more clearly what kinds of foundational problems may at all be amenable to empirical resolution. And since we wouldn't want to see precious bullets wasted on a minor target, our individual choice of experiment will likely contain a reflection of what each of us considers the most pressing questions in quantum foundations (see also Question 2, *Big Issues*).

GUIDO BACCIAGALUPPI · Either a very large-scale interference experiment, or a really small-scale test of Born's rule (to test for collapse theories and pilot-wave theories, respectively).

ČASLAV BRUKNER · What exactly *is* a measurement? Wigner reformulated the Schrödinger-cat thought experiment as the "Wigner's friend" experiment and proposed that the consciousness of an observer is the demarcation line that defines what a measurement is and when it happens. There is nothing in the quantum formalism, however, that demands that a state of consciousness cannot be in a superposition. Still, experiments with humans seem interesting, because in a "von Neumann chain," at the latest the observer herself should know if and when the measurement takes place.

In the Wigner's-friend thought experiment, an experimenter (the friend) is performing a measurement on a decaying atomic state in a sealed laboratory. The atom has a one-half probability of decaying from an excited state to the ground state after one hour. If the atom decays, it will emit a visible photon into the eye of the experimenter, resulting in the perceptual state "I see the photon." If the atom does not decay, no photon will be emitted, and the state of the observer's perception will be "I see no photon." A second experimenter (Wigner) is stationed outside the sealed laboratory. To him on the outside, and on the basis of all the information that is in principle available to him, the physical description of the state in the laboratory will be a superposition of the two scenarios.

What will Wigner's friend inside the sealed laboratory perceive after one hour? Will she be definite about whether she has observed a photon? One is tempted to answer these questions with the quantum-mechanical resolution that inside the laboratory, the friend's act of observation will collapse the quantum state into one of the two outcomes, and so the friend will either observe the photon or not. But then, if Wigner had "all information that is in principle available to him," wouldn't that include knowing the friend's measurement outcome? Shouldn't the mere availability of the information about the outcome somewhere in the universe—specifically, in the environment inside the laboratory, which includes the friend's consciousness—collapse the wave function that Wigner assigns? Or does Wigner's friend observe some kind of blurred reality, while Wigner keeps describing the situation in terms of the superposition state?

The most interesting point is that these questions could be answered experimentally, at least in principle. I think it might have been Deutsch who once suggested a gedankenexperiment along the following lines. Wigner could learn whether his friend has observed a definite outcome, without himself learning which outcome she has observed. It is enough for the friend to communicate "I see a definite outcome" to Wigner. This message contains no information about which outcome has occurred and thus shouldn't lead to a collapse of the wave function assigned by Wigner. Three different results of the experiment are possible:

(1)   The wave function collapses due to a breakdown of the quantum-mechanical laws when applied to states of consciousness or to systems of sufficiently large size, mass, complexity, and the like. Wigner then concludes that although he could exclude all known effects caused by conventional decoherence, the wave function still collapses.

(2)   The quantum formalism is unmodified; Wigner's state assignment is the superposition state, and his friend perceives a blurred reality that she cannot associate with either seeing or not seeing the photon. I cannot make much sense out of this option.

(3)   The quantum formalism is unmodified; Wigner's state assignment is the superposition state, and yet the friend observes a definite outcome.

In the last case, the two observers have complementary pieces of information. Taken together, they would violate the Heisenberg uncertainty relation. The point is that they cannot be taken together. They are redundantly imprinted in two complementary environments ("in two worlds," one is almost tempted to say): the sealed laboratory and the outside, respectively. They will remain separate as long as there is no communication between them on the *relevant* information. If we respect that there should be no preferred observers, then both Wigner and his friend have evidence that the records—such as a click in a photodetector, a certain position of a pointer device, a printout of a computer, or a definite human brain state—*are* created. But these records of each of the observers individually cannot be comprised as "facts of the world" independently of specifying in which "environment" they have happened. For me, this dramatic departure from naive realism would be the final proof of the validity of Bohr's dictum: "It is wrong to think that the task of physics is to find out how nature is. Physics concerns what we can say about nature."

JEFFREY BUB · There have been various proposals in the literature that wave-function "collapse" is a real dynamical process, not primarily associated with measurement or decoherence. The Ghirardi–Rimini–Weber theory, in several variants, is one such proposal. Penrose has argued that this is a gravitational effect. Briefly, he considers a tiny crystal that is in a superposition of two quantum stationary states, $|\psi\rangle$ and $|\phi\rangle$, at two different locations. The superposition is also a stationary state, with same energy. Taking account of the gravitational field of the crystal, the two different space-times of the two states $|\psi\rangle$ and $|\phi\rangle$ are associated with different Killing vectors. As a consequence, Penrose argues, the superposition would be unstable and collapse to $|\psi\rangle$ or $|\phi\rangle$ in a decay time of the order of $\hbar/E_G$, where E_G is the gravitational self-energy of the difference between the expectation values of the mass distributions in the two locations of the crystal. Penrose has suggested several versions of an experiment to test this. For a large system like Schrödinger's cat, the collapse would be virtually instantaneous; but for a tiny crystal, the decay time would be measurable, provided the experiment could be conducted in such a way that the hypothetical dynamical collapse could be distinguished from environmentally induced decoherence.

It would be nice to do an experiment to check whether wave-function collapse is a real dynamical process. If that turned out to be the case, the current version of quantum mechanics with unitary evolution would be an approximation to a structurally different theory, and the conceptual situation would be radically altered.

ARTHUR FINE · I am not an experimentalist and cannot hope to describe an experiment in any decent detail. But there are two *kinds* of experiment I would like to see. One addresses the interpretation of superpositions, which is arguably the most fundamental question of interpretation for any quantum theory. The experiment would be to fulfill Leggett's hope and to make something—beyond SQUIDs or buckyballs—that would correspond unquestionably to a genuine superposition of macroscopic observables (or, to show that this is just not possible).

The second experiment relates to my answer to Question 8 (see page 169) and the Bell inequalities. The photon-correlation experiments that test Bell-like constraints have generally had suboptimal rates of detection, which allow certain local hidden-variables models for the experiment, my "prism" models. Other correlation experiments have sufficient play in synchronizing the coincidence events that other sorts of local models, what I call "synchronization" models, are possible. These models of the experiments introduce selective resources, according to which the detected pairs form a subensemble that violates the Bell inequalities, whereas in the larger ensemble of all emitted pairs the Bell inequalities are satisfied. Still other experiments, those with very high detection rates (like trapped-ion experiments), allow for subluminal signaling between the measurement events. All these experiments, then, allow a thoroughly classical simulation of the outcomes. There are proposals for experiments that claim to eliminate all three sources of classicality together. I'd like to see whether they can be carried out, and be able to look carefully at the results.

CHRISTOPHER FUCHS · I can think of two experiments I would like to see with this outlandish proviso! (Actually, they're connected, as you'll see.) Anton Zeilinger can be our guinea pig. First, we contract his lab to do a double-slit experiment on him—you know, prepare his center of mass in an approximate momentum eigenstate and let it scatter off two small slits in a wall. I'd then wait somewhere behind the wall (at a second wall) with my eyes closed until I expect it overwhelmingly likely to see him. Upon opening my eyes and seeing where he is, I'd ask him which slit he went through. My guess is he'd say that he doesn't remember a thing between walking into the preparation chamber and the conversation we had—as if he had been anesthetized—but I might be wrong. In any case, I wouldn't expect him to be qualitatively different from any other physical system.

For the second experiment, we'd need a computer far more advanced than the present-day pride-and-joy of IBM Corporation—the one they are training to compete on-air against two former champions of the television game show *Jeopardy!* It should be a computer that would pass any number of Turing tests with any number of people, one that would be able to obtain a high-school diploma and then enroll in college and obtain a physics degree as well. Furthermore, it'd be nice to fit it into a human-size robotic housing, with enough control and flexibility of its limbs and phalanges that its manipulation of small optical components would be on par with one of Anton's best graduate students. Suppose we had that. (Since IBM named their computer Watson, we might name ours de Finetti.) For the actual experiment, we would contract Anton to assign de Finetti some experimental project in his lab—perhaps something like preparing an exotic entangled state of five photons that had never been prepared before, and then checking the Bell-inequality violations it gives rise to. My guess is that de Finetti, after a proper training in laboratory technique, would be able to pass the test with flying colors, but I might be wrong. In any case, I wouldn't expect him to be qualitatively different from any other agent.

GIANCARLO GHIRARDI · Even though the dynamical-reduction models are fundamentally phenomenological, once one adds nonlinear and stochastic terms to the standard quantum evolution, not much space remains for the candidate theories and for the parameters appearing in them. This is why I attribute great importance to the work of scientists like Stephen Adler, who try to identify possible experimental tests of such theories against standard quantum mechanics. So far, no feasible test has been found, but there surely are areas involving mesoscopic systems where one might hope to be able to soon perform *experimenta crucis* for the dynamical-reduction theories. I look with keen interest to this area of experimental research. An analogous line of thought has been pursued by Roger Penrose, although with a different theoretical perspective. As is well known, Penrose aims to make quantum gravity responsible for breaking the linear nature of the standard theory.

SHELLY GOLDSTEIN · A variety of experiments have been proposed to distinguish quantum mechanics from collapse theories, such as the GRW theory and other theories that involve spontaneous localization of the wave function. These theories make predictions different from those of quantum mechanics for the results of suitable experiments probing macroscopic quantum interference. I shall not describe any detailed experiments here—I assume that others will.

But with regard to the foundational question of trying to choose the correct version of quantum mechanics—which involves trying to decide between theories that make exactly the same experimental predictions—no experiment will be of any help for this. For this sort of task it is not the experimental facts that are at issue but the explanation of those facts. That theories can be different—and describe rather different sorts of physical realities—and yet be empirically equivalent is perhaps not a happy thought. But it is a fact. We'll have to come to terms with it.

DANIEL GREENBERGER · In line with the ideas on a fundamental length (which might be macroscopic) given in my answer to Question 15 (see page 265), I would like to see a search for some form of quantization at the astrophysical or cosmological level. I would expect this to be different in kind from our usual quantization, a sort of gravitational quantization in strong gravity fields, which would then have to be hooked up to our usual quantum-mechanical quantization. I think this would change our whole conception of how to link up gravity with quantum theory.

LUCIEN HARDY · I would choose an experiment to probe interference with macroscopic objects. The best version of this is Roger Penrose's proposal, which I will describe below. But first let me discuss a gedankenexperiment which, I believe, goes back to Yakir Aharonov.

Consider sending a particle with charge q through an interferometer. If this particle is an electron, then we know we will see interference (such experiments have been done). If we increase the charge, however, then there has to be a point at which the interference is largely wiped out. This is because an experimentalist could come along some time T afterward and introduce a test charge at a distance of about T/c from the interference experiment and deduce from the force on this charge which path the charge q took through the interferometer. Even if we do not introduce a test charge later, the fact that we could implies that the information as to which path the charge took is imprinted on the electromagnetic field, and so we will still not see interference. But why do we see interference for an electron? Clearly, the continuous classical theory of electromagnetism cannot be operating at the level of an electron charge. This implies that we need a theory of quantum electromagnetism of some sort.

A similar argument goes through for mass. We know that we see interference for particles with sufficiently small mass (in fact, interference has been seen by Anton Zeilinger's group with buckyballs, so the mass can be quite big). For a sufficiently large mass, however, interference would have to be wiped out for the same reason as given above. We could introduce a test mass at a suitable distance some time later and deduce from the force on the test mass which path the particle took through the interferometer. Hence, we need a theory of quantum gravity. The theory of quantum gravity will probably be much more radical than the theory of quantum electromagnetism. In quantum electromagnetism, we can assume a fixed background space-time on which to evolve our fields. This will fail in quantum gravity. The theory of quantum gravity still has to explain, however, why we see interference for small masses but not for sufficiently large masses.

Penrose argues in the following way. If a sufficiently large mass (around about the Planck mass, 10^{-6} grams) were to go into a superposition, then it would cause space-time to curve in two different ways according to which term in the superposition one looks at. There is no unique way, however, to map two space-time manifolds onto each other, and so there would be a confusion as to what was the forward direction in time in this region. Such a direction is needed so that the wave function can be evolved. Hence, Penrose argues, something has to give to prevent this inconsistency. He supposes that the wave function would have to collapse. Penrose provides a more detailed model predicting a timescale for this collapse.

Regardless of whether one accepts the particular argumentation given by Penrose, it seems that an experiment of this nature is exactly what is required to probe the experimental domain in which the conceptual structures of quantum theory and general relativity clash. The detailed data collected from such an experiment would provide some insight into how to construct a theory of quantum gravity.

Now, a straightforward interference experiment where a Planck-scale mass goes through an interferometer is very difficult to perform. Penrose therefore proposes an ingenious interference experiment, in which a photon is used to put a small mirror of Planck mass into a superposition. The small mirror is attached to a spring and intersects one path of a photon as it passes through a Michelson–Morley-type interferometer. Whenever the photon goes along the path with the small mirror, it causes a recoil. This recoil is canceled by the photon on its return journey. According to standard unitary quantum theory, the mirror will go into a quantum superposition at the intermediate time. If phases are chosen appropriately, the return trip of the photon will reverse the unitary evolution associated with the outgoing trip, and we see interference of the photon. On the other hand, if there is effective collapse of the state of the small mirror, then we will not see interference of the photon.

Such an experiment is not far beyond the reach of today's technology. Some experimentalists are already working in this direction. This experiment is interesting because it probes nature exactly where we expect an interplay to happen between quantum theory and general relativity.

ANTHONY LEGGETT · I'm not entirely clear how to interpret "technical feasibility." Am I allowed to contemplate a genuine "Wigner's friend" experiment—that is, Schrödinger's cat with the cat replaced by a sentient human, who will later be able to report on his or her experience while in the box—*and* assume that the friend can be isolated from the outside world and from "intrinsic" decoherence in such a way that a correct application of quantum mechanics would lead me, outside the box, to predict interference of the friend's two macroscopically (and subjectively) different states?

If that is too much to ask, then I would settle for an EPR–Bell experiment done with immediate readout by human agents spatially separated by, say, five to ten light seconds, so that each agent consciously registers an outcome before a signal has time to arrive from the distant station.

TIM MAUDLIN · I would perform a two-slit interference experiment with macroscopic objects, where the slits are at macroscopic scale (using bowling balls, for example). The persistence of interference in such a regime would settle whether there is any collapse of the wave function that plays a role in solving the measurement problem.

DAVID MERMIN · The foundational issues about quantum mechanics that perplex me are all predicated on the assumption that the theory is correct. I would like to be able to make better sense of what it says. I am not persuaded that my perplexity is so acute that I should seek the answer in a breakdown of the theory. Therefore I would not expect any experimental test to shed light on a foundational question.

I exclude here the possibility that a crucial foundational issue might be associated with an application of the theory so intricate that the relevant calculation might be too difficult to perform, thereby requiring an experimental test. It does seem to me that all the puzzling features of the theory emerge full-blown in its most calculationally elementary applications.

This is not to say that the breakdowns of quantum mechanics suggested by some interpretations are not worth exploring through experiment. (See also my answer to Question 7, page 156.) I expect quantum mechanics to break down at some scale. Indeed, I find it amazing, in view of the body of data that gave rise to it, that it seems to be working perfectly well within the atomic nucleus and even within the nucleon. This lends support to viewing quantum mechanics as a "mode of thought," as Chris Fuchs and Rüdiger Schack once put it, rather than as a description of the world.

So I would be surprised (and rather disappointed) if foundational issues were settled by observing a breakdown of quantum mechanics. I would expect them to be settled by our acquiring a deeper understanding of the existing theory, within its domain of validity.

LEE SMOLIN · Good question. Any experiment that got a result that disagreed with a prediction of quantum mechanics.

ANTONY VALENTINI · There are at least five different experiments that I would be keen to do, all of them involving tests of the Born probability rule.

First, I would like to test the Born rule for particles that have been emitted by an evaporating black hole. This could be done at least in principle, should we discover primordial black holes—left over from the early universe—that are currently evaporating. I would also like to test the Born rule for particles that are entangled with partners that have fallen behind the event horizon of a black hole. This might be possible if we find appropriate atomic-cascade emissions taking place naturally in the neighborhood of a supermassive black hole. I would also be keen on testing the Born rule for any kind of particle at the Planck scale. The motivation for these three experiments is my suggestion that the quantum equilibrium state might become unstable in the presence of gravity.

Another worthwhile place to look, in my view, is in the neighborhood of nodes of the wave function, where the de Broglie–Bohm dynamics breaks down.

The fifth experiment I'm keen on is to test the Born rule for relic cosmological particles that decoupled (at early times) when their wavelengths were larger than the instantaneous Hubble radius. By analysis of the relaxation process on expanding space, I've shown that relaxation can be suppressed at super-Hubble wavelengths. So it's possible that such particles never underwent relaxation—they could still exist in our universe today and violate the usual rules of quantum mechanics. Specifically,

I would suggest testing their polarization probabilities, for the particles themselves or for their decay products, and to search for violations of Malus's law. Since I have to choose only one experiment, let it be this last one.

DAVID WALLACE · I'd do the two-slit experiment, but using gravity waves.

Of course, that's taking the "regardless of its current technical feasibility" clause *really* seriously! We haven't yet succeeded in detecting classical gravity waves, still less seen if they're quantized. But it would be one of those experiments that would be astounding whatever happened. If we see what effective field theory predicts—quantization of detection events, interference continuing even when the wave amplitude is so low that gravitons pass through only every few seconds—that would be an incredible triumph for quantum theory and quantum field theory. And of course, if we *didn't* see that, it would be unambiguous evidence that not only general relativity stands in need of modification, but quantum mechanics too.

ANTON ZEILINGER · To put forward a true challenge: independent of current technical feasibility, the most important experiment would, in my eyes, be a real Schrödinger-cat experiment. I have been saying tongue-in-cheek that I call an entity a cat only if it actually meows. Clearly, using explosives would not be very kind, but obviously there are versions of the Schrödinger-cat experiment that are much more benevolent to the poor feline. More seriously, a quantum-superposition experiment—or even a quantum-entanglement experiment—with real living beings would be a major step forward. It would indicate that we have experimental tools of unprecedented power at our hands. It is clear that this will open up applications we presently cannot even imagine. As an experimentalist, I read all arguments against the possibility of such an experiment as a challenge to be creative and find an experimental way to work around the difficulties.

WOJCIECH ZUREK · A careful test of the symmetries of entanglement—that is, of "entanglement-assisted invariance," or "envariance"—would be on top of my list. There is a strong immediate motivation behind it: if we knew that nature really abides by such nonlocal symmetries, we would understand where probabilities in physics come from. But there is a deeper motivation here as well: entanglement is how quantum systems know about one another, and it is also how they become, via decoherence, classical. And finally, there is also a practical angle: many of the suggested applications of quantum mechanics take for granted subsets of symmetries of states of composite quantum systems.

There are several variants of such an experiment. The simplest one would involve local manipulations of a Bell state to demonstrate that one can really take the state through all of its accessible composite Hilbert space using local operations. As I have discussed in my answer to Question 5 (see page 122), this would be also enough to establish equiprobability in Bell states, which is the basis for the envariant derivation of Born's rule.

A more advanced version would deal with states that do not have the symmetry of Bell states, so that probabilities of distinct outcomes are not equal. In this case, one can test the interplay between the relative quantum phases and the amplitudes, an interplay that in the end results in Born's rule for probabilities.

The most advanced version would entangle a microsystem with a macrosystem and test whether the symmetries are still there. In a sense, this is a bit like the original version of Schrödinger's cat (gedanken)experiment, where it was the entanglement (and not just the superposition) that was the focus. The objective would be to see whether various unitary transformations applied at one end can be successfully undone by the appropriate countertransformations at the other end.

The tricky part of the experiment is that one would like to be able to transform the composite system using local unitaries but test whether the *global* state is indeed following the prescribed trajectory in the composite Hilbert space. Ideally, this would involve a global measurement.

The good news is that such transformations (when everything goes right) are deterministic, so that one knows what observable to measure in order to confirm that the composite system has not, in some way, strayed off the track prescribed by quantum theory. (It is this certainty about the evolution of the global state that implies local indeterminacy, leading to probabilities and, eventually, Born's rule.) So if everything goes well, one does not really need to do quantum tomography, which would be resource-intensive.

The bad news is that the good news above applies to the case of the appropriate *global* measurement. Global measurements are, in any case, what one should attempt to do—envariance in its essence is a symmetry of entangled states, so its testing necessarily involves global measurements—but they are generally difficult to carry out.

SWITCHING SIDES

If you have a preferred interpretation of quantum mechanics,
what would it take to make you switch sides?

THE FIELD OF QUANTUM FOUNDATIONS is a genre-bender that feasts on a smorgasbord of influences ranging from physics and mathematics to the philosophy and history of science. In this sense, it has open-mindedness and lateral thinking hardwired into it. At the same time, it is often portrayed as a battleground where different factions are pitted against each other in a relentless ideological trench war—a stalemate in which each side refuses to cede territory but is unable to produce a defining argument that would change the hearts and minds of the opponents. And while for some it is the dissimilarities between the various interpretations that hinders reconciliation and progress, others have pointed to a lack of any tangible difference as the real problem.

For philosophers, intellectual turf wars have always been the modus operandi of their field. It is simply de rigueur to publicly proclaim and vociferously defend one's individual philosophical position. Scientists, on the other hand, are much less familiar with such battles of opinion as part of their discipline. The subject of quantum foundations, by virtue of its mixed background, seems caught in the middle.

In these times of culture wars and foundational skirmishes, there's often nothing more instructive than putting yourself in someone else's shoes. So I figured it would be an illuminating exercise to ask our interviewees what, if anything, would make them desert their own interpretive camp—assuming they count themselves as belonging to one (see Question 3, *My Favorite Interpretation*)—and seek refuge elsewhere, or perhaps go off to a deserted island to do some soul-searching for a while.

GUIDO BACCIAGALUPPI · I do not have a preferred interpretation of quantum mechanics. If I had one, then I believe it would only be empirical evidence, or very persistent failure to find it, that would make me switch sides (other criteria, such as conceptual and aesthetic issues, point in conflicting directions, and thus I find them inconclusive).

Specifically, evidence for macroscopic interference would turn me from a collapse theorist into a Bohmian or Everettian, while persistent failure to find other physical effects of an environment supposedly responsible for the apparent collapse would turn me from a Bohmian or Everettian into a collapse theorist. And if the evidence militates in favor of no-collapse positions, evidence for exceptions to the Born rule would turn me into a Bohmian, while persistent failure to find such exceptions would turn me into an Everettian.

Of course, it is difficult to define persistent failure, and it depends crucially on which kinds of innovative experiments are actually made. In the absence of significant experiments, there is just not enough evidence one way or another.

ČASLAV BRUKNER · It is my belief that all one can ever expect from a good interpretation is for it to be helpful in extending the present knowledge to new theoretical avenues or new phenomenological domains. In this sense, there can hardly be a more convincing argument for judging the Everett interpretation a good interpretation than Deutsch's recognition that he profited from the intuition of this interpretation when laying the groundwork for the idea of quantum computation. And yet there are other researchers, myself included, who find the Everett interpretation bizarre. To me, this supports a view defended by Svozil: that one should not accept claims of the absolute truth of any particular interpretation. After all, no interpretation can experimentally be distinguished from any other.

This call for an open and tolerant attitude toward the variety of interpretations does not impose restrictions on our recognition that at a certain stage in the development of a theory, one interpretation may foster research progress while others don't. Quantum information science serves as the best example. It is indisputable that the perspective on quantum states as "states of an agent's knowledge"—as opposed to "states of reality"—has prompted researchers to find practical solutions across all subdisciplines of quantum information science (cryptography, communication, and computation). Despite such progress, a large number of highly interesting and technically challenging problems in quantum information science, and in quantum foundations in general, are still unsolved. Let me give just two examples. First, it is known that the maximal number of mutually unbiased bases sets exist in Hilbert-space dimensions that are integers of a power prime. Do they exist for arbitrary dimensions? Second, it is known that the vacuum state of a massive quantum field can maximally violate the Bell inequalities for suitable spacelike-separated observables. Is the violation for a given massive field still possible as the two localization regions are moved arbitrarily far apart?

Let me get back to the interview question. If novel techniques or insights coming from a competing interpretation of quantum mechanics enable us to attack previously unsolved physical problems that are at the level of concreteness of the aforementioned examples, I will not hesitate to rethink my position regarding this interpretation. Likewise, I will do so if the physical and mathematical framework of a competing interpretation turns out to be more conducive to a merger of quantum mechanics with general relativity into a quantum theory of gravity. In the meantime, I retain my preference for the tradition of the Copenhagen interpretation.

JEFFREY BUB · My preferred interpretation is the information-theoretic interpretation that I sketched in my answer to Question 3 (see page 67). On this view, the "big" measurement problem is a pseudoproblem, but there remains the "small" measurement problem: the problem of accounting for our familiar experience of a classical, or Boolean, macroworld. I characterized this problem as a consistency problem: the problem of explaining the dynamical emergence of an effectively classical probability space of macroscopic measurement outcomes in a quantum measurement process—much like the problem in special relativity of explaining Lorentz contraction dynamically in terms of a dynamics consistent with the kinematic structure of Minkowski space-time.

If someone convinced me that there was no possibility of a plausible solution to the consistency problem, or that the role of decoherence in resolving the problem was somehow problematic (as it is in the case of the "big" measurement problem), then I would probably want to think harder about the Everett interpretation, and about the role of decoherence there in selecting a preferred basis and securing a coherent interpretation of quantum probabilities.

ARTHUR FINE · As explained in my reply to Question 2 (see page 44), I do not have a preferred interpretation, although I do have a rough ordering of the ones I know (with some minimal version of the standard interpretation on top and, maybe, many worlds near the bottom). I regard some interpretations simply as nonsensical. A demonstration of how they might make sense could help improve their rank. Some I regard as too easy or trivial. Here, considerations of heuristics would be important; for instance, showing that the interpretation suggests new sorts of experiments, or measurements, or new lines of generalization. I suspect, however, that firm adherents of one school of interpretation or another may be more moved by temperament (perhaps under the guise of "philosophy") than by argument. Even so, if one is able to listen openly to competing ideas and intuitions, then attitudes can be tempered. My experience with Wenzel is a wonderful example (see my response to Question 1, page 26). Still, in many cases, switching sides is probably not really an option.

CHRISTOPHER FUCHS · Switch sides to what? The premise of the question is that there is something coherent on the other side—I no longer think there is. Of course, I toyed with all kinds of crazy ideas as a boy—from hidden variables, to collapse models, to there being no space-time "underneath" entangled quantum states, and so forth. I can promise you I started as no Bayesian about probabilities, quantum or otherwise, and certainly no personalist Bayesian about quantum states. Like most students of quantum mechanics, when the textbook said, "Suppose a hydrogen atom is in its ground state, blah, blah," I thought the ground state was something the atom could actually *be in* ... all by itself and without the aid of any agent contemplating it. But the years went by, and I slowly, painfully, came to the opinion that I have today: that those nonpersonalist ideas about quantum states and the outcomes of quantum measurements *just don't fit* the actual structure of quantum theory. They are fairy tales from some fantasyland, not the world we actually have.

Still, I can certainly give a list of things that would have deterred my pursuing a Bayesian account of quantum states if they had been true of quantum theory: If a single instance of an unknown quantum state could be identified by measurement. If an unknown quantum state could be cloned. If collapsing a state on one system could cause an instantaneous, detectable signal on another. If the time evolution equation of quantum theory were nonlinear. All these things, if they had been true of quantum theory, would have indicated that quantum states do not have the character of epistemic states. (Remember the discussion in my answer to Question 3, page 70.) But, of course, all these things are not true: the structure of quantum theory allows none of them. And that's the point.

OK, then: granting quantum states to be epistemic, what would it take to deter me from a personalist account of quantum measurements? Under what conditions would I believe it fruitful to pursue a hidden-variables reconstruction of quantum theory? If Bell's theorem were not violated. If Gleason's theorem were not true. If Kochen and Specker could not have found a noncolorable set. If the ontological baggage required of a hidden-variables account weren't every bit as large as the space of epistemic states, as shown by Alberto Montina in his paper "Exponential Complexity and Ontological Theories of Quantum Mechanics." (What would it mean to draw a distinction between the epistemic and ontic states then anyway?) But the structure of quantum theory allows for none of these things. And again, that's the point.

GianCarlo Ghirardi · My preferred position with respect to the standard theory corresponds to accepting an empirically testable alternative to quantum mechanics, and I realize that not much wiggle room remains for those who want to dynamically and consistently induce the macro-objectification of the position of the objects of our common experience. So I'm ready to switch sides—and, in fact, completely abandon the view I presently hold—if some experimental test will yield an explicit proof of the occurrence and persistence in nature of quantum superpositions of states corresponding to different macroscopic positions. Put differently, I think it's very important to try to test the superposition principle in those mesoscopic situations in which the dynamical-reduction models claim that it is almost immediately violated.

Shelly Goldstein · I have switched sides, several times in fact. I began as a proponent of textbook quantum theory. But the more I learned about it, the more implausible it seemed. At some point long ago, I was attracted to many worlds and to the idea, associated with many worlds, that a theory can determine its own interpretation. For me, this was connected with certain materialist approaches to the philosophy of mind. Once I came to recognize that materialism is hopeless—that consciousness, which can't coherently be denied, transcends physics as we know it—I gave up on many worlds. I later learned about Nelson's stochastic mechanics and was surprised that quantum probabilities and quantum measurements could be understood in such a straightforward way. But then I learned that stochastic mechanics, as simple as it had seemed, was still far more complicated than necessary, and that with Bohmian mechanics, the quantum mysteries can be eliminated in an almost trivial way.

I'll switch again if I can be persuaded that there is, or if I can find, an even simpler approach. Or if experiments demonstrate that quantum mechanics is, in fact, wrong.

DANIEL GREENBERGER · What would it take to get me to switch sides in my interpretation? As I said in my response to Question 3 (see page 75), I like the Copenhagen interpretation because I believe it better describes the experimental reality, and because it blends in beautifully with the superposition principle. But there are two things that could get me to switch. One is an interpretation that is really closer to experimental reality. You would think this wouldn't be hard to do, since with the standard interpretation you have to swallow so many nonintuitive ideas, like the existence of state vectors in Hilbert space.

The problem of the best interpretation is sort of like choosing the best explanation of the Kennedy assassination. The official story is so shot full of holes as to be totally implausible. But all the conspiracy theories are even more implausible, so one keeps an open mind on the subject, hoping that something will come out that will clarify the situation. So too with Copenhagen. Maybe something will come along that fits as well, or better, with the experiments that one does. Not too likely. The second possibility I mentioned above is that I would settle for something that is much easier to explain to my students. That would be a big plus. Then I could perhaps alternate between the interpretations, or use the one to help clarify the other. The only definitive victory would be if one interpretation could help explain something that was very poorly done in Copenhagen, such as give a lot of insight into the gravitational problem. Then it couldn't be ignored.

Maybe there is a better interpretation out there. Maybe if you kiss enough frogs, one will turn into a prince. The problem is, they all look like plain old frogs.

LUCIEN HARDY · I do not have a preferred interpretation of quantum theory. So let me say, instead, what it would take for me to adopt one. The following qualities are essential:

- That it provides an account of why the world appears as it does. This implies providing a solution to the measurement problem.
- That the interpretation is not ad hoc. This would be the case particularly if all the equations of the interpretation could be derived from some simple compelling ideas.
- That it is a theory of quantum gravity. Strictly speaking, a theory of quantum gravity may be a different and deeper theory than quantum theory. But we should really seek an interpretation of the most fundamental theory.

The following features would make the interpretation more compelling:

- That it provides insight into the nonlocality predicted by Bell's theorem.

- That it enables us to do calculations we could not previously do, or at least that it suggests new calculations we would not previously have thought of doing.
- That it leads to new experimental predictions that are subsequently confirmed.
- That it leads to progress in other fields of physics.

I am hopeful that a theory of quantum gravity, when we eventually get our hands on it, will do all of these things. Then, and only then, do I expect to embrace a particular interpretation. I hope it will be the correct one.

ANTHONY LEGGETT · I'm not sure whether the minimalist point of view that I have above labeled the "statistical interpretation" can really be called an interpretation. Like the so-called Copenhagen interpretation, it feels to me to be rather a *refusal* to interpret the quantum formalism as anything other than a recipe, pure and simple. Some prima facie alternative "interpretations," such as the consistent-histories approach, seem to me to be merely adding soothing words to this point of view. Others, such as the Everett–Wheeler approach, as explicated by DeWitt and Graham (the "many worlds" interpretation), are quite literally meaningless: I simply do not understand what, as applied to the unobserved universes, the words "equally real," ostensibly English, are supposed to mean. So if we exclude the possibility that experiments will find a spectacular breakdown of the predictions of quantum mechanics within my lifetime (which would produce a whole new ball game), I don't see myself changing my point of view anytime soon.

TIM MAUDLIN · The most critical consideration is, of course, empirical evidence. In principle, a theory that postulates collapse of the wave function will make different predictions than one that does not, so one could experimentally decide between, for example, a Bohmian approach and a GRW approach.

Beyond that sort of straightforward empirical test, the amenability of a theory to treat gravity might be convincing. It is possible that some clear and plausible approach could only be implemented in one foundational framework.

DAVID MERMIN · My intuitions about the nature of quantum mechanics are not coherent enough to add up to anything I would dignify with the term "interpretation." Admittedly, shortly after turning sixty, I did write a few papers setting out what I called the Ithaca interpretation (see also my answer to Question 6, page 136). But I was young then, innocent, and overly willing to sacrifice an accurate phrase for an entertaining one.

One of those papers made its argument under the banner of Bohr's statement that the purpose of our description of nature is "only to track down, so far as it is possible, relations between the manifold aspects of our experience." When I wrote the paper, the crucial word for me was *relations*. My motto was *correlations without correlata*. What led me to stop giving physics colloquia on the IIQM after only a year was the obvious question: "Correlations between what?" Abner Shimony aptly complained that the Ithaca interpretation "had no foreign policy." Exchanges with Chris Fuchs persuaded me that just as important as *relations* was *our experience*, which I was too ready to hide beneath the same rug under which I had (correctly) swept the problem of consciousness.

So insofar as I had a preferred interpretation in 1998, what persuaded me that it was, at best, insufficiently developed was somebody making me aware of some interesting ideas that hadn't occurred to me. It remains entirely possible that some wise, imaginative, and readable person may in the future lure me away from the position I am trying to sketch in my answers to these questions.

To make me switch to some interpretations I now reject would require a breakdown of quantum mechanics along lines suggested by these currently unpalatable points of view. Of course, at that point they would no longer be interpretations of existing theory, but alternative theories. To convert me to Bohmian mechanics, for example, I would have to see clear evidence of particles that were not in "quantum equilibrium." Without that breakdown of orthodox quantum mechanics, the reintroduction of particle trajectories seems an unnecessary complication that raises questions at least as vexing as those raised by the orthodox theory. To convert me to the view that "wave-function collapse" was a real physical process and not just an updating of expectations on the basis of new information, I would have to see convincing evidence of deviations from quantum probabilities produced by Ghirardi–Rimini–Weber–Pearle "hits."

A simple nontrivial example of a history containing many different times that exactly satisfied the consistency conditions might persuade me to take another look at consistent histories (see my answer to Question 16, page 279).

LEE SMOLIN · I've been wondering this lately, and I've been examining the assumptions that have motivated my own research on quantum foundations over many years. I've asked myself whether looking for some novel point of view might lead to the solution faster than sticking to what is, after all, by now a very old point of view.

I can't imagine, however, switching to take on one of the standard views according to which quantum mechanics is a fundamental theory. I admire the cleverness of both Bohr's interpretation and the Everett interpretation, but I have too vivid an understanding of their inadequacies to imagine switching to them. What would impress me greatly would be a new perspective. Indeed, when a problem has remained unsolved for so long, it is likely that the resolution will come from taking a novel point of view that leads to unexpected theoretical or experimental results.

ANTONY VALENTINI · The observation of spontaneous collapse would, of course, make me switch to collapse theories. If nonequilibrium violations of quantum theory, as I envisage them, are *not* observed, say over the next one hundred years, then I would start to have serious doubts about hidden variables (were I still alive). I don't think the equilibrium de Broglie–Bohm theory by itself is scientifically satisfactory, even if logically it is a possibility, because the details of the trajectories can never be tested even in principle. I might consider Everett's theory if it can be shown to be conceptually coherent.

I might also change my views if someone made important progress in theoretical physics—for example, about information loss in black holes, or about quantum gravity—for which a particular interpretation played a crucial role.

DAVID WALLACE · I do have a preferred interpretation of quantum mechanics: the Everett interpretation. I take Everett's basic insight to be that we don't have to treat quantum mechanics differently from any other physical theory: we can just regard the theory's mathematical models (in quantum mechanics, unitarily evolving quantum states) as representing physical states of affairs—just as in classical mechanics, or classical field theory, or general relativity. (From that point of view, the "measurement problem" arose because we erroneously thought quantum mechanics couldn't be understood that way.)

So to "change sides," I'd have to be convinced either that (1) something was wrong with the theory itself, or (2) that for some reason Everett's insight is wrong, and that after all we can't just take quantum mechanics as a straightforward physical theory like classical mechanics.

I'm pretty clear what it would take to persuade me of option 1: empirical evidence in contradiction with quantum mechanics. In particular, if we find a violation of the superposition principle, that would be pretty good reason to reject the Everett interpretation of quantum mechanics—but we'd be rejecting quantum mechanics (and the Everett interpretation along with it), rather than rejecting an interpretation but keeping the theory.

I'm less sure what it would take to persuade me of option 2. At one point I'd have said, "Strong philosophical reasons to think that probability doesn't make sense in the Everett interpretation." But I've become more and more convinced that probability doesn't make any more sense in non-Everettian contexts (and indeed, that probably it makes less sense). And certainly, *mathematically* probability works fine in Everettian quantum mechanics, at least where decoherence is applicable.

Overall, I think in most cases I'd be more willing to revise a philosophical principle that was in conflict with the Everett interpretation than revise my interpretation of quantum mechanics. I guess I'm just not that confident that we'd have reasons to believe some given philosophical principle that were so persuasive that they'd require us to modify quantum mechanics or to interpret it completely differently from the way in which we normally interpret scientific theories. Perhaps that's just my lack of imagination, though.

ANTON ZEILINGER · Switching from one interpretation to another is not the important issue. What's important is to clearly identify the specific basic notions (explicit and implicit) from which each specific interpretation starts. The present coexistence of various interpretations that all agree with the formalism is in itself an important lesson. For example, why is it that very different fundamental interpretive positions lead to the same experimental results? What remains invariant when going from one interpretation to another? From such an analysis, I hope to arrive at a deeper understanding of the really indispensable elements of any interpretation. I should note that here I mean interpretations of strictly the same formalism. Personally, I expect approaches that modify the mathematical formalism (for example, GRW)—as fruitful as they are in exploring alternative theories—to be ruled out by experiment in the future.

WOJCIECH ZUREK · I do not have a preferred interpretation. I feel inspired when reading Bohr and Everett, and I believe (with Wheeler) that much of the antagonism between these two camps is based on a mutual misunderstanding.

BELIEFS AND VALUES

—

*How do personal beliefs and values influence
one's choice of interpretation?*

A GOOD FRIEND OF MINE, who grew up in a large Irish Catholic family, once told me that she wouldn't vote for John Kerry for the sole reason that he didn't oppose abortion. After heated discussion, mixing the taboo subjects of politics and religion into a strong cocktail, the term "value voter" was no longer abstract media jargon to me.

Well, science is not politics or religion, of course. Scientists rarely talk about the influence of values and beliefs—whether philosophical, political, spiritual, or otherwise—on their work. What counts are hard facts, and the very mention of the term "belief" in the context of a scientifically amenable question sends shivers down the practitioner's spine. On the other hand, there are many areas at the frontiers of science that allow for different views of what the central problems are, and of what approaches one may reasonably pursue or forgo. The foundations of quantum mechanics certainly fall into this category, as this book hopefully demonstrates in copious amounts (as if any further evidence was ever needed). And in making choices, wouldn't it be natural to assume that one cannot help but be informed by personal values and beliefs—not in the shape and name of religious dogma or blind faith or political ideology, but in the form of certain, say, methodological or philosophical dispositions? In *The Structure of Scientific Revolutions*, Thomas Kuhn forcefully argued that this indeed must be the case, that subjective factors—what Kuhn calls in one place "personal and inarticulate aesthetic considerations"—inevitably play a decisive role in the grand battle of competing scientific ideas and theories.

As the discussion of the Bell inequalities in Question 8 has exemplified, even if hard facts are brought to the table—namely, a concrete mathematical model tested by experiment—there remains plenty of leeway in accommodating rather disparate conclusions. And once we take a good look at the various interpretations of quantum theory, we may expect the motivation for pursuing any one program to be rooted even more so in beliefs-and-values terrain. Deterministic or indeterministic? Epistemic or ontic? Realist or nonrealist? Local or nonlocal? In the absence of empirical distinguishability, it would appear that a good part of such preferences must surely, if only subconsciously, be rooted in one's temperament and deepest philosophical convictions and intuitions.

Or perhaps not? Is this an overly psychological, anthropocentric view of quantum foundations? To what extent do beliefs and values really influence the kind of foundational questions and interpretive programs one focuses on, the kind of strategies one pursues, the kind of solutions one proposes? Is it inevitable for beliefs and values to play a role in this regard? Is it acceptable?

GUIDO BACCIAGALUPPI · This is a rather tricky question: are these meant to be beliefs of a general philosophical nature, or also moral, political, or religious beliefs? It apparently used to be the case that hidden-variables theories appealed to communists—because more "materialist" than the "idealist" Copenhagen view—but that is definitely a matter of the past.

General philosophical views can shape one's sympathies in quantum foundations. For instance, a dislike of positivism will turn one against Copenhagen-style views (whether or not positivism had a role in shaping those views). Or one's views in philosophy of mind might influence one's choice of approach in the philosophy of quantum mechanics.

Looking at my own personal case, I'm extremely open as regards questions of interpretation, but I'm not sure I can identify any personal reasons for this. As an undergraduate, I was taught for several years by Paul Feyerabend, who certainly instilled in me a sense of tolerance and an appreciation of opposing viewpoints. But I'm equally certainly not promoting conflicting views just for the sake of diversity. I'm genuinely interested in many different approaches to the foundations of quantum mechanics. I rather think I picked up a sense of fairness and an openness of outlook working in the philosophy of physics group at Cambridge.

Quite generally, there is very little, if any, dogmatism in the philosophy of physics community, and no big egos to hamper research. That is one of the reasons why I love working in this field.

ČASLAV BRUKNER · Totalitarianism was the biggest tragedy of the twentieth century. With lasting danger of an increase in the influence of collectivist ideologies, it is important for us to continue to study them so we can learn how to avoid them, or offer resistance to them when they are on the rise, or diminish their consequences when they get to power. Thus far, I've had the opportunity to be exposed to three ideologically different social structures: Tito's socialism, with "workers' self-management" as a propaganda façade for continuing a one-party political monopoly; Milosevic's brutal and manipulative nationalism; and finally, Austria's liberal democracy, with its everyday latent-but-pretty-obvious xenophobic political reality. In reaction to these experiences, I have developed a firm conviction about the importance of independence and self-reliance, and about the importance of opposing external interference with one's own beliefs and desires and with the beliefs and desires of those we love and care about.

Therefore, I definitely do not like to associate any interpretation with "truth." My attitude is that only "the rule of reason and the law of liberty shall prevail throughout the realms of science," as Arthur S. Eve ended his (and the only) reply to Johannes Stark's pamphlet "The Pragmatic and the Dogmatic Spirit in Physics," published by invitation in the April 30, 1938, issue of *Nature*.

The request for tolerance and openness toward new ideas is at the core of my belief in an open future—a future that is not fixed but remains to be determined in a process where both the decisions and the experience that precedes these decisions participate. An open future is not possible in classical physics, where the idea of a "coming into being" or "becoming" and of a "free will" are fundamentally illusory. Weizsäcker's thesis of probability as the only available scientific description of the open future, and his understanding of the past as the realization of facts through documents, remain iconic reference points to me—just as Whitehead's guiding idea that natural existence consists of, and is best understood in terms of, processes rather than things. Convinced that an ontological picture of our world has to include both possibilities and facts, I am naturally left with the tradition of the Copenhagen interpretation.

JEFFREY BUB · If by "personal beliefs and values" one includes ontological and metaphysical assumptions outside physics, then I think the answer is, a great deal—as I indicated in my answer to Question 10 (see page 203). Lorentz's assumptions about space and time induced him to "regard the aether, which can be the seat of an electromagnetic field with its energy and its vibrations, as endowed with a certain degree of substantiality, however different it may be from all ordinary matter." So for Lorentz, distances and time intervals should be measured "by means of rods and clocks having a fixed position relative to the aether," and it makes a difference whether a body moves through the aether. Since this difference does not seem to be detectable, one then has to show via a dynamical argument why this difference can never be detected in measurement—just as Bohm's theory involves a dynamical explanation for the inaccessibility of the hidden variables via any conceivable measurement, once the equilibrium distribution of hidden variables has been achieved.

By contrast, the influence of personal beliefs and values is potentially less toxic in the program of reconstructing quantum mechanics from general principles, information-theoretic or otherwise. Here, one begins with a broad class of theories, for example, the class of no-signaling theories, and the task is to come up with principles, like information causality, that are general enough to provide an interesting starting point for physics, and to show that these principles exclude all theories in the class except quantum mechanics. Of course, one's personal beliefs and values might be relevant to one's appraisal of the suitability of a particular principle as a starting point for physics, but here the principle is explicit as a premise in the derivation of some salient feature of quantum mechanics, rather than being implicit in the motivation for a particular interpretation.

ARTHUR FINE · Please see my responses to Question 12 (page 229) and Question 14 (page 252).

CHRISTOPHER FUCHS · You know by now that I like to quote William James. I do it because he writes better than I do. In any case, there is no better way to answer your question than to quote him again:

> The history of philosophy is to a great extent that of a certain clash of human temperaments. Undignified as such a treatment may seem to some of my colleagues, I shall have to take account of this clash and explain a good many of the divergencies of philosophies by it. Of whatever temperament a professional philosopher is, he tries, when philosophizing, to sink the fact of his temperament. Temperament is no conventionally recognized reason, so he urges impersonal reasons only for his conclusions. Yet his temperament really gives him a stronger bias than any of his more strictly objective premises. It loads the evidence for him one way or the other, making a more sentimental or more hard-hearted view of the universe, just as this fact or that principle would. He *trusts* his temperament. Wanting a universe that suits it, he believes in any representation of the universe that does suit it. He feels men of opposite temper to be out of key with the world's character, and in his heart considers them incompetent and "not in it," in the philosophic business, even though they may far excel him in dialectical ability.
>
> Yet in the forum he can make no claim, on the bare ground of his temperament, to superior discernment or authority. There arises thus a certain insincerity in our philosophic discussions: the potentest of all our premises is never mentioned. I am sure it would contribute to clearness if in these lectures we should break this rule and mention it, and I accordingly feel free to do so.

I think that says it all.

The state of New Hampshire has a motto, "Live Free or Die." Quantum theory, I would say, is the first physical theory to indicate that we might live again (like George Bailey) and live free. It is the first physical theory to expose with technical beauty all the cracks in the block-universe conception. I bank my career on that value: science, like Darwin, will eventually make its natural selection. To be let live in this other sense is the most any scientist can hope for.

GIANCARLO GHIRARDI · I'm of the opinion that personal beliefs can and must play a crucial role not only in the elucidation of the meaning of existing and well-established theoretical schemes that do not admit a clear and simple interpretation, but also in the formation of scientific alternatives to such schemes and in the discovery of unexpected aspects of natural processes. There is no doubt that adhering to a realistic view of the world has played a decisive role for Einstein—particularly with respect to his development of the EPR puzzle—and for Bell in the derivation of his famous inequality. At a completely different level, and without attaching any particular relevance to it: there is no doubt that my prejudice of not being willing to attribute a fundamental role to conscious observers, and my belief that a fundamental theory should account for our definite perceptions in a logically clear way, has been the main motivation for my long-standing efforts to work out collapse models.

SHELLY GOLDSTEIN · Physicists are people, and people are not fundamentally rational. So which theories physicists end up believing, including which versions of quantum mechanics ("one's choice of interpretation"), will depend upon a host of historical, emotional, and maybe even some rational factors. Sometimes we believe what we first learned, or what we first thought we really understood, or what got us excited at the time.

If we value stability and tradition—and there's something to be said for that, of course—we will resist departing from our views, often in the face of compelling arguments against them. And if we value authority—and there's something to be said for that as well—we will, of course, be more inclined to accept the official view of things. If we value what is fashionable, we will, of course, be inclined accordingly. And we will often use our rationality to justify conclusions to which we have already been led by nonrational considerations.

If we believe in the centrality of humanity in the vast scheme of things, we might be more inclined toward versions of quantum mechanics in which observation plays a fundamental role, such as the Copenhagen interpretation. We are also then more likely to believe that quantum information theory has something deep to convey about the nature of quantum reality. If we are realists, and believe that there is a physical reality outside of us and that it's the purpose of physics to grasp that reality, we'll be inclined differently than if we're positivists or if we believe that the purpose of physics is to facilitate prediction.

DANIEL GREENBERGER · Scientists like to believe that science is an objective enterprise, and physicists generally see scientific truth as the goal of their research, which they credit as being objective. My own experience tends to the opposite conclusion, in that people get very worked up over their favorite interpretation, which often depends on their psychological inclinations. For example, some people cannot abide by an interpretation that is not realist, and they cannot believe that probabilities are an underlying part of reality. Others are hung up on causality, in the classical sense. Often there is a religious basis to these strong feelings.

My own inclinations lead me in the opposite direction. I don't think we know much about the "real world," and I tend to accept interpretations that remind us of that. So I believe that probabilities are here to stay, and that when quantum theory breaks down, it will be because it is not weird enough.

I tend to have a spiritual view of the world, but not in the sense of organized religion. I think that whoever designed the world, it was not with us in mind, and I doubt that we will ever understand it. I think looking for the order in the universe is a noble enterprise, and I like to be part of it, but I am highly skeptical of the outcome. That makes the achievements of a Newton or an Einstein seem even more remarkable. But finding the "theory of everything" is a pretty tall order for creatures who understand almost nothing, and I'm afraid I'm pretty skeptical about that enterprise.

LUCIEN HARDY · The relationship between personal beliefs and values and the output of creative people, whether they be artists or scientists, is a complicated one. While a little self-knowledge is a good thing, I do not think it is terribly conducive to the creative process to be overly concerned with how, for example, our spiritual and political beliefs influence our research direction. Further, there is a danger that such arguments be used to reduce (by way of apparent explanation) rather than celebrate the creativity inherent in what we do.

So let me instead use this opportunity to say what a few of my (not especially personal) beliefs are as pertinent to the field of quantum foundations.

First, workers in foundations have to engage in the great project of physics. Our job is not simply to sit back and tidy up the mess left behind after the physics has been done, but rather to be involved in developing new theories and new experiments. We should bring foundational thinking to the heart of the process of pushing forward the frontiers of knowledge about the physical world. This is good for us. But it is also the case, as has been argued by Lee Smolin, that foundational thinking is essential to get fundamental physics out of the rut it has been in for the last thirty years or so. We have a duty to engage. We have got to change the course of physics. Nothing less will do.

Second, what marks out foundational workers is a belief that it is profitable in physics to ask conceptual questions and to try to understand a physical theory (whatever "understand" means). Once a physical theory has been formulated, there is a pressure to move one's attention to more practical questions. How do we use the theory to do calculations? How do we apply the mathematical tools of the theory to other situations? It is only by thinking about the conceptual issues, however, that we (1) really understand what the new theory is telling us about reality, and (2) have a chance of understanding how to construct the next physical theory. My main criticism of the program of loop quantum gravity, for example, is that it is blindly applying the mathematical tools of quantization to canonical formulations of general relativity without doing sufficient conceptual work up front. Conceptual thinking needs to be involved in the genesis of a new physical theory. Otherwise, it runs the chance of importing the conceptual inadequacies of earlier theories hidden in the mathematical structure of those theories.

Third, we should seek to construct general mathematical frameworks from suitable conceptual ideas, and then attempt to gain insight by seeing how our physical theories can (or cannot) be formulated in such frameworks. This sort of structural work is, I anticipate, the best way of moving beyond our current theories. Further, it can help to move foundational debate beyond the realm of words into the realm of mathematical theorems and such like.

Fourth, we should not be overly seduced by impressive mathematics. That a mathematical structure is beautiful is not reason enough for it to play a role in physics. It is better to let physical thinking determine the structure of the mathematics we use than vice versa. Relatedly, we should seek out axioms for our physical theories that can be expressed in a physical rather than mathematical language. From the point of view of the physicist, mathematics is there to serve physics, and not the other way round.

Fifth, we should aim to account for the appearance of the physical world in terms of a realist interpretation of our ultimate physical theory. This is a rather sorry task since we probably will never get to the ultimate physical theory. But we can expect major clarifications every few centuries or so. We have to bear in mind that reality is probably a lot more weird than we are currently capable of understanding, and that most likely, some of the things we currently take for granted as being true of the world are completely wrong. But we should not give up on seeking out the ultimate understanding of reality.

These are the core beliefs that drive my approach to physics.

ANTHONY LEGGETT · At least in my own case—and that of most of the colleagues I know well enough for the question to be relevant—I don't think they do. Religious believers and assertive atheists can easily gravitate to the same interpretation.

TIM MAUDLIN · The main belief that seems to influence these sorts of interpretational questions is the belief that we are capable of formulating a true, objective account of the physical world. Many physicists deny either that this is possible or that it is their own personal concern: they are satisfied with an algorithm for making predictions that is clear enough to be used in everyday circumstances and accurate in its output. Many of John Bell's later writings were arguments that the "received view" about quantum theory did not meet the standards of precision and exactitude that a physicist should seek. He acknowledged that the standard approach is just fine For All Practical Purposes, but insisted that one should want more than this. One should seek a theory with a clear ontology defined at all scales, where the dynamics is all in equations rather than in surrounding talk couched in vague terms (such as "measurement"). The largest gap is not between the various "interpretations" (that is, between the various competing exact theories), but between those who are seeking such an exact theory and those who just don't care as long as the present practice works. Since this is a fundamental judgment about the whole point of developing physics, it cannot be a matter of proof. For myself, I am only interested in physics insofar as it attempts to provide a clear and comprehensible account of what the physical world is. If I were convinced that it can only provide practical guidance to engineers but no ultimate account of what exists, then I would not be interested in studying it.

DAVID MERMIN · The belief that physics is, or ought to be, the whole story surely plays a role. Those who believe that physics describes the external world *as it relates to us* have an interpretive flexibility unavailable to those who insist that "we" have no place in the story except as complex physical systems. (See also the third of my answers to Question 2, page 53.)

I have the impression that those physicists who believe in God tend, perhaps unsurprisingly, to take a more strongly realistic view of the abstractions that make up the quantum formalism than do many of us who take an atheistic view of the world.

There are also those who maintain that while God does not exist, Physical Law does. Since I agree with the first half of this proposition, I would not call them idolatrous. But others might.

Values (as opposed to beliefs) are harder to identify. I sometimes detect them in the attitudes of those who believe in, or search for, hidden-variables models of quantum mechanics. I have heard ringing declarations about the nature of science, exhortations not to give up the good fight, and expressions of scorn for contemporary obscurity.

See also my answer to Question 15, page 268.

LEE SMOLIN · I don't think that my view is different from those I disagree with because we have different values. Most of those who work on the foundations of quantum mechanics do so because we believe strongly in getting the foundations of science right, because this is where the deep truths science teaches about nature are to be found. We believe that our understanding of the laws of physics captures true aspects of nature, and we all care very deeply about discovering those truths.

Having said this, it is curious that there appear to be bundles of ideas that go together. I've noticed that some people who believe in the many-worlds interpretation also believe in the possibility of strong artificial intelligence and the anthropic principle. Denial of all three of these speculative ideas also seems to go together. Perhaps what plays a role is how optimistic one is about how close we are to a final physical theory. Those who believe we may be close try to string together a metaphysical picture based on what we know now. I have perhaps a more tragic view of the state of science: my sense is that there are great mysteries that science still does not have a toehold on. These mysteries include the reason why nature appears quantum, the nature of consciousness, and the problem of why certain laws are satisfied rather than others.

ANTONY VALENTINI · It can be interesting and insightful to ask where people's ideas and preferences come from, but only up to a point. In the end, what counts is how the ideas stand up to theoretical and experimental scrutiny.

What I find more interesting is why you ask this question. It's a peculiar question to ask a scientist. The whole point of science is to arrive at objective truth by a combination of reason and experiment, and to leave personal beliefs by the wayside. Not to say that it's easy. But imagine asking a biologist how "personal beliefs and values" affect the interpretation of fossils. Or a condensed-matter physicist regarding the interpretation of superconductivity. The question would seem peculiar, and an insinuation that the person being questioned was or might be behaving unscientifically by allowing personal beliefs to cloud their judgement.

Dennis Sciama used to say that when it comes to the interpretation of quantum mechanics, "the standard of argument suddenly drops to zero." It's still a field that is often short on argument and long on prejudice. It's as if the usual rules of rational, scientific argument tend to be suspended in this area. What have "personal beliefs and values" got to do with a scientific discussion?

It might be claimed that the question is reasonable in the context of quantum foundations, where there are different and radically divergent interpretations. But our present uncertainty is no reason for compromising basic standards. We should focus on arguments and evidence, not on beliefs. The whole point of science is to get away from mere belief.

Some will say that realism in physics amounts to a "belief" or "value," but that is confused and mistaken.

It is true that personal beliefs, values, and inclinations can provide motivation and inspiration for new ideas or for following a certain road. In principle, there's nothing wrong with that. Perhaps it's even necessary if one is going to make a serious effort exploring an idea. But again, once ideas or research directions have been thought of and decided on, one has to find out if there is any theoretical or experimental support for them. I don't think it's scientifically healthy to give much weight to why someone thought of something or why they're attracted to exploring a certain idea, as it distracts from the scientific heart of the matter—which is whether or not the idea itself makes sense and is correct.

I would say that the widespread neglect of objectivity and realism in quantum physics has contributed to an erosion of the idea of science as finding out what the world is really made of and how it works. The objective world is oblivious to our personal beliefs. Doubts about the existence of the former tend to foster an emphasis on the importance of the latter. Actually, the roots of this go back to the late eighteenth and early nineteenth centuries, which saw the rise of German idealism as a force in philosophy. The development of quantum mechanics itself, in the 1920s, shows traces of its influence, with an emphasis on the subjective knowledge of the human observer as opposed to objective reality.

Some people are uncomfortable with the objective and rational world of science, which is sometimes seen as an impersonal world devoid of human meaning, with no room for religious or spiritual belief. Sometimes, this is what lies behind an attraction to subjective, nonontological interpretations of quantum physics. This seems to have been the case for some of the founding quantum physicists, certainly in the case of Wolfgang Pauli. But the trouble with subjective, nonontological interpretations is not so much the motivation that sometimes lies behind them, but the fact that they don't make sense (see my discussion of the measurement problem in my answer to Question 7, page 157).

DAVID WALLACE · I'd like to say that they don't, but what I really mean is that they *shouldn't*. This is an objective question, even a scientific one: what does our best theory of the microscopic tell us about the physical world? That might not be a question directly answerable by experiment—though experiment bears on it quite a lot—but it shouldn't be a matter of taste. My beliefs and values shouldn't influence my take on the quantum measurement problem any more than they should influence my take on global warming or gamma-ray bursts.

Dynamical-collapse theories are a really strong example here, of course, because they really are testable. But even in the case of the pilot-wave theory, which makes the same predictions as quantum mechanics in normal circumstances, it's still a different theory with a very different formalism, and adopting it would have pretty major consequences for how we go beyond the Standard Model or how we quantize gravity. And even something like a "pure" interpretation, such as the approaches based on information, are mainly explored by people who think this is really telling us something important about the objective structure of the world, maybe even something with experimental consequences sooner or later. In every case, the choice of "interpretation"—"theory" would be a better word—is actually influencing the science that people are doing.

I haven't mentioned the Everett interpretation—which is the interpretation I think is correct—because oddly enough I think it's the conservative option, the one that doesn't really require any (well, much!) change in how we do, or think about, quantum physics. That's because the Everett interpretation, at least as I see it, basically just tells us to take quantum mechanics literally, and reassures us that there's no immediate paradox in doing so, that macroscopic superpositions aren't in contradiction with our observations. And in day-to-day physics, we basically do take quantum mechanics literally—we regard the quantum state of a system as something about that system, something we can prepare and modify and interact with. We use "collapse of the wave function" as a shorthand, but when pushed we quickly retreat to saying that decoherence makes the superposition unobservable, not to regarding the collapse as some objective nonunitarity.

Now, it's true that we don't always *believe* what we think our theories *say*. So even if someone acknowledges that quantum mechanics *says* that after a measurement the world is still in a superposition but the superposition is unobservable, they might not believe that that's *true*, that the theory really can be trusted when it says that. And maybe that is a matter of our "beliefs and values," maybe there isn't any knockdown argument to convince somebody who uses quantum mechanics as a predictive tool that he ought to believe what the theory says about the physical world. But that isn't anything specific about quantum mechanics—any scientist is at liberty to carry on using his theory but not really believe its claims, if that's what he wants to do. (It seems to me a pretty strange thing to want to do, but maybe that's *my* beliefs and values talking!)

ANTON ZEILINGER · This is a very important question, and I do not know the precise answer. In my personal case, I have a strong feeling that having grown up in Vienna—with its history of Ernst Mach and the Vienna Circle—was important in formulating my own position in the foundations of quantum mechanics.

WOJCIECH ZUREK · I do not have a preferred interpretation. Indeed, I think personal beliefs should be allowed to evolve with scientific evidence. Trying to do it the other way around is dangerous, as the examples of Copernicus and Galileo demonstrate. Therefore, I cannot really answer this question. But I do believe—and this has become a part of my personal beliefs!—that one should trust firmly established facts of science, and this includes the really quantum part of quantum theory.

(Mind you, this does not include any specific interpretation, and my confidence in the less quantum parts of quantum textbook credo increases as its key elements, like Hermiticity of observables, emerge from the uncontroversial postulates.)

In my interpretation-oriented thinking—for instance, in the "existential interpretation," which emphasizes persistence in time as the key attribute of existence—I follow an operational approach to the interpretational issues and take the quantum postulates very seriously. But I also think it is important not to declare victory prematurely. This is an important virtue of decoherence: it is not attached to a specific interpretation. And clearly there is more to be done to put it on a firmer and, above all, more fundamental foundation, and to go beyond what it has already revealed about our quantum universe.

THE ROLE OF PHILOSOPHY

*What is the role of philosophy in advancing our understanding
of the foundations of quantum mechanics?*

EINSTEIN, IN HIS INCOMPARABLE WAY, once called philosophy a "mother who gave birth to and endowed all the other sciences." And therefore, he went on, "one should not scorn her in her nakedness and poverty, but should hope, rather, that part of her Don Quixote ideal will live on in her children so that they do not sink into philistinism."

The sheer existence of the field of quantum foundations is a testament to Einstein's words. By its very nature, this field is all about contemplating the bigger picture, the fundamentals, the conceptual nuggets, the meaning of this and that—something that is often associated (sometimes scornfully) with a philosophical spirit or mindset. In this most general sense, the subject is certainly deeply philosophical. But one can, of course, be philosophically minded without having any knowledge of philosophy as practiced as a proper academic discipline, with all its idiosyncratic codes and doctrines and methods and ways of reasoning that can feel both utterly charming and oddly foreign to anyone trying to break in from the outside.

But what exactly can academic philosophy and its practitioners do for quantum foundations? Are there services that only philosophy can provide—things that make quantum foundations more resilient at the core, things that aid us in separating the wheat from the chaff? And what's the nature of this import? Is it mainly a general mode of analytical thinking? A bag of useful tools and methods? Is it conceptual clarification, applied a posteriori to the theories devised by the scientists? Or are there genuinely new insights and discoveries to be had, fueled by a symbiosis where philosophy informs and helps create the very ideas and theories relevant to quantum foundations and to physics as a whole? Should every aspiring quantum foundationalist enroll in a double major in physics and philosophy? Or is it not science that needs help from philosophy, but rather the other way around?

Here, at last, John Wheeler's immortal words come to mind: "Maybe philosophy is too important to be left to the philosophers!"

GUIDO BACCIAGALUPPI · I have always believed that the philosophy of physics and the foundations of physics are not two distinct disciplines but two institutional labels for one single discipline (used in philosophy and physics departments, respectively). Of course, work done within philosophy might on average be less technical, but these are differences of degree.

Nevertheless, there are foundational questions that relate directly to questions encountered elsewhere in philosophy, and to which philosophers can contribute in a distinctive way. I'm thinking in particular of questions of emergence and reduction, of the nature of probability, and of the nature of mentality. These come up in various places, but they have been of special importance in recent developments within the Everett interpretation. And indeed, philosophers such as Saunders and Wallace have been crucial in working out a convincing version of Everett in recent years.

None of this, of course, precludes physicists from making contributions that are distinctly philosophical, as with David Deutsch originating the decision-theoretic approach in Everett, with Caves, Fuchs, and Schack applying subjective probabilities to quantum mechanics, or with Dieter Zeh developing the many-minds aspect of the Everett interpretation.

ČASLAV BRUKNER · My colleague and friend Marek Żukowski likes to say that philosophical propositions may be defined as those that are not observationally or experimentally falsifiable at a given moment in the development of human knowledge, or as those that are, in pure mathematical theory, not logically derivable. While I agree with him that philosophical positions are not subjects of proof, I would like to add that traditional characteristics of philosophical work—such as rigor of thought, clarity of expression, and, above all, questioning beliefs—can have great value. Philosophy can and should aid, to use Dirac's words, "in overcoming a prejudice" and in prompting alterations to theoretical ideas that have proved unfruitful in extending physical phenomenology. Does today's philosophy of quantum physics meet this demand?

I've often had the chance of observing the following situation. One of our philosophy fellows listens to a quantum-mechanical talk devoted to decoherence. After the talk, he shakes his head disparagingly and remarks, "But your argumentation is only for all practical purposes!" He seems disappointed. His objection seems strong: no matter how small the interference terms become in the process of decoherence, they are still there. So how can the question of what exists be settled by an approximation?

I think that besides being reminiscent of classical realism, such questions reflect ignorance of the simple fact that quantum theory cannot be both universal and not just "for all practical purposes" (or "FAPP," to use Bell's acronym). On the one hand, measurements must lead to irreversible facts, for otherwise the notion of measurement itself would become meaningless, as no measurement would ever be conclusive. On the other hand, this irreversibility must be only FAPP if quantum theory is in principle applicable to any system, including to the measurement apparatus itself. The key point is that the subject can, of course, be described as an object, but then only as an object for another subject. I would like to see more studies in the philosophy of physics explaining the fundamental nature of FAPP—the philosophy of

FAPP, if you like—rather than regular attempts to expel this term from the foundations of physics on the basis of a presupposed philosophical doctrine.

The conceptual grounds for such philosophical studies can be found in the works
of Bohr, von Weizsäcker, and Wheeler, and in Kant's understanding that the foundations of empirical science are based on the preconditions of experience. These authors did not take the laws discovered by us to exist as "laws of nature" in the outside
world. Rather, these laws appeared to them as necessities of the mind for making any
sense whatsoever of the data of our experience. As von Weizsäcker put it succinctly,
"Nature is earlier than man, but man is earlier than natural science."

JEFFREY BUB · One perception of this role, to which I do not subscribe, is that
physicists come up with interesting results but are not sufficiently literate as philosophers to articulate the broader significance of their discoveries for our conception of
physical reality, so the philosophers come in afterward as a sort of cleanup crew to
sanitize the messy metaphysics of the physicists. This division of labor relegates the
philosopher to a rather boring and sterile role.

I think that seminal work in the foundations of quantum mechanics originates
with individuals or groups who combine—in some measure depending on interests—physical intuition, mathematical ability, and a nose for a philosophical issue.
Take Bell's investigation that led to his inequality, for example. After showing that
various no-go proofs for hidden variables underlying the quantum statistics all de-
pended on unwarranted assumptions, Bell noted that Bohm's theory, which evades
these results, introduces "an explicit causal mechanism ... whereby the disposition of
one piece of apparatus affects the results obtained with a distant piece," so that "the
Einstein–Podolsky–Rosen paradox is resolved in a way which Einstein would have
liked least," to quote from Bell's 1966 *Reviews of Modern Physics* paper. The crucial
question he asked was whether it could be proved that "*any* hidden variable account
of quantum mechanics *must* have this extraordinary character." This question renewed the completeness issue raised by EPR—the core of the foundational debate
between Einstein and Bohr—as a serious topic in physics after being moribund for
thirty years. No doubt, physicists were originally interested because Bell's inequality
could be tested experimentally, but this led eventually to the theoretical investigation of quantum information and a rigorous reconsideration of the foundations of
quantum mechanics, both theoretically and experimentally.

So where do philosophers fit in? After Bell's result, physicists became interested
in how to exploit entanglement in information-processing tasks, that is, in what you
could do with Bell's insight. What philosophers focused on was Bell's assumptions,
just as Bell—in philosophical mode—had focused on the assumptions of prior no-
go theorems. In so doing, they kept alive issues that were initially disregarded by
physicists. For example, one of Bell's implicit assumptions was the freedom of the
observer in a Bell experiment to choose the measurement setting. Philosophers referred to this as the "no conspiracy" assumption, alluding to the absence of a conspiracy of nature in tying the choice of measurement setting to the values of the hidden
variables. Initially, there was no particular interest in this assumption, which was

regarded as noncontroversial, but as I mentioned in my answer to Question 9 (see page 183), recently Barrett and Gisin proved the intriguing result that if one party's choice of measurement setting is constrained by one bit, then the correlations can be reproduced by local resources.

So I think the role of the philosopher is to come up with new questions, or to sustain interest in questions of a fundamental nature that might otherwise be ignored by physicists, and to work on answering these sorts of questions. The questions posed and answered by EPR and Bell are examples of such questions. Of course, EPR and Bell were physicists, but in working on these questions, they were wearing their philosophical hats, so to speak. One could cite examples of philosophers who have raised and answered such questions. Maudlin, a philosopher of physics, was the first to consider the question of how much communication it would require for Alice and Bob to simulate the correlations of a Bell state for projective measurements, and he provided the first of a succession of partial answers before Toner and Bacon's one-bit result. Pitowsky's work on the geometry and complexity of correlation polytopes in his 1989 book *Quantum Probability, Quantum Logic* was crucial in bringing this topic to the attention of physicists. Simon Saunders and David Wallace, Oxford philosophers, have breathed new life into the Everett interpretation by asking and answering questions about decoherence and probability. Specker, a logician, considered a prediction game in which the player is required to come up with a noncontextual assignment of values but loses because the constraints of the game require a contextual assignment. These considerations led to the Kochen–Specker theorem and to current work on contextuality in quantum mechanics.

ARTHUR FINE · In the context of foundations, it is useful to think of philosophy under two hats. One is analytical. It is concerned with analysis of concepts, clarity of thought, and rigor of argument. Under this hat, a great deal of foundations is thoroughly philosophical, and (perhaps unwittingly) it works every day with the analytical tools that philosophical investigations have sharpened. The other hat is synthetic. Here philosophy opens up the repository of ideas embedded in different philosophical traditions and explores the connections of these ideas and traditions. Thus, ideas associated with realism, pragmatism (instrumentalism), or positivism are recognized players in quantum foundations. More recently, ideas developed in philosophical discussions of criteria of identity, decision theory, and Bayesian or subjective probability have been prominent in foundational work on Born's rule, the Everett theory, and the whole informational approach.

Unfortunately, in the physics literature "philosophy" is often used as a term of abuse, as a way of marginalizing a point of view. Usually, this is a sign that the author (or speaker) has run out of good arguments. It can also signal the view that we have come down to personal beliefs, or a conflict of values. In this case, I think more philosophy, in the analytical sense, would be useful. For personal beliefs and values themselves are not outside the realm of rational thought. Intelligence functions here as well in as elsewhere. Nor should one buy into the implicit contrast being drawn between facts and values. For what is factual is always also laden with values, and what might be presented as a value judgment will always involve a host of factual elements as well. To put it in quantum terms, facts and values are entangled. Just so are philosophy and science. No shallow relativism—"*this* is my personal belief, *that* is yours"; "*that* is philosophy, *this* is science"—does justice to how things stand as we actually lead our personal and professional lives.

CHRISTOPHER FUCHS · If you catch me on a bad day, I'd say "no role." But that's on a bad day; the truth is my troubles are much more narrow, and I shouldn't portray them otherwise. The real culprit is that a large fraction of the philosophers of science who work on quantum foundations have never seemed to me to bring much to the table that might help move us forward. Except for their willingness to engage in foundational discussions in a way most physicists will not, they almost represent an impediment. There is no doubt that my stance would not be what it is today if I had not had a sustained interaction with this community, but their role has always been a negative and resistive one; what I have gotten out of the deal is that it has been a kind of whetstone for sharpening my presentations, not my substance. I'd rather say that I've learned something directly from them—that my eyes were opened by this or that consideration that only a philosopher would see naturally—but it hasn't been so.

One trouble is that they advertise their role as one of checking the consistency and logic of what physics presents to them—checking the plumbing, as Allen Stairs says—but it has been my experience that it is most often a game they use in the service of *their own* prejudices. The manipulations of logic work just as well on false values as they do on truth values. What logic cannot reveal are prejudices, predispositions, and assumptions. If you read my answers to Question 2 (see page 45) and Question 3 (see page 70), you'll know some of the prejudices and predispositions I mean.

On the other hand, I have been affected very deeply by some dead philosophers of a certain strain, ones who knew not a thing of quantum mechanics. These are the turn-of-the-century American Pragmatists: William James, John Dewey, Ferdinand Schiller, and some of their disciples. A more modern-day pragmatist for who I have significant sympathy for parts of his thought (though he is dead now too) is Richard Rorty.

The role these guys have played in my life is that they give me examples of what the world would be like if it were thought of in terms antithetical to a block-universe conception of things. I then go to the quantum formalism and ask myself, "Can I see something similar there?" When I can, I further ask myself, "Can I expose the essential point more convincingly than they ever could with the aid of this formalism?" It has been a great technique for me and has carried us really very far down the technical path of QBism.

The story of how this technique came about is worth telling—for the relationship between the pragmatists and me is really very accidental. In July 1999 I gave a talk at Cambridge University on our then freshly proven quantum de Finetti theorem (a purely technical result in Quantum Bayesianism). At the end of the talk, in the question-and-answer session, Matthew Donald boomed out from the back of the audience, "You're an American Pragmatist!" Well, I didn't know what he meant by that, and I didn't get a chance to talk to him afterward (until two years later). But the thought hounded me from time to time, "What did he mean that I'm an American Pragmatist?" As luck would turn out, I ran across a copy of Martin Gardner's book *The Whys of a Philosophical Scrivener* at a hospital charity sale a month before that later meeting with Donald and bought it for fifty cents. I did so because it contained an essay titled, "Why I Am Not a Pragmatist," and I read it in the car while my wife did some more browsing. In that little half hour, it was like a flash of enlightenment! Every time Gardner would give a reason for eschewing a "linguistic preference" of the pragmatists, I would find myself thinking, "Well, you just don't understand quantum mechanics." By the end of the article, the adrenaline was surging through my body, "I am an American Pragmatist!"

Now there are nearly seven hundred books on the subject sitting on my bookshelves at home and in my mind, and if you were to ask me on a good day, I would say, "Philosophy can indeed play *quite* some role in advancing our understanding of quantum mechanics."

GianCarlo Ghirardi · Let me answer this question by resorting to an illuminating historical example. At the time the EPR paper appeared, it was judged, almost unanimously, as exclusively dictated by Einstein's philosophical prejudices about standard quantum mechanics. Here, for instance, are the words of Pauli, made in direct reference to the paper:

> As O. [Otto] Stern said recently, one should no more rack one's brain about the problem of whether something one cannot know anything about exists all the same, than about the ancient question of how many angels are able to sit on the point of a needle. But it seems to me that Einstein's questions are ultimately always of this kind.

Similarly, Abraham Pais wrote—in 1982!—about the EPR paper:

> The only part of this article that will ultimately survive, I believe, is this last phrase, which so poignantly summarizes Einstein's views on quantum mechanics in his later years. ... [Einstein's] conclusion has not affected subsequent developments in physics, and it is doubtful that it ever will.

I invite lucid readers to compare these statements with the fact that—besides Bell's discovery of the nonlocal nature of natural processes, which stems directly from a critical consideration of Einstein's conclusions—all recent and promising technological developments, such as quantum teleportation, quantum cryptography, and the advances in the field of quantum computation, are fundamentally based on the consideration of an EPR-like situation.

Shelly Goldstein · The project of physics, or at least one of its projects, is to understand physical reality at the deepest, most fundamental level, and to formulate theories, expressed in mathematical terms, that embody that understanding. Almost everyone would agree that mathematics plays an important role in this project. But so does philosophy.

Philosophy is important for the foundations of quantum mechanics not because the foundations of quantum mechanics is what you are doing when you go beyond physics, to questions of interpretation that transcend physics. Rather, philosophy is important for the foundations of quantum mechanics precisely because the foundations of quantum mechanics—finding a precise version of quantum mechanics, a precise physical theory that accounts for quantum phenomena—is simply physics.

And I doubt that there can be a clean separation between doing philosophy and being engaged in physics at a fundamental level. Physics of this sort often will involve the replacement of familiar concepts—for example, simultaneity or absolute time—with new concepts. Until the proposal of their replacement, the familiar concepts may have been regarded as essential for physics and physical understanding. If the new theory involving the new concepts is ultimately to be successful, the proponents of the new theory will have to explain how it is that, appearances to the contrary notwithstanding, the new concepts can accomplish what almost everyone had believed required the old. This sort of activity involves physics at its best, and it is obviously also essentially philosophical.

Physical theories often involve statements about probability, either in their formulation or in their implications. Now, while probability theory is clear as mathematics, the meaning of real-world probability is not at all clear. Philosophers have, of course, addressed the question of the meaning of probability. Physicists have as well—as they must if they are to know what they mean when they use the notion.

DANIEL GREENBERGER · It is true that most physicists have a positive disdain for philosophy. Eugene Wigner, a rather philosophical physicist, nonetheless repeats the wonderful quote that "philosophy is the systematic misuse of a terminology that was invented for just that purpose!" But like the fellow who discovers that he has been using prose all his life, I think that everyone, whether they know it or not, is guided by a philosophy. But philosophy has had an overt effect on quantum physicists in general, in that a particular philosophy has largely taken over the field, probably mostly for the worse: the philosophy of the strong positivists. Most physicists believe that one must strive to talk about what can be measured, and to avoid concepts that refer to unmeasurable things. And they tend to come down hard on concepts that are clearly and overtly unmeasurable, which they tend to dismiss as nonobjective.

And yet many of the everyday concepts of quantum physics refer to unmeasurable quantities, such as the wave function, the wavelength, the phase, and so on. I don't think we are hypocrites; we just don't tend to think about these things. The most damaging aspect of all this is that if one is applying for a grant, the mentioning of the philosophical aspect of something is the kiss of death on the application. Might as well hope to win the lottery. I think that this attitude does little credit to us as a profession, but it isn't about to change soon. We are still in the "don't ask, don't tell" stage when it comes to philosophy.

Nonetheless, many of the really great physicists were philosophically minded, and the early great debates were strongly colored by it.

LUCIEN HARDY · The divide between physics and philosophy in the foundations of quantum theory is not a sharp one. Certainly, there is a great deal of overlap in terms of the kind of quantum-foundational research people in physics and philosophy departments do. In part, this is probably because the job situation for people interested in quantum foundations in physics departments has in the past been rather difficult (the situation is now much improved), and consequently many people from physics backgrounds have gone on to be employed in philosophy departments. A typical conference in quantum foundations will, generally, have a healthy mix of physicists and philosophers (along with a few mathematicians and computer scientists as well).

Question 14 asks about the role of philosophy, however, rather than the role of philosophers. I think that philosophy is crucial for the foundations of physics, but it sometimes brings with it a dangerous attitude we have to guard against. It is a common insult for a physicist to describe something as "mere philosophy." But it is philosophical questions such as, "What is reality like?" that drive work in quantum foundations and, perhaps less explicitly, much of the rest of physics. All physicists interested in fundamental questions are, in truth, motivated by philosophical questions. The danger posed by philosophy to physics is that there is an attitude that the task of academic philosophy is to provide interpretation of existing theories, rather than to be engaged in the process of discovering new physics. This attitude is bad for philosophy, but it is really bad for physics.

ANTHONY LEGGETT · As elsewhere in the foundations of physics, I would regard the main role of philosophy—and particularly of "philosophy of physics" as practiced, for instance, by the Oxford school—as being simply to keep the bulk of the practitioners (mostly professional physicists without any substantial philosophical background) intellectually honest. I am probably unusual in having gone into physics only after an initial training in philosophy. I would regard the dominant effect of this training as being a tendency to scrutinize harmless-sounding phrases to see if they are really meaningful (compare my answer to Question 12, page 232).

TIM MAUDLIN · "Philosophy" in this context should not be considered a body of doctrine, but rather an intellectual attitude. Philosophy demands clarity of expression and explicit formulation of arguments and theories. One could say that it takes a philosophical temperament to appreciate the clarity of the EPR argument and the obscurity of Bohr's response to that argument, for example. In this sense, Einstein, Schrödinger, and Bell are philosophical, while physicists who are content with the fundamental unclarity of the Copenhagen interpretation (or with the uncurious attitude of "shut up and calculate") are not. But although the philosophical attitude is essential for advancing our understanding—because it is the attitude that seeks understanding in the first place—the doctrines of professional philosophers have little to contribute.

DAVID MERMIN · If quantum mechanics is correct, or even if it is only correct to a high degree of accuracy in some yet-to-be-delimited domain, then everything in quantum foundations counts as philosophy. Let me rephrase the question: what role have professional philosophers played in advancing our understanding of the foundations of quantum mechanics? I do not count as "philosophers" professional philosophers who are also professional physicists, and I count as "professional" anybody with a Ph.D. in the field.

When I got into this business thirty years ago, I had hoped that philosophers would bring to the conversation their historical expertise in the Big Questions. What is the nature of human knowledge? How do people construct a model of the world external to themselves? How does our mental organization limit our ability to picture phenomena? How does our need to communicate with each other constrain the kinds of science we can develop? Those kinds of questions.

To my disappointment, it seems to me that professional philosophers prefer to behave as amateur physicists. They don't try to view the formalism as part of a Bigger Picture. On the contrary, they seem to prefer to interpret it more literally and less imaginatively than many professional physicists. Because they are less proficient than physicists in using the tools of physics, they tend not to do as good a job on these narrower matters. They often come through as naive and unsophisticated.

So I would say that up to now, professional philosophers have not played a significant role in advancing our understanding of quantum foundations. I would not (and could not) discourage them from working in quantum foundations. But I would urge them to keep their eyes on the Big Questions.

See also my answer to Question 15, page 268.

LEE SMOLIN · The questions in the foundations of quantum mechanics are philosophical as well as scientific. They rest on the most basic questions asked by philosophers: what is space, what is time, what is matter, what is the relationship of knowledge to the world? Yet if a novel perspective on these questions emerges from efforts to solve the foundational problems, it will certainly have implications for experiment.

Philosophical preconceptions are also behind the different strategies that people take toward solving the foundational problems. This is to say that to be successful at solving them, we will likely have to weaken the disciplinary boundaries between science and philosophy and revive the tradition of natural philosophy, by which experimental questions are approached with the clear thinking that a grounding in the philosophical tradition brings.

ANTONY VALENTINI · I would say that our thinking about quantum physics became muddled in the 1920s, under the influence of certain incorrect philosophical ideas that were fashionable in some circles at the time. I think physicists need to unlearn some of those wrong ideas in order to return to clear scientific thinking about quantum theory. In this, some exposure to analytical philosophy can be helpful. For example, any graduate student in the subject knows that, if anything, indeterminism makes free will even harder to explain, since our actions would be occurring for no reason, and yet one often hears physicists citing our apparent free will as a reason for abandoning determinism.

On the other hand, if people in quantum foundations would start thinking like other physicists and scientists do—in terms of an objective reality that we need to discover—there would be little need for philosophy. For example, spectroscopic analysis enables us to deduce the chemical composition of the stars, despite Auguste Comte's infamous claim in the nineteenth century that this would never be possible. Yet astrophysicists don't need to study philosophy. Similarly, biologists and geologists have deduced that certain events occurred on earth millions of years ago, without worrying about philosophical questions.

I've already mentioned, in my answer to Question 13 (see page 244), the unfortunate effects of German idealism on scientific thought. That philosophical movement was deeply influential in Denmark, as well as in the German-speaking world. It was sparked off by the publication in 1781 of Kant's *Critique of Pure Reason*, a work that was widely interpreted as having undone the Copernican revolution and having restored human beings to a central place in the order of things, a claim that was later repeated in 1927 with the rise of quantum mechanics. Kant, like many others of his time, believed that Newton had discovered the true physics. On the other hand, David Hume argued that certain knowledge about the world was impossible to acquire. Kant was faced with a paradox. How had Newton done it? Kant's ingenious answer, as least as widely interpreted, was that Newtonian physics reflected the structure of human thought, not the structure of the world, and that the world "in itself" was unknowable.

Now, a lot happened between 1781 and 1927. But if we ask why Bohr and Heisenberg took seriously the absurd idea that experiments *must* be described in terms of classical physics—a claim that is easily refuted by describing experiments in terms of, for example, de Broglie's nonclassical pilot-wave dynamics of 1927—then the answer is that they took seriously the bizarre Kantian claim that classical physics is essential to the structure of human thought. In order to clear away such wrong-headed ideas, it helps to know where they came from and why they were proposed. For this purpose, a knowledge of the history of philosophy can be helpful.

Another wrong idea that needs clearing away is operationalism. This has roots in the philosophy of Ernst Mach and entered physics with the publication of Einstein's first relativity paper in 1905. For the first time, the human observer seemed to play a central role in physics. More substantially, as I've already said in my answer to Question 10 (see page 210), Einstein made the fatal mistake of treating macroscopic equipment—in Einstein's case, rods and clocks—as if they were fundamental objects, when in fact they are emergent and approximate objects built out of more elementary things such as particles and fields. The widespread respect for Einstein's approach seemed to justify Bohr's subsequent belief that classical apparatus played a fundamental role in quantum theory, even though any apparatus is built out of nonclassical atoms.

Another mistake propagated by Einstein's first paper—at least as it was widely interpreted—is the idea that physics should be based only on what is observable, when in fact, as Einstein later explained to Heisenberg in a private conversation in 1926, some body of theory is required before we can even know how to make reliable observations. What we can observe is determined by the theory, not the other way round. To correct these kinds of mistakes, analytical philosophy can be helpful.

But at the end of the day, what is needed, essentially, is clear and objective scientific thinking, of the sort exemplified by Bell—what was once normal practice, among people like Maxwell, Boltzmann, and de Broglie, for example, and what to this day remains normal practice in almost every branch of science, including physics. The essential point, shared by almost all scientists, is that there is a real world, and that it is the task of science to find out about it. In the narrow context of quantum foundations I would say that, for the most part, we need to *un*learn some bad philosophical ideas that have become associated with the subject and that scientists in other areas would never take seriously.

DAVID WALLACE · I guess I ought to be in the ideal place to answer this question, since my original training was in physics and I moved into philosophy after my doctorate. But actually, I don't have a systematic answer to give. Ultimately, you make progress with problems by applying whatever the needed techniques and tools are. Whether those tools, or the departmental affiliation of the tool-user, count as "physics" or as "philosophy" isn't that important.

That said, what a philosophy training tends to give you is not so much a body of relevant knowledge, so much as a certain way of analyzing a problem. Philosophy teaches you to be very careful, very attentive to whether the steps of your argument really do follow from one another, very concerned about what the conceptual assumptions are in your reasoning, very worried about whether ideas you're using have been properly defined.

Now, quite often the style of reasoning in physics—even theoretical physics—is a lot more freewheeling than that. There's generally an impatience to get to the point at which concrete calculations can be done, and a willingness to cut corners—mathematically and conceptually—in doing so. As a rule, the proof of the pudding is in the eating: if you've managed to calculate something accurately, you must have been doing things right.

That might sound as if I'm building up to criticize the physics way of doing things, but in general I'm not—in fact, I think philosophers sometimes both underestimate its power, and confuse lack of mathematical rigor with lack of conceptual clarity. (For example, a lot of people in philosophy of physics seriously underestimate how much conceptual progress quantum field theory has made with issues of renormalization, just because that progress doesn't lend itself to rigorous axiomatization.) But when you're trying to understand the foundational structure of a subject and not just do calculations with it, the philosophy style of reasoning can be a useful complement to the physics style.

(I'm generalizing, of course. Plenty of physicists—Einstein, most famously—can reason in both styles, according to what's needed at the time. But I think it's often the case that some formal philosophy training can help develop a more conceptually careful style.)

That's not to say that there aren't places in foundations of physics where philosophical knowledge, and not just philosophical technique, can come in handy. In particular, physicists—or some physicists, at any rate—can end up saying very silly things about some philosophical issues. Free will and the problem of consciousness are pretty key examples: these are topics that have been thought about for a very long time, and while there's not a consensus on the *right* way to think about them either, there's a lot that's been learned about superficially plausible but actually plain *wrong* ways of thinking about them. So sometimes, unfortunately, you get situations where someone claims that quantum theory has profound implications for (say) freedom of will, where actually they're just working with a philosophically naive and uninteresting notion of free will. You get the same problem in discussions of operationalism in quantum mechanics (to a lesser extent, though: physicists tend to be better informed on these closer-to-home topics).

As always when work gets interdisciplinary, the solution is to find a cooperative colleague in the other discipline and talk to them. That happens less in physics–philosophy interdisciplinary work than in, say, the physics–biology case. I think that's partly because philosophy has a bit of a bad reputation among scientists, and partly because—I'm sorry to say—that bad reputation is often deserved: too many philosophers end up saying really silly things about science in general and physics in particular, because they haven't done their homework and haven't consulted a colleague. But that's not true for everyone in philosophy, any more than it's true that everyone in physics is ignorant of relevant philosophical ideas.

ANTON ZEILINGER · I am convinced that any really fundamental progress has to involve a very careful analysis of the basic notions we use. This reminds me a lot of the situation in relativity theory, where Ernst Mach's analysis of the notion of space and time was crucial for Albert Einstein. It is to be noted that most of the founders of modern quantum mechanics were very knowledgeable in philosophy. This has been lost in the second half of the twentieth century, and I am convinced that it has to be regained.

WOJCIECH ZUREK · Philosophers often ask good, incisive questions. They have to be taken seriously. But more often than not, these questions are posed in everyday language. As I have said earlier, I do not trust everyday language in discussions of quantum foundations. Nevertheless, philosophically motivated questions can (and have) been an inspiration for questions that one can then formulate more precisely using the quantum formalism.

UNIFICATION

———

What new input and perspectives for the foundations of quantum
mechanics may come from the interplay between quantum theory
and gravity/relativity, and from the search for a unified theory?

T HE GRAND PROJECT of quantum foundations is caught in a dilemma. On
the one hand, everybody keeps singing the praises of quantum mechanics:
how there's never been a physical theory whose strange predictions have been
so stunningly verified by experiment. Which means that we're usually inclined, at
least in our daily dealings, to think of quantum mechanics as some kind of finality,
worth our undivided attention—much in the spirit, by the way, of Born and Heisenberg's declaration, at the 1927 Solvay meeting, that quantum mechanics is a "closed
theory, whose fundamental physical and mathematical assumptions are no longer
susceptible of any modification."

On the other hand, we all know too well that quantum mechanics isn't very comfortable around our other best friend, general relativity, the theory of choice for describing gravity. To be sure, the incompatibility is not much of a problem in practice:
for the most part, quantum effects become relevant only in the realm of the microscopic and mesoscopic, while general-relativistic phenomena dominantly appear in
the cosmological regime of black holes and other supermassive objects. There's a relatively safe zone separating the quantum and general-relativistic regimes—a range
covering dozens of orders of magnitude in size and mass, a territory over which classical physics still rules like a fat, aging godfather who knows his ultimate limits but
can rest assured that his wisdom will continue to be in demand in most situations of
practical interest.

But no self-respecting physicist will want to stop here. All of us realize that something will ultimately have to give. And sure enough, there's been a beehive of activity
aimed at finding ways of squaring quantum mechanics and general relativity. The
buzzword is quantum gravity: the promise of some self-consistent über-theory from
which quantum mechanics and general relativity fall out as limiting cases at opposite
ends of the mass scale. And because scientists cannot help thinking big—and because
once they've thought big, they want to think even bigger—they have already set their
eyes on the next target, on what many see as the holy grail of physics: a fabled, and
usually unabashedly capitalized, Theory of Everything that would someday, so it is
hoped, unify all the known forces.

How does all of this affect how we rank and think about and approach problems in the foundations of quantum mechanics? What kind of new foundational questions suggest themselves? And which of the well-known questions get a makeover? Which ones get boosted to the top of the priority list? Which ones are rendered less acute, or become altogether obsolete? How, and where, can the field of quantum foundations productively intersect with ongoing efforts toward unification, and what are the sort of promising future research directions that come up once we look beyond the confines of quantum theory per se? Are there concrete problems and questions that can, and should, be worked on immediately, or do we need to sit idle and wait around until some day the final theory is put into our lap?

And furthermore, what are modern-day practitioners of quantum foundations to do? Can they meaningfully continue to focus on quantum mechanics alone? Or would they, in doing so, run the danger of wasting time and mental sanity in trying to read too much into a theory whose fundamental status might someday evaporate once the larger endeavor of unification reduces the formalism to a mere approximation? Must all quantum foundationalists worth their name become literate and skilled in the work done on the quantum-gravity and unification fronts—and perhaps even eagerly throw their hats in the ring and make themselves available as active contributors to these research areas?

GUIDO BACCIAGALUPPI · I'm not sure I can say much that is definite on this. It could be that a quantum theory of gravity (or unified theory) is sufficiently different from standard quantum mechanics that it would dissolve traditional puzzles like the measurement problem. I remember a paper by Joy Christian from perhaps fifteen years ago, in which he was able to quantize Newton–Cartan theory and obtain a superselection rule for different space-time geometries (or sufficiently different ones—I cannot remember exactly). Alternatively, it could be that some foundational approach to quantum mechanics provides key insights for arriving at a quantum theory of gravity. Or again, a quantum theory of gravity might suffer from the same foundational problems as standard quantum mechanics, as seems to be the case, by and large, in quantum field theory. Anything is conceivable.

ČASLAV BRUKNER · In general relativity, the space-time metric is not a fixed background structure but rather a dynamical physical entity. This suggests that the conceptual and structural framework of quantum theory should be applicable also to space and time, provided quantum theory applies to this domain of physical reality. But the difficulties that arise when attempting to merge quantum theory and general relativity are so large and have lasted for so long that one suspects that they are not merely technical and mathematical but rather conceptual and fundamental.

What goes wrong in these attempts? There is no easy answer to this question, and certainly no consensus among the specialists. The lack of experimental guidance has made it extremely hard to narrow down the problem. In such a situation, it is useful to remind ourselves of the pivotal role that the thought-experiment technique played in deepening the conceptual understanding of quantum theory at its early stage of development in the 1930s. Not only did these thought experiments confirm the consistency of the theory in every detail, but they also pushed research into new experimental directions that eventually led to *actual* experiments. That the culture of discussing thought experiments is largely nonexistent in the quantum-gravity literature is to be regretted and might be one of the key methodological failures that prevent a proliferation of expectations of how the theory might be used.

So perhaps a simple thought experiment is in order to illustrate how quantum gravity challenges our conventional ideas of space and time. Take a massive lump of material—like a Bose–Einstein condensate or a nanomechanical oscillator—and put it into a quantum superposition of states corresponding to two spatially separated locations. The lump is in a quantum superposition; the gravitational field is in a quantum superposition; and so is the space-time, according to a fundamental lesson of general relativity. Now take a probe particle and scatter it in the gravitational field. What is the quantum state of the particle? In nonrelativistic quantum theory, but also in relativistic particle dynamics or quantum field theory, quantum states are specified at some given "instant of time," where "time" is treated as a fixed background parameter. Yet in our thought experiment, the background metric is not well-defined, and thus it is difficult to see what "time" could mean.

The point of this simple example is that the subject of quantum gravity generates a number of conceptual problems for quantum foundations over and above those that are already present in quantum gravity in general. These problems cannot be dismissed on technical grounds by suggesting that the low-energy scattering results in our example could be correctly calculated from a weak-field perturbative expansion of the Einstein Lagrangian in the flat space-time background. At best, such a calculation is likely to yield only an approximation to the complete theory. And it could be extremely misleading, because another fundamental lesson of general relativity is that the presence of mass curves space-time. What our little example also shows is that the deficiency of a classical description of space-time arises far away from the so-called Planck scale, despite repeated claims in the literature to the contrary.

Let me conclude with a remark. The long story of the quest for a theory of quantum gravity reminds me of some physicists telling us that the end of physics is just around the corner. When I hear this, I can't help being reminded of the medieval description of heaven as a curtain on which the stars hang. I'm convinced that our contemporary concepts of space and time will appear to future generations as naive and silly as this picture looks naive and silly to us today.

JEFFREY BUB · General relativity allows the possibility of closed timelike curves (CTCs). An interesting thing to consider is quantum-information processing with circuits that implement unitary interactions between "causality respecting" qubits and qubits traveling along closed timelike curves. There are two models for such interactions: one proposed by Deutsch (D-CTCs), and another based on entanglement and postselection proposed independently by Svetlichny, and by Bennett and Schumacher, and developed further by Lloyd and coworkers (P-CTCs). It turns out that with D-CTCs you could distinguish arbitrary sets of nonorthogonal quantum states, cloning an arbitrary unknown pure state would be possible, and different mixtures of pure states represented by the same mixed state would become distinguishable. Aaronson and Watrous showed that classical and quantum computation would become equivalent, with the power of the complexity class **PSPACE**. This is the class of all decision problems solvable by a deterministic Turing machine using a polynomial amount of memory—presumably, but not yet provably, very much larger than the class **P** of problems solvable by a deterministic Turing machine in polynomial time, the class of "tractable" or "efficiently solvable" problems, or the class **NP** of problems for which a proof of a positive answer can be verified in **P**. The power of P-CTCs is less, but they are closely related to information loss from black holes. CTC interactions provide a fascinating arena to study the possible implications of a theory of quantum gravity for the foundations of quantum mechanics.

ARTHUR FINE · I can only understand this question as asking what future developments in physics might have to say about today's foundational questions. My response is to ask: how could we know, or even make reasonable conjectures, without knowing at least the limits of future developments? Of course, one might explore things speculatively, for example, like this. Suppose one thought that gravity, as in the dynamics of black holes, entailed (à la Penrose) that nature has a way of converting a pure state to a mixture. Then one might hold that today's unitary dynamics is only an approximation to an appropriate nonunitary theory. Then one might well expect a solution to the measurement problem in terms of something like dynamical collapse. On the other hand, dynamics of that kind (that is, where the evolution corresponds to a completely positive linear map on the density operators) can be represented as the reduced dynamics associated with the unitary evolution of states in an enlarged state space. Thus, a supposed resolution of the measurement problem provided by an admissible nonunitary theory can be considered as perspectival, just like the resolution suggested by decoherence today.

Instead of such speculation, and argument, I would suggest that we take Dewey's words to heart (see the quote in my answer to Question 10, page 204). With a really new theoretical advance, we should not expect to find answers to many of today's questions, for those questions depend on background assumptions that the new advance may well undo. Rather, like old soldiers, we should expect old questions to fade away, and to be replaced by new, more fruitful issues of foundational concern.

CHRISTOPHER FUCHS · Honestly, my feeling is that it's too early to answer this question in any sensible way. All I will commit is that I think the flow of the question is backward. Maybe the reverse would be better: what new perspectives on gravity will we get from thinking deeply about the foundations of quantum mechanics? Lucien Hardy sometimes says half-jokingly that he is looking for a Copenhagen interpretation of general relativity. That strikes me as being closer to the right consideration.

GIANCARLO GHIRARDI · For many years, serious attempts have been made—in particular by Philip Pearle—to develop a relativistic generalization of the GRW theory along the lines of a quantum field–theoretical model. These attempts, however, have not yet led to a satisfactory solution, due to the appearance of intractable divergences. I'm now starting to believe that a radically different approach might be called for. In this sense, I look with particular interest to proposals by Roger Penrose that aim to relate the collapse process to gravitational effects. Although Penrose has made repeated references to such a program, I should mention that he has not yet formulated an explicit model of a theory with the desired features.

SHELLY GOLDSTEIN · Who knows?

When thinking about a quantum theory of gravity, however, we tend to be pushed to seriously consider the wave function of the universe and its meaning. And this wave function has a very different feel to it than the wave functions of small subsystems used in more down-to-earth physics. This different feel can naturally lead to a different perspective. In fact, I think it leads us to want to regard the wave function as nomological, as a representation of the law governing the structure of space-time and the behavior of its fundamental occupants.

This change of perspective may utterly transform the way we think about quantum theory: from being a theory involving novel entities—some (if not all) of which are wave functions, and with the behavior of which the theory sometimes seems primarily concerned—to being a theory of a different sort, for which the wave function is not a fundamental entity at all.

But for such a change of perspective to be successful, the wave function of the universe must have features compatible with its being regarded as nomological. As part of a Bohmian theory, the equations of motion that it generates—governing the behavior of the occupants of the universe—would have to be sufficiently simple. Whether this can be so depends, of course, on the choice of wave function of the universe. But it depends as well as on the theory of which it is a part. Perhaps if that theory is sufficiently unified—has sufficient symmetry and structural integrity—it will support a wave function of the universe that is clearly nomological.

DANIEL GREENBERGER · There are people who believe that gravity will help solve the measurement problem. I am skeptical of this, but I won't comment on it. My own belief is that relativity and gravity have extremely important things to say about quantum mechanics, but that their role is much more important than in measurement.

If one takes a wave packet representing a free particle, one sees that if one follows its phase at the center of mass of the packet, one gets, if we use $p = m_0 v \gamma$, where $\gamma = \left(1 - v^2/c^2\right)^{-1/2}$, $E = m_0 c^2 \gamma$, and along the center of the wave packet $r = vt$,

$$\exp i\varphi = \exp\left[i\left(p \cdot r - Et\right)/\hbar\right] = \exp\left[im_0\left(\nu\gamma \cdot \nu t - c^2\gamma t\right)/\hbar\right]$$
$$= \exp\left[-im_0 c^2 \gamma \left(1 - \nu^2/c^2\right) t/\hbar\right] = \exp\left[-im_0 c^2 \tau/\hbar\right],$$

where $\tau = \left(1 - v^2/c^2\right)^{1/2} t$. So the phase of the wave measures the passage of proper time, τ, along the path of the particle. Even in the nonrelativistic limit this is true, and so the proper time leaves a residue in this limit that shows up in many problems, even though the concept of proper time is not recognized in nonrelativistic physics. (One particular example is when one makes a nonrelativistic boost to a system moving with constant velocity, called a Galilean transformation. If one later makes a boost back into the original system, there is a mysterious phase factor that shows up. This phase factor is just the residue of the proper-time difference between the two systems, the accelerated one and the original one—it is the nonrelativistic residue of the twin paradox, which isn't recognized as such nonrelativistically but cannot suddenly disappear, since it causes a real phase shift.)

In classical physics, one can ignore the concept of proper time without paying a price for it. But quantum-mechanically it shows up as a phase, which has many manifestations in the nonrelativistic limit, and in my opinion one should incorporate the concept into nonrelativistic quantum theory.

Similar problems arise with the concept of mass. The problem here is that mass plays no dynamical role in the formalism of quantum theory; it is included as a passive parameter. But if two particles interact, their binding energy changes, and this should automatically show up as a change in the mass of the system. One can put this in by hand, but mass really should appear as an operator in the theory. I suspect that it is actually inconsistent to treat it as a mere parameter. For example, when we create a particle, we give it a definite mass state. But actually, this is EPR thinking (the kind of thinking that led to the Einstein–Bohr debate, which was not resolved until the Bell theorems). The result of that debate was our realization that a particle does not have a definite spin state until we measure it, but we claim it has a definite mass state from the moment it was created. Isn't there an inconsistency there? I suspect there is a Bell theorem for this situation.

I think this is related to our problems with gravity. If you put a particle into a really strong external gravitational field, it should behave like any other particle in that field and its mass should drop out, according to the weak equivalence principle. But quantum mechanically, that doesn't happen. We give the particle a mass at the beginning, and the mass enters the calculation, and this violates weak equivalence. So I think that there are parts of quantum theory that we do not understand at a very simple level. And to use our makeshift solutions, which are good enough for some of the simple things we do but probably conceptually wrong—to try to use these incomplete concepts that we have, to try to solve really difficult problems like quantum gravity—seems to me like putting the cart before the horse.

LUCIEN HARDY · I think that people working in quantum foundations need to *wake up*! The issue of quantum gravity is completely central to our subject and affords us a great opportunity to apply the methods we have developed over the past decades.

The problem of quantum gravity is to find a theory that reduces, in appropriate limits, to quantum theory on the one hand and to general relativity on the other. Of course, we also require that the theory is consistent with subsequent experiments to test it. We are very fortunate to live in a time when such a juicy problem remains open.

We have, with the problem of quantum gravity, a clash between two worlds—two entirely different ways of formulating a physical theory. Quantum theory represents a radical departure from the physics that went before it, in that it is inherently probabilistic. There is no way to formulate standard quantum theory without reference to probabilities. But it is deeply conservative in that it operates on a fixed causal structure. In particular, we need a background time to evolve the state. General relativity, on the other hand, is conservative in that it is deterministic, but it represents a radical departure from earlier physics, in that it has nonfixed causal structure. Space-time curves in response to the presence of matter. Although general relativity does not challenge our classical notions of reality as deeply as quantum theory, we should not underestimate just how radical a departure from Newtonian physics it is.

Now, when we combine quantum theory and general relativity, we expect to get a theory that is radical in both respects. In fact, we expect to get a theory that is even more radical still. This is because quantities that are not fixed are, in quantum theory, subject to fundamental indefiniteness. (In quantum theory this corresponds to a linear superposition of terms, but in a theory of quantum gravity it may correspond to something else.) Since causal structure is not fixed in general relativity, we therefore expect to have indefinite causal structure in a theory of quantum gravity. There will be situations in which there is no matter of fact as to whether two events are spacelike or timelike separated, unless an actual measurement is made—for example, by trying to send a particle between them. This is inconsistent with the notion of an evolving state, since we will not be able to foliate space-time into a sequence of spacelike hypersurfaces. Hence, the usual formalism of quantum theory is undermined. Indefinite causal structure is also inconsistent with local field equations, since then we cannot define a local region around an event in the usual way. Hence, the usual formalism of general relativity is undermined.

The challenge of finding a way to deal with indefinite causal structure is likely to impact on the two deepest problems of quantum foundations: the measurement problem and the issue of nonlocality (Bell's theorem). I think that a theory of quantum gravity will look very different from both quantum theory (as formulated in terms of Hilbert space) and general relativity (as formulated in terms of tensor densities).

My work in this direction has been to construct a general operational framework for probabilistic theories having indefinite causal structure. I call this the causaloid framework. It is possible to formulate quantum theory in this framework, and I am currently working on enhancing the framework to allow me to formulate probabilistic general relativity. Once I have formulated these two theories, I hope that the insights provided will suggest a way to formulate a theory of quantum gravity in the same framework.

The more general goal of finding a unified theory—one that accommodates all the fundamental forces—is, of course, a worthy one. It might be the case that the search for a theory of quantum gravity will naturally do this anyway. But my feeling is that it is better to concentrate on the conceptual clash between quantum theory and general relativity for the time being, because it is in the midst of a clash of this nature that foundational thinking is most likely to be useful.

ANTHONY LEGGETT · I suspect none. The most worrying paradoxes in the foundations of quantum mechanics come not from the realm of the ultrasmall, but precisely from the opposite end—that is, from the direction of our own experience. And while some people (for example, Steve Adler) have proposed that there may be a deep connection between the two realms, I would personally bet against it. If there should be new input, I suspect it will have to do with a reorientation of our views concerning the arrow of time (see my answer to Question 2, page 268).

TIM MAUDLIN · The clearest way for a physical theory to make contact with observation is through its predictions concerning local beables: the disposition of local matter in space-time. These are the sorts of facts that we take ourselves to have firm access to on a macroscopic scale. This basic picture employs a division of the physical world into a spatiotemporal "stage" and the material "players." The stage might be fixed and unchanging, as in Newton's theory, or dynamically affected by the matter, as in general relativity. So far, the quantum treatment of matter and the general-relativistic treatment of space-time have not been reconciled in one foundational package. Finding a way forward puts pressure on the foundations of both theories.

There are many fascinating questions here. One is whether the nonlocality of quantum theory can be reconciled with a fundamentally relativistic account of space-time. One can also ask whether the deep relationship between gravity and space-time structure postulated by general relativity can survive the attempt to treat gravity with the same theoretical tools that are applied to the other forces. And most fundamentally, we must ask whether the distinction between spatiotemporal structure and "matter" can even be drawn at a foundational level.

The interesting thing is that any answer to these questions will be surprising. If gravity is unlike the other forces because of its connection with space-time, then the attempt to model a theory of gravity along the lines of the theories of electromagnetism and the weak and strong nuclear forces may be misplaced. But if gravity isn't special in this way, then the apparent central insight of general relativity is lost. And if the distinction between the spatiotemporal and the material breaks down, then we need an entirely new framework of physical structure.

DAVID MERMIN · My guess is that an understanding of the connection between gravity and quantum mechanics will have to await new input and perspectives from the foundations of both disciplines. Space and time in quantum field theory are classical parameters. They're on *our* side of the subject–object boundary. Extrapolating them down to sub-nucleon levels—let alone to the Planck scale—strikes me as un-warranted and even arrogant. (I note with interest a hint of some personal values here. Compare my answer to Question 13, page 243.) Spatial and temporal coordinates describe the readings of our instruments.

I'm just as skeptical about quantum cosmologists applying quantum mechanics to the universe as a whole. For quantum mechanics to make sense, there has to be an inside ("the system") and an outside ("us").

So insofar as gravity is a theory of the structure of space-time, I'd be surprised if real progress were made in incorporating it into quantum theory without a more thoughtful and (dare I say it?) *philosophical* examination of the foundations of both fields.

LEE SMOLIN · I have always believed that the problems in the foundations of quantum mechanics would play a necessary role in resolving the problem of the relationship between quantum mechanics and the dynamics of space-time. I went into quantum gravity in the hope that the search for a theory of quantum gravity would fail, and that in that failure we would see where quantum mechanics had to be superseded or modified. I think we have reached this point and the key issue is the role of time. I don't think a complete unification of quantum theory and space-time physics is possible in a cosmological setting without a framework in which there is a real global time.

ANTONY VALENTINI · There has, of course, already been a lot of input. Work in quantum gravity often has a cosmological setting, where in the very early universe the lack of an external classical background makes textbook quantum theory inadequate. This was, historically, one of the motivations for the Everett interpretation. Today, according to inflationary cosmology, the remnants of primordial quantum fluctuations are imprinted on the cosmic microwave background, and a proper understanding of the quantum-to-classical transition during the inflationary era again forces us to think beyond the textbooks. In the context of a theory like inflation, which is currently being tested experimentally, the Copenhagen interpretation can't be taken seriously. Quantum foundations needs to catch up with what has been going on elsewhere in fundamental physics and cosmology. How can we, for example, return to something like "operational quantum theory," which relies on a classical background containing macroscopic apparatus, when there is an experimental need to discuss a quantum-to-classical transition that took place in the earliest moments after the big bang?

I'd also like to point out that there is currently a great opportunity to use cosmology as a testing ground for quantum theory under new and extreme conditions, namely, at very short distances and very high energies. Inflationary cosmology, in particular, is being used as a laboratory to test almost every modification of high-energy physics that theorists are able to think of, and yet hardly anyone is using it to test quantum theory itself. A handful of people, such as Daniel Sudarsky, have considered how to use it to test collapse theories, and I have studied how to use it to test for quantum nonequilibrium in the early universe. But there is a vast amount of further work that could and should be done.

On the subject of gravity proper, there is the puzzle of black-hole information loss, and the alarming possibility that a closed system can evolve from a pure to a mixed state. This problem has fueled an immense amount of work in high-energy physics and string theory, where it is hoped that ideas like the AdS/CFT correspondence will provide a fundamentally unitary description of black-hole formation and evaporation. This is an important problem in quantum foundations, and I've speculated, in a hidden-variables context, that Hawking radiation may consist of nonequilibrium particles that violate the Born rule—where such states can carry more information than conventional quantum states. That's a line of thought I hope to develop further.

The nonlocality of de Broglie–Bohm theory, and of hidden-variables theories generally, points to the existence of an absolute time. This might help with solving the notorious "problem of time" in quantum gravity. But already at the level of standard quantum gauge theories, in Minkowski space-time, there is a tension with manifest Lorentz covariance, which requires the introduction of bosonic "ghost" states with negative norm. I've always thought that the simplicity of noncovariant and ghost-free gauges, such as the temporal gauge, in theories such as quantum chromodynamics, points to the existence of an underlying preferred state of rest, and I find the pilot-wave version of gauge field theory—at least as I formulate it, with three-vector gauge fields on an Aristotelian space-time—to be more elegant than the standard version. I suspect that this line of thought may be worth developing further.

I find it interesting that the AdS/CFT correspondence might be interpreted as saying that physics is really based on a Yang–Mills gauge theory on flat space-time. It would be straightforward to make a de Broglie–Bohm version of the latter, and one has to wonder how the underlying preferred frame would relate to the emergent gravitational description.

Finally, it wouldn't be surprising if one particular interpretation of quantum theory proved to be crucial in developing a unified theory. But we won't find out for as long as quantum foundations remains so removed from the rest of physics.

DAVID WALLACE · If I knew that, I think I'd probably be most of the way to having that unified theory myself!

Seriously, I think it's very interesting how *little* modification either string theory (as the current leading candidate for a quantum theory of gravity) or loop quantum gravity (as the current runner-up) make to the basic conceptual structure of quantum theory. In both cases, we basically hold on to unitary dynamics, transition amplitudes, Hilbert spaces, and the like.

I don't think we should be terribly surprised by that. Most of the great advances in theoretical physics come from a kind of radical conservatism: we try to push the basic principles of our extant theories as far as we can and see where that leads us. String theory and loop quantum gravity adopt that kind of conservatism toward quantum mechanics. (One too-glib way of characterizing the difference between them is that string theory also adopts it toward particle physics, and loop quantum gravity also adopts it toward general relativity).

Whether it's sensible to bet on that strategy is going to depend pretty strongly on your take on the measurement problem. It's hard to make any sense of quantum gravity unless you understand it in an observer-independent way—that is, in Everett's way. If you think that notions of observation and measurement play an essential role in quantum theory—or if you think that quantum theory doesn't really make sense as a theory, and needs to be supplemented with hidden variables or modified to introduce a collapse of the wave function—you should probably be skeptical about mainstream quantum-gravity research. (Roger Penrose is probably the most famous example of someone who accepts this way of thinking: he sees dynamical collapse as something that we should expect to be caused by trying to create superpositions of space-time geometries.)

On the other hand, if you think Everett's approach to quantum mechanics is basically satisfactory—which I do—then we don't have any reason to expect the foundations of quantum mechanics to be particularly illuminated by the search for quantum gravity. And if string theory or loop quantum gravity turns out to be basically correct, the general structure of quantum theory won't really be modified at all by the incorporation of gravity. (The specific quantum mechanics in question, of course, will be modified a lot.)

Does that mean I'd bet on those programs succeeding? Not especially. Making progress so far ahead of the experimental data is bound to be chancy at best. But at any rate, I don't think we have much positive reason to reject their shared assumption that quantum theory continues to be applicable even in the general-relativistic regime—nor, if that shared assumption fails, much of a clue as to what will take its place.

Incidentally, this is a way in which the Everett interpretation is almost disappointing, at least compared to strategies like dynamical collapse that change the quantum formalism. If we really did expect some failure of quantum theory in the vicinity of the measurement process, that would be an amazing experimental regime to probe—hard, but way easier than quantum-gravity experiments—and might give us the experimental clues we need to make progress on quantum gravity. But the universe isn't designed for our convenience, and the fact that it would be useful for dynamical collapse to occur doesn't give us any reason to think it does occur.

ANTON ZEILINGER · Personally, I feel that the reverse is true. The process of understanding the connections between quantum mechanics and gravity on a fundamental level necessitates a deeper understanding of the foundations of quantum mechanics. In my eyes, the pictures we have of the notions of space and time are still too realistic to this date.

WOJCIECH ZUREK · The fact that nonrelativistic quantum theory "knows" about special relativity—as illustrated by, for example, the no-cloning theorem—is a hint that quantum states and space-time may have intertwined origins at some deep level, presumably deeper than relativistic field theory. Problems with quantizing gravity, as well as black-hole thermodynamics, support this suspicion.

I note that all of these subjects have "information" as a part of the underpinnings. This is consistent with my hope that whatever insights we might gain, they may be also useful in the central problem of the relation between information and existence. And vice versa: I think that starting with fundamental primitives of information acquisition and transfer may shed a new light on physics.

THE NEXT BIG BANG

Where would you put your money when it comes
to predicting the next major development in
the foundations of quantum mechanics?

POPULAR WISDOM HOLDS that fundamental change and new discoveries tend to happen precisely when they're least expected—and when everyone is vehemently denying their possibility. In this context, it has become almost a cliché to quote James Clerk Maxwell's famous statement, made during his inaugural lecture at the University of Cambridge in 1871:

> The opinion seems to have got abroad, that in a few years all the great physical constants will have been approximately estimated, and that the only occupation which will then be left to men of science will be to carry on these measurements to another place of decimals.

Whether this "opinion" Maxwell refers to also reflected his own personal sentiment is not easy to tell from his speech, as he first derides a sole focus on "careful measurement" as "out of place in the University," but reappraises it a little later on. At any rate, he certainly wanted to capture the mood of boundless scientific optimism prevalent among many of his contemporaries, at a time when classical mechanics, thermodynamics, and electrodynamics appeared to be the great consummation of the project of science.

Of course, not long after Maxwell's remark, the supposedly sturdy edifice of classical physics came crashing down. Blackbody radiation and the ultraviolet catastrophe, the photoelectric effect, the null result of the Michelson–Morley experiment, the problem of explaining the stability of atoms: physics was suddenly plunged into crisis mode. But almost as quickly, in the true spirit of a Kuhnian revolution, the predicament was miraculously and productively turned into radically new theories that took science and our picture of nature further than even the most progressive and unconventional thinkers at the time of Maxwell's address could have foreseen.

To this day, we're still struggling to comprehend the repercussions of these new theories—a situation that is, of course, responsible for the existence of books like the one you're holding in your hands. Minor conceptual and theoretical earthquakes, like those created by Bell's results and quantum information theory, have helped hustle us toward new insights and points of view. But the thirst of quantum foundationalists remains far from quenched. Sometimes it feels as if we're all holding out for some seismic event that will decisively change the course of quantum foundations, and as if, in the meantime, we're just whiling away the long afternoons by putting another coat of paint onto our personal foundational pet project.

But what form could such a profound event possibly take? And what kind of event is most likely to occur in the foreseeable future? It goes without saying that these are tough questions; to come up with answers will be little more than a shot in the dark. Still, what would science be without dreaming big and speculating wildly?

So let's turn to our interviewees for hints of what might constitute the next sweeping breakthrough in the field of quantum foundations. No matter how the story of quantum mechanics ultimately pans out, it'll be illuminating to hear some educated guesses on what the future might hold. Should one of these predictions turn out to be right on the money—and I think chances are good—whoever made it will be able to point, with a beaming told-ya-so smile, to his interview response as proof of infallible clairvoyance. And should quantum mechanics end up taking a completely unforeseen course, well, then no harm will have been done, and we may perhaps take some solace in the fact that, as Paul Milo recounts in *Your Flying Car Awaits*, the twentieth century was filled with predictions that never came true. Baby factories, anyone? Lunar vacations? Knowledge pills? Algae as the main source of human nutrition? And where's my four-day workweek?

GUIDO BACCIAGALUPPI · A lot of effort is currently going into understanding the exact nature of quantum-mechanical nonlocality (for instance, the $2\sqrt{2}$ bound in the Bell inequalities). These are also relatively new questions, so there should be plenty of scope for new discoveries. I think it's fairly likely that the next significant advance in foundations will be in this field. (Indeed, we may have seen one such breakthrough already with Wehner and Oppenheim's recent work on the trade-off between nonlocality and uncertainty.)

ČASLAV BRUKNER · In 1935 Schrödinger attempted to illustrate the bizarreness of quantum mechanics through a thought experiment in which a cat is put into a quantum superposition of alive and dead states. A cat is placed in a box, together with a vial of poison gas, a radioactive atom, and a hammer hooked to a Geiger counter. The radioactive atom has a one-half probability of decaying after one hour. If the atom decays, the Geiger counter will detect the radiation, break the vial, and kill the cat. If the atom does not decay, the vial will remain intact and the cat will be spared. After opening the box and repeating the experiment many times, in about one-half of the cases the cat is found alive, and in the other one-half dead. What happens to the cat in the box after one hour in a single run of the experiment? Our intuition says that it must be dead or alive, yet quantum theory claims that it can be in a superposition of both.

The idea of this gedankenexperiment remained a mere academic curiosity until very recently, when newest technological developments have increased the level of control over individual quantum systems to a degree that hitherto inaccessible parameter and resolution regimes can now be reached. What was widely considered bold fantasy just ten years ago is now likely to become experimental reality in the foreseeable future: the demonstration of superpositions of macroscopically distinct states, that is, of "Schrödinger kittens" if you like. Here's just one candidate for the realization of such states: nano- and micromechanical devices, which contain up to 10^{20} atoms and may even allow for superposition states where the relevant displacements—hundreds of nanometers—are of the order of the physical size of the object itself.

When it comes to predicting the next major development in the foundations of quantum mechanics, I have little doubt that these experiments—which will potentially also include the smallest living organisms, such as viruses—will play one of the key roles. For most researchers, the motivation for this experimental program is the speculation that due to collapse models and the like, the superposition principle may break down at some stage between the level of an atom and that of human consciousness. I am not very enthusiastic about such an outcome. To me, the main motivation for forging ahead with research on Schrödinger-cat states is that these states are truly new states of matter not familiar to us. Let me explain my point. Popular-science books, and even research articles in quantum foundations, are filled with statements such as, "A cat in a Schrödinger-cat state is both alive and dead at the same time." That assessment is about as wrong as could be. For example, when we talk about the spin of an electron in the state "spin up" along the x axis, $|x+\rangle = \frac{1}{\sqrt{2}}(|z+\rangle + |z-\rangle)$, we never say that the electron has "spin parallel and antiparallel to the z axis at the same time." The point is that the state $|x+\rangle$ represents a completely new "quality" that has little to do with the property of spin along the z axis; it is a state of definite spin along the x axis. Likewise, the original Schrödinger-cat state represents a "new quality" of a cat not known to us, but one that is definitely different from alive and dead. The new experiments will give us a major insight into what it is.

JEFFREY BUB · There are indications that issues of communication complexity and computational complexity are relevant to the foundations of quantum mechanics.

It is strongly believed, but not yet proved, that quantum computers are more powerful than classical computers. More precisely, we don't have a proof that **BQP**—the class of all problems that can be solved efficiently, i.e., in polynomial time, with bounded probability of error using a quantum computer—is strictly greater than **P**, or strictly greater than **BPP**, the class of problems solvable with a probabilistic Turing machine in polynomial time with bounded probability of error. Since **BQP** is known to be in **PSPACE**, such a proof would also be a demonstration of something else that is strongly believed but for which we don't yet have a proof: that **P** is distinct from **PSPACE**.

Communication complexity has to do with the amount of information that must be transmitted between two parties to compute some distributed Boolean function f: $\{0,1\}^n \times \{0,1\}^n \longrightarrow \{0,1\}$ of private inputs $\vec{x}, \vec{y} \in \{0,1\}^n$, where Alice possesses the $\vec{x}$ string and Bob possesses the $\vec{y}$ string. Van Dam has shown that if we lived in a world with PR-box correlations rather than the correlations of entangled quantum states, distributed computation would be trivial: any computation could be performed with just one bit of communication (no bits would violate "no signaling").

I think the next major development in the foundations of quantum mechanics will involve some new insight about complexity and physics.

ARTHUR FINE · The strongest sources of "quantumicity" seem to be entanglement and noncommutativity. There seem to be significant connections, not yet well understood, between them. For example, entanglement for bipartite systems produces joint probabilities that violate the Bell inequalities for many quadruples of observables, provided pairs defined on the same subsystem do not commute. On the other hand, if a pair of observables does not commute, then we can specify states for which the assumption of a random-variables representation with respect to those states contradicts the noncommutativity of the pair. Thus, the background from which one might derive Bell inequalities for an entangled bipartite system (i.e., treating both single and joint probabilities as random variables on a common space) is itself in conflict with noncommutativity. Currently, connections like this are being triangulated by a number of experimental and theoretical investigations. Advances there would help us understand the roots of quantumicity. More generally, then, I expect some important advances in understanding the quantum limits on classicality.

CHRISTOPHER FUCHS · I don't know "on what" I'd put my money, but I do know "on where." I'd put it on the Perimeter Institute for Theoretical Physics in Waterloo, Canada!

GIANCARLO GHIRARDI · I would definitely put my money on the fact that it will turn out that the linear nature of quantum theory does not possess universal validity. Nonlinear elements must be added to the quantum dynamics. I realize that this bet is quite risky and that it might lead to my losing my money. But something in my mind tells me that nonlinearity is the unavoidable conclusion.

Whether the violation of linearity will emerge from schemes like collapse theories, or from quantum gravity, or from wormhole theories, I do not dare to say. I'm aware of the difficulties encountered by attempts toward an adequate relativistic generalization of dynamical-reduction theories. Also, no precise quantum-gravity model breaking the linearity at the Planck-mass level (the obvious level for our purposes) has been worked out so far. And wormhole theories lead to collapse models only through a series of approximations that are mathematically not completely satisfactory. All of this makes me cautious about trying to predict which, if any, route will lead us out of the contradictory situation in which we now find ourselves with respect to our most fundamental theory.

SHELLY GOLDSTEIN · I don't know. If the question concerns some substantive big idea, how can anyone anticipate such a development? If the question refers to fashion, I really have no idea. But I do wonder whether the next major developments will be in the direction of a more rational and coherent physics, or in the opposite direction. The history of the foundations of quantum mechanics does not, it seems to me, provide strong grounds for optimism. Nonetheless, the very fact that a book such as this one is in the works should be taken as a distinctly positive development.

DANIEL GREENBERGER · Predicting tomorrow's weather is a dangerous thing to do, even though we have all sorts of instruments to record wind, pressure, humidity, clouds, approaching storms, and so on. I notice that weather forecasters, who are wrong about a third of the time, never apologize for past mistakes. Economic- and technological-innovation forecasters, who are almost never right, also don't apologize. But I am happy to apologize in advance for a projection into the future that is based purely on my own feelings. No input here from any experts. I will not predict the next major development but something even wilder. I will give you my ideas on how quantum theory will break down—but not when. At the moment, there is not the slightest indication from experiment that quantum theory will ever break down, and many physicists believe that it never will. Many even believe that we are approaching a "Theory of Everything."

I confess that I consider such thoughts to be sort of absurd. Given where we are along our evolutionary path—just a few million years into some sort of humanity, and maybe fifty thousand or so into some sort of rationality—I would guess we were near the beginning of understanding things. We don't understand anything about consciousness, which would be step number one toward understanding nature. We don't even know how to ask why anything works, the interesting question—only how it works, the dull question. And even there our tools are restricted. I suspect that in a hundred years or so, people will look back on where we are today in sort of the same way we now look back on the ancient Greeks: very bright, but hopelessly naive. So all our theories will break down, because today we don't even know what a good theory is. Time will tell us that—a long time.

So given that quantum theory will break down, and we don't know how, are there nonetheless any clues as to how it might do so? Well, back in my answer to Question 7 (see page 152), I sort of indicated how some of my thoughts are going. I think we don't understand elementary relativity and the meaning of mass very well. But let me concentrate here on one specific problem, the weak equivalence principle. Galileo noted that a particle in an external gravitational field behaves independently of its mass. That is, if you drop it from a fixed position with a definite velocity, it will follow a certain trajectory. If you drop a different particle with a different mass (say, it is twice as heavy) from the same position and give it the same initial velocity, it will follow the exact same trajectory. Einstein then noted that this means the motion of the particle depends only on its external environment and not on the particle at all. The external field determines the geometry of space, and just as a free particle in an inertial system with no forces moves in a straight line, so the particle in free fall in this external gravitational field is taking the shortest path it can in this new geometry: it is moving along a "geodesic." I think this beautiful observation should be one of the fundamental laws of nature.

But in quantum theory, this is not true. A free particle has a Compton wavelength $\lambda_{Compton} = h/mc$ (h = Planck's constant, c = the speed of light, m = the mass of the particle) and a de Broglie wavelength $\lambda_{deBroglie} = h/mv$ (h = Planck's constant again, and m = the mass again, and v = the speed of the particle). Both of these depend on the mass of the particle, and so the motion of the particle depends on its mass. For example, if a light particle m is placed in the neighborhood of a very heavy particle of mass M, the heavy particle produces a gravitational field that acts as an external field for the light particle. If one solves for the circular gravitational Bohr orbits in this field (assuming a potential of the form $V = GMm/r$), one finds that

$$r_n = \frac{n^2 \hbar^2}{GMm^2}.$$

So one sees that one could find the mass of the particle by measuring the lowest Bohr orbit ($n = 1$). The velocity also depends on the mass. This is not at all what one expects from the equivalence principle, classically, where neither of these quantities should depend on the mass. What is going wrong here?

The problem is that the quantum rules have the mass built in. So dimensional arguments tell us that r will be a function of m. In fact, $r_n = f(n\hbar/m)$, and you can see that the gravitational Bohr radius above follows this formula. The same is true for the velocity. So quantum theory by its very nature conflicts with the weak equivalence principle.

This leads to an interesting question: if the mass dependence is built in, how does it disappear in the classical limit? The answer is that if a second particle has K times the mass of the first, then in the limit of high quantum numbers, the number n must be K times greater. So there is a kind of scaling in phase space (i.e., p plotted against x), such that if the velocity is to be the same, then the momentum, $p = mv$, must be K times greater. Then we will have for the second particle $r(m_2) = f(Kn_1\hbar/Km_1) = f(n_1\hbar/m_1) = r(m_1)$. So it is the ratio of masses that drops out in the classical limit, where the states are packed close together. But for low-lying states this scaling—and equivalence—breaks down.

This whole problem occurs because there is no natural length scale in quantum theory. I would imagine that in a very strong gravitational field, equivalence would hold, and the fundamental, natural scale would exert itself. Then one would have a different quantization law based on this scale, instead of Planck's constant, and the two quantization schemes would have to reconcile themselves. But at least one can predict that the place to look for a fundamental scale and evidence of a different type of quantization is in a strong gravitational field. By the way, there is no reason to believe that this scale has anything to do with the Planck length, $r_{\text{Planck}} = g\hbar/c^3 =$ 10^{-34} cm. The Planck length is where the theory would be expected to break down if nothing new enters. But if something new does enter, like a fundamental length, all bets are off. Historically, such guesses as to where a theory will break down have always been wrong. New phenomena always enter. For example, classical physics broke down when quantum phenomena entered totally unexpectedly.

Gravitationally, there is every reason to believe something new must enter. The sizes of phenomena, based on the numbers we know, are all wrong. For example, if one asked for the gravitational Bohr radius of two neutrons orbiting each other, it comes out to be about 10^{27} cm, about the size of the universe, and clearly nonsensical. Similarly, an example I like to use is that if we knew about gravity and quantum theory but not about electricity, and if we wanted to predict what nature's fundamental velocity c^* would be, we might try $c^* = Gm^2/\hbar$, where m is the mass of the neutron. This comes out to about 10^{-30} cm/sec, an idiotic result much closer to the speed of darkness than to that of light. So I don't place much store in the Planck length as a predictor of future phenomena. For that, you have to believe that we are close to a theory of everything and that there is nothing new to be discovered. For my part, I would expect that with the discovery of dark matter and dark energy, we are much closer to seeing physics turned on its head than to seeing its mysteries solved. My ideas for the future of physics may be totally wrong, but I doubt if anyone else's are any more likely to be true.

LUCIEN HARDY · The great thing about major developments is that they usually come as a surprise. That consideration aside, here are the things I am hoping for. First, a satisfactory reconstruction of quantum theory from really compelling axioms. Second, a reconstruction of (probabilistic) general relativity from really compelling axioms. Third, a construction of quantum gravity from a set of really compelling axioms.

I hope that these three things are related, in that the reconstructions of quantum theory and general relativity aid the construction of a theory of quantum gravity, and that under appropriate limits, the axioms of the theory of quantum gravity relax to those of quantum theory on the one hand and general relativity on the other hand. I also hope that the theory of quantum gravity leads us to a natural realist interpretation.

ANTHONY LEGGETT · The experimental demonstration that quantum mechanics is not the whole truth about the physical world—though I don't know if that counts as a "development in its foundations"!

TIM MAUDLIN · The first successful complete quantum theory of gravity may push foundational research in a new direction. But I wouldn't bet the bank on it.

DAVID MERMIN · Let me put the question in a more manageable form: what was the last major development in the foundations of quantum mechanics? (It remains basically the same question, since none of the developments that follow have been broadly accepted as the most illuminating way to look at the subject.)

I would nominate for the most important recent development the application of quantum mechanics to the processing of information, starting with the invention of quantum cryptography by Bennett and Brassard in 1984, continuing with the development of quantum computation, and the fascinating efforts of Chris Fuchs to make a coherent whole out of it all. As runner-up, I would cite the study of pre- and postselected ensembles by Aharanov and his collaborators, and (perhaps—I still lack a good feeling for it) the ensuing notion of weak measurement. In third place, I would put the consistent-histories point of view, as put forth by Bob Griffiths.

What all three of these developments have in common is that they are standard quantum mechanics applied in highly nonstandard settings. In this respect, they are all conservative approaches to quantum foundations. They use the orthodox theory to answer simple questions that it had never before occurred to anybody to ask. The answers provide intriguing new perspectives on the theory.

Because the last of the three seems to have been widely ignored in the quantum-foundations community and is unrepresented among the authors of this volume, I'll say a little about it. (My old friend Pierre Hohenberg has tried valiantly to get me to take this stuff seriously. Pierre and I were in both college and graduate school together, but in all those years nobody ever warned me to stay away from him; see my answer to Question 1, page 31. Maybe somebody should have.)

Consistent historians offer an unusual fusion of collapse and no-collapse points of view. Underlying their weltanschauung is an old formula of Aharonov, Bergmann, and Lebowitz (ABL), which compactly gives the probabilities of the outcomes of a whole sequence of (von Neumann) measurements carried out at different times on a system in a given initial state. Prior to its reinterpretation by consistent historians, the ABL formula was understood to be an expression of the fact that immediately after any particular measurement, the state of the system collapses according to the standard Born rule; this postcollapse state then evolves under the unitary dynamics until the next measurement in the sequence produces another collapse. Unitary evolution, followed by measurement and collapse, followed by more unitary evolution, followed by more measurement and collapse, and so on.

Consistent historians eliminate measurement and collapse from the story by reinterpreting these probabilities to be probabilities of what I would call *actual states of being*—called histories. These histories (or, more accurately, the subset of them deemed "consistent," as noted below) have nothing to do with measurement outcomes. For consistent historians the ABL formula is thus more fundamental and broader in scope than the Born rule. The Born rule can be extracted from the consistent historians' version of ABL in some very special cases, but measurement vanishes from ABL in the general case, which according to consistent historians gives probabilities not of measurement outcomes but of actual states of being.

How can they get away with this vast extension of actuality to entities whose nonexistence lies at the very heart of conventional quantum mechanics? Easily! They do it by forbidding the extension whenever it gets you into trouble; they impose stringent consistency conditions on the probabilities appearing in any candidate for a valid history. Any history that meets these consistency conditions can describe the probabilities of an actual state of being, and not the mere outcomes of a set of piddling laboratory operations. Any history that violates the consistency conditions is utter nonsense—not a history at all, and certainly not a description of actual states of being.

As one might expect, there can be many distinct histories, all of which meet their own internal consistency conditions, although the state of being that combines the actual states of being associated with more than one of those histories need not satisfy its own internal consistency conditions. When this happens, the combination of the two actual states of being is not an actual state of being.

Rather than concluding from this that the project is dead in the water, the consistent historians elevate it to a fundamental ontological principle. Reality is multifaceted. There can be this reality or there can be that reality, and provided you refrain from combining actualities from mutually inconsistent realities, all of the incompatible realities have an equally valid claim to actuality. This tangle of mutually incompatible candidates for actuality (associated with different "frameworks") constitutes the no-collapse side of consistent histories. The collapse side lies in the fact that each of these peacefully coexisting mutually exclusive actualities is associated with what from the orthodox point of view (which consistent historians reject) would be a sequence of measurements and Born-rule collapses.

This multiplicity of incompatible realities reminds me of special relativity, where there is time in this frame of reference and time in that frame of reference, and provided only that you do not combine temporal statements valid in two different frames of reference, one set of temporal statements is as valid a description of reality as the other.

But I am disconcerted by the reluctance of some consistent historians to acknowledge the utterly radical nature of what they are proposing. The relativity of time was a pretty big pill to swallow, but the relativity of reality itself is to the relativity of time as an elephant is to a gnat. (Murray Gell-Mann, in his talk of "demon worlds," comes close to acknowledging this, yet he dismisses much less extravagant examples of quantum mysteries as so much "flapdoodle.")

LEE SMOLIN · I would put money on supporting a few people with the imagination and courage to invent new approaches to resolving the problem. I would also bet that any approach that has been on the table for more than twenty years is unlikely to be the answer. And so, as much as I like my friends who work on fifty-year-old approaches, I would encourage people who want to solve the problem to look for new ideas.

ANTONY VALENTINI · I suspect that the field of quantum foundations will develop properly only when it starts attracting people who are fully conversant with modern theoretical physics and its problems at an advanced level. There has been too much work on elementary quantum mechanics—for example, for simple entangled systems. As I said in my reply to Question 15 (see page 269), there are important theoretical problems concerning the early universe, black holes, and quantum gravity, and exploring these further might lead to something important. But if I was asked to make one guess, then if the history of physics is anything to go by, I'd say it's most likely that an experimental breakthrough will be needed to make further progress.

Some areas of theoretical physics have lost contact with physics as an experimental science, not only in the trivial sense that we don't have experimental anomalies that defy explanation with current theories, but also in the more sinister sense that some theorists grossly underestimate the crucial role that experiment has played in the historical development of our theories. Contrary to folklore, no really major advance in fundamental physics has ever occurred without guidance from experiment. Theorists like to believe that pure thought can suffice, and they often cite the example of general relativity in 1915, where textbooks and popular accounts often give the impression that the correct perihelion motion of Mercury came out of the theory as an unexpected bonus. In truth, Einstein used the observed perihelion motion to rule out his own pre-1915 theory of gravity. Once he found the field equations that gave the correct perihelion motion, only then could he be confident in the other predictions coming from those same equations. And there are plenty of other examples. Schrödinger's original wave equation was beautiful and Lorentz-invariant—and it gave the wrong energy levels for hydrogen. His nonrelativistic version seemed less elegant, but gave the right energy levels. And so on. If we look at the examples that theorists often cite, and if we examine what really happened historically (as opposed to the folklore in theoretical textbooks), we always find that experiment played a much bigger role than theorists like to believe.

So it's likely that we need an experimental clue, an empirical window. To that end, we should be trying harder to test quantum mechanics in genuinely new and extreme domains. My "prediction," for what it's worth, is that we will find an experimental breakdown of quantum theory. My guess is that quantum theory will turn out to be an equilibrium case of a broader theory based on hidden variables, and this motivates me to suggest experiments searching for nonequilibrium violations of quantum theory.

DAVID WALLACE · I wouldn't—not much of it, anyway. Major changes in a field are by their nature pretty much impossible to predict in advance, and I'm not close enough to the detailed work in quantum information and computation to predict the relevant next steps there.

That said, I might wager a small sum on our making some fairly substantial breakthrough before *too* long in how to think about quantum computation and information flow in quantum systems—something that would give us a better handle on why quantum computers seem almost-but-not-quite equivalent to their classical counterparts. That's no more than a hunch, though, and it's largely driven by the exciting progress in recent years on diagrammatic ways to think about quantum mechanics. (Never underestimate the power of a new notation!)

I'd also put quite a bit of money on our *not* finding any experimental failure of unitarity, or any other evidence that quantum theory breaks down (anywhere outside the general-relativistic regime, at any rate). Given the coherence of the Everett interpretation as a solution to the measurement problem, and given the problems with relativity and with field theory involved in changing the quantum formalism, I strongly suspect that unitarity, and the universality of the superposition principle, are here to stay.

Come to think of it, though, it probably makes sense for me to hedge, and to put my money on finding violations of unitarity after all. If my preferred approach to quantum mechanics were to be empirically falsified, at least I'd be rich.

ANTON ZEILINGER · Here I have two answers.

Experimentally speaking, I am convinced that the next major development in the foundations of quantum mechanics will come from investigating phenomena in higher-dimensional Hilbert spaces. Just remember the fact that whenever we increased the dimension of Hilbert space—either for a single particle, or by adding more particles—we discovered something fundamentally new. For example, by going from two-state systems to three-state systems, we discovered the Kochen–Specker theorem. Going from one particle to two particles, Bell's theorem arose, and going from two particles to three, the GHZ theorem. I am convinced that similar phenomena are hidden in higher dimensions. The real challenge will be to formulate the right questions.

On the level of our basic notions, we should realize that some of the most important developments in physics happened when we abandoned distinctions we had previously made. For example, Newton gave up the distinction between laws governing how apples fall on earth, and laws governing the motion of heavenly bodies. Likewise, in my eyes, we have to give up the distinction between information and reality. The reason is that there cannot be any operational procedure through which we could investigate anything about any possible reality without actually employing our current (or possible future) information. This is not to say that everything is just information or knowledge. What I mean is that we need a new fundamental concept unifying the notions of information and reality.

WOJCIECH ZUREK · I think it would be a surprise. But the themes brought up in my take on Question 15 (see page 271)—including in particular black-hole thermodynamics—are on the list of usual suspects. And that list naturally includes the relation between existence and information: the "epiontic" nature of quantum states, and the primitive notions of information I have discussed in my answers to Question 5 (see page 122) and Question 9 (see page 194).

Thermodynamics has been often prescient in the past about the connections between physics and information. There are significant analogies between spontaneous symmetry breaking and quantum jumps, namely, the selection of preferred states (see my reply to Question 9, page 194). In fact, this is one of the reasons why I'm interested in the dynamics of symmetry breaking in second-order phase transitions. So this may be a good place to look for more than just analogies.

DEAR ORACLE

*What single question about the foundations of quantum
mechanics would you put to an omniscient being?*

GUIDO BACCIAGALUPPI · Why should I want to spoil the fun?

ČASLAV BRUKNER · Who cares about the foundations of quantum mechanics
when offered an exclusive opportunity for posing a single question to an omniscient
being? I would rather like to ask her or him whether death is really worse than the
chicken at Tresky's restaurant (see Woody Allen's *Love and Death*).

JEFFREY BUB · Is wave-function "collapse" a real dynamical process, not primarily associated with measurement or decoherence, or is it a kinematical effect connected with the loss of information on measurement, just as Lorentz contraction is
a kinematical effect of motion?

ARTHUR FINE · Here's the question, dear being. If I have a system in a superposition of two states localized far away from one another (say, here and on the moon),
is the system somewhere definite and, if so, tell me where, please, is it? (For position,
oh being, you may substitute your own favorite observable.)

CHRISTOPHER FUCHS · There are no omniscient beings—I believe this is one
of the greatest lessons of quantum theory. For there to be an omniscient being, the
world would have to be written from beginning to end like a completed book. But if
there is no such thing as *the* universe in any completed and waiting-to-be-discovered
sense, then there is no completed book to be read, no omniscient being. I find the
message in this tremendously exciting. In a QBist understanding of quantum theory,
it is not that nature is hidden from us. It is that it is not all there yet and never will
be; nature is being hammered out as we speak.

But in honor of John Archibald Wheeler, I will repeat one of *his* questions to our
finite physics community. With him, I deem that there is a chance we can answer it
(or at least part of it) in our lifetimes:

It is difficult to escape asking a challenging question. Is the entirety of existence, rather than being built on particles or fields of force or multidimensional geometry, built upon billions upon billions of elementary quantum phenomena, those elementary acts of "observer-participancy," those most ethereal of all the entities that have been forced upon us by the progress of science?

Wheeler, who brought me into quantum theory, should have the last word anyway.

GIANCARLO GHIRARDI · Has the superposition principle universal validity? Or is it progressively violated in passing from genuinely microscopic to mesoscopic and macroscopic situations? Do linear superpositions of macroscopically and perceptually different states actually occur, so that we could claim that we are fundamentally deceived by our perceptions concerning the universe? Or is, as I sincerely hope, the emergence of precise properties at the macrolevel a genuinely physical process, rather than simply an illusion of our perceptual apparatuses?

SHELLY GOLDSTEIN · I would ask God the following: what is the final fundamental physical theory? I would also request that the theory be expressed in comprehensible terms—and not the way more standard versions of quantum mechanics have traditionally been formulated.

That final theory may in fact have nothing to do with quantum mechanics. But once I knew that, I would no longer much care about the foundations of quantum mechanics. And if the final theory were a quantum theory, then, with God's formulation, the theory would be so clear that any foundational questions about it would seem irrelevant.

DANIEL GREENBERGER · There's the famous joke about Pauli dying and going to heaven and asking God to explain how the universe actually works. And when God starts to explain, Pauli interrupts and says, "No, no, that's not the way to do it!"

My fears are exactly the opposite. I'm afraid that when I ask the same question, I won't understand what he is talking about. But I don't think any human being could understand. I think he will look at me and say, "Look, why don't you ask me that in fifty thousand years, after you've had some time to grow up." If I were to ask him a question about quantum theory, he would probably burst out laughing: "Why not ask about epicycles? That would be equally relevant!" I doubt that we know enough about the universe to ask any intelligent question. And when you think about it, would you really want to live in a universe that was so simple that you could understand it, even if God himself tried to explain it to you?

LUCIEN HARDY · Of course, the question we would really like to have an answer for is, "What is the ultimate theory of reality?" But that would be cheating, and in any case, there is a fair chance we would not understand the answer anyway.

So it is better to ask a question that has a yes/no answer that would guide us in theory construction. Even then, there is a risk our question will contain implicit assumptions that render any meaningful answer impossible. Let us then hope that this omniscient being is also benevolent and gives us the answer that will most help us to make the next important development in physics (which I take to be developing a theory of quantum gravity).

So, without further ado, here is my question: "Is there definite causal structure?" Or, more precisely, "Is there always a matter of fact as to whether it would be possible for a particle to pass between any two given events?"

If the answer to this question is yes, then my assertion about quantum gravity leading to indefinite causal structure is ultimately wrong. On the other hand, if (as I would bet) the answer is no, then we have to rethink how we go about formulating our physical theories. The idea of a state evolving in time would simply be wrong. The idea of local field equations as usually formulated in terms of a metric would be wrong. My hope is that coming to terms with indefinite causal structure will ultimately lead us to a deeper understanding of reality.

ANTHONY LEGGETT · Is quantum mechanics the whole truth?

TIM MAUDLIN · Is the violation of Bell's inequality accomplished by means of a preferred foliation of space-time?

DAVID MERMIN · I'd ask, "How has the uncertainty principle altered the 'omni' of your omniscience?"

Joking aside—but it wasn't really a joke—I have trouble imagining an omniscient being. Let me rephrase the question: if you could be frozen for 150 years and revived intact, what question would you ask physicists when you woke up?

I'd ask something like this:

Is the fundamental physics of a system still described in terms of quantum states that evolve linearly in time and that specify probabilities of the outcomes of tests that we can perform on that system? If so, is anybody puzzled by the meaning of this conceptual structure? If not, is there general agreement on the meaning of the structure that replaced it?

In early twenty-first-century terms: has the structure of quantum mechanics survived intact for a century and a half? If so, are there still foundational problems? If not, are there still foundational problems?

I chose 150 years because a century might not be long enough to get an interesting answer. But I also worry that physicists two centuries from now, no matter how I phrased the question, might not understand it. It might elicit only polite bewilderment, just as a pressing aether-theoretic query at the end of the nineteenth century might seem not only irrelevant but downright incomprehensible to a physicist of the early twenty-first.

There are two possible grounds for future bewilderment at my question. One is that quantum mechanics will have been discovered, as Einstein always hoped, to be a phenomenology based on a more fundamental view of the world, which is more detailed and more intuitively accessible. This strikes me as unlikely, because John Bell showed that any theory detailed enough to satisfy certain common-sense yearnings would also have to contain instantaneous action-at-a-distance. (See my answer to Question 8, page 176.) So while the discovery of a more fundamental view of the world during the next century and a half seems entirely possible, I'd be surprised if the new theory turned out to be more intuitive than our current understanding.

An appropriate timescale for the survival of quantum mechanics is set by the fact that its basic conceptual machinery has suffered no alterations whatever, beyond a little tidying-up, for over eighty years. Not a bad run when you compare what happened to fundamental knowledge between 1860 and 1940, though not close to the more than two centuries that classical mechanics remained the fundamental theory. So the persistence of the same basic formalism for another 150 years seems at least plausible.

Even so, my question might elicit mid-twenty-second-century bewilderment, be-cause after several more generations of physicists, chemists, biologists, engineers, and computer scientists had worked with the theory, it might finally, in Feynman's words, have become obvious to everybody that there's no real problem. We early twenty-first-century people, who believed there ought to be a better way to under-stand the theory, will then have been consigned to the same dustbin of history as the early twentieth-century aether theorists.

I hope that's not how it works out. It is, for example, now possible to articulate the nature of the wrong thinking that made relativity seem shockingly counterintu-itive to many people during its early years. People had simply deluded themselves into believing that there was something called "time" that clocks recorded, rather than recognizing that "time" was a remarkably convenient abstraction—I would say an ingenious abstraction, except that nobody set out deliberately to invent it—that enables us to talk efficiently and even-handedly about the correlation among many different kinds of clocks.

There is now no generally agreed-upon key to dissolving the puzzlement that quantum mechanics engenders today in many of us. (For that matter, I have en-countered otherwise sensible physicists who disagree with the above resolution of the puzzles of relativity.) I would hope that within the next 150 years, such a key might be found that almost everybody would agree clarifies the character of the the-ory, in contrast to today's state of affairs, where no school of thought commands more than ten percent of the population, except for those who maintain—but can they really mean it?—that there is nothing to be puzzled about.

LEE SMOLIN · If I believed in such a being, I wouldn't be working on theoretical physics. But if you force me to reply, I would ask to have a dream in which I run into Einstein on the street and invite him home to have a drink. I would see if I could get a laugh out of him by bringing him up to date on the questions he worked on.

ANTONY VALENTINI · My question would be: "Is it possible in principle to use entangled systems for superluminal signaling?"

If the answer was yes, I would hope that I was on the right track (though non-linearities might yield similar effects). If the answer was no, I would abandon deterministic hidden-variables theories.

DAVID WALLACE · I'm going to cheat and offer him the choice of two.

First question: "Is quantum theory—not quantum field theory, or any other *particular* quantum theory, but the general dynamical framework of quantum theory—ultimately correct? Or is the quantum framework, like the classical framework, just something to be superseded in due course?"

A quick comment on this one. It's fashionable to say that scientific theories are always being superseded and replaced, and at some level that's true. But at a deeper level, we've only *ever* had two dynamical frameworks that were sufficiently well developed to actually do any proper calculations: classical physics and quantum physics. (And the problem with quantum gravity is that we only know how to write a relativistic theory of gravity in the classical framework.) It's a live option—the option string theory bets on—to suppose that the quantum framework really is the ultimate dynamical framework. But either answer would be fascinating.

Second question: "Do the unobserved branches in macroscopic superpositions represent physically real states of affairs, as real to their inhabitants as our surroundings are to us? If not, why not?"

In a way, the first question would be more sensible, as I'm more confident I know the answer to the second already. On the other hand, the second would settle a lot of arguments! (And it sneakily combines the question, "Is the Everett interpretation the right way to understand *unitary quantum mechanics*?" with the question, "Is unitary quantum mechanics *true*?")

ANTON ZEILINGER · My question to the omniscient being would be, "Which of the questions and answers in this book do you find interesting?"

WOJCIECH ZUREK · I think I would simply ask the omniscient being whether it is governed by the quantum laws. Answer to that question alone would tell me a lot about our universe. Like whether the being can be a part of our universe or, because it is omniscient, necessarily has to be on the outside.

I could also ask whether the being can predict the future, and about the nature of that future. For instance, is there just a single "branch" for the being's future—a branch like I perceive, although perhaps defined with more resolution—or does the future consist of all the branches? (In that case, we would have to define more precisely whose future we're talking about!)

There are other good questions: the theory of everything, and whether quantum theory "as we know it" is compatible with general relativity, and so on. But that begins to sound like a longer conversation.

EPILOGUE

SEVENTEEN QUESTIONS and close to three hundred responses later, what have we learned? Trying to draw objective conclusions would be like trying to propose a definitive interpretation of David Lynch's *Mulholland Drive*. With both the film and this interview book, everyone will take away something different. It's a freedom as deliberate as desired.

Neither do I intend to launch into a tedious question-by-question summary, nor a grand analysis complete with pie charts. Instead, let me focus on one particular observation. In my introductions to Questions 3 and 12, I talked a lot about warring interpretive factions. But perhaps that's an outdated image. I think the interviews make it overwhelmingly clear that what's happening today is more accurately described as a sharp contrast, in mindset and approach, between an interpretation-focused, realist, ontological camp on the one hand, and a reconstruction-focused, epistemic-informational camp on the other.

The people in the first camp are wedded to the idea that we ought to exorcise observers from the picture and make quantum mechanics, as Bub (page 67) puts it, "conform to some ideal of classical comprehensibility," by embedding quantum mechanics into a realist interpretive framework with an explicit ontology. The people in the second camp pursue some form of reconstructive approach infused with the spirit of quantum information, heeding Wheeler's why-the-quantum call and taking an epistemic view of the formalism.

Let me expand these characterizations a little. The interpretation-focused, realist, ontological camp roughly thinks like this. Let's take standard quantum mechanics as our starting point, because we already know that its (statistical) predictions match our observations. In particular, we won't attempt to rederive the formalism from deeper principles. But we cannot accept the standard textbook presentation of quantum mechanics: it makes quantum mechanics into a ragtag creature, studded with severe deformities and clinically deluded in its talk of "observers" and "measurements." We are appalled that hardly anyone else seems to notice or care. And so we take it upon ourselves to fashion some new clothes for quantum mechanics, such that it may better match our expectations and join the ranks of what we consider proper physical theories.

Take the interview responses from people partial to de Broglie–Bohm and collapse approaches as an example. Many of these responses display an outspoken disdain for textbook quantum mechanics, which is denounced as "internally inconsistent" (Ghirardi, page 47) and "unprofessionally vague and ambiguous" (Goldstein, quoting Bell, page 49). The goal becomes to lift the "smokescreen of the Copenhagen interpretation" (Valentini, page 54). This entails, among other things, a firm commitment to "beables," that is, to an ontology, because "accounts that are vague or noncommittal about their beables are not precise physical theories at all" (Maudlin, page 52). So building a satisfactory quantum theory requires, first of all, the specification of a definite ontology—particles in de Broglie–Bohm, mass densities (or "flashes") in collapse theories—and the specification of its dynamics. Crucially, statements referring to observers and measurements are to be purged from the formulation of the theory, based on the reductionist argument that such structures cannot have fundamental status but must instead be understood in terms of the beables of the theory. The pre-Socratic philosopher Democritus believed that "in truth, there is nothing but atoms and the void." For a Bohmian, then, "in truth," there is nothing but particles and guiding fields. For a collapsist, "in truth," there is nothing but mass densities and their nonlinear evolution. For an Everettian, "in truth," there is nothing but the global wave function and the Schrödinger equation. Quantum theory is thus made to largely feel like a classical, materialistic theory, save for some interpretation-specific idiosyncrasies, such as the abstract, nonlocal ontology of the Everett picture.

The reconstruction-focused, epistemic-informational camp, in contrast, thinks like this. Let's start neither from the ready-made quantum formalism, nor from some kind of prejudice about a prerequisite ontology that's to be mounted onto quantum mechanics like a luggage rack on a car. Crucially, we see the prominent, fundamental role of observers and measurements in quantum theory not as a critical flaw to be remedied at all costs, but as a constructive starting point. We see it as something suggestive of fundamentally new ways of thinking about physics, about nature, about the role and status of physical theories, and about the relationship between subject and object. The fact that observers and measurements prominently appear in the axioms of quantum mechanics doesn't mean that they're to be regarded as entities *physically* different from other objects in our world, as is often suggested (usually by critics of Copenhagen-style quantum mechanics). Rather, it is the *application* of the quantum formalism that requires a split between observed and observer, because this formalism is essentially a kind of map for observers navigating the world; it is not the world itself. Our goal, then, is to pick out the features of the world that inform the structure of the formalism—the features that make quantum theory such an excellent map. On this reading, we consider it misguided to try to turn quantum theory into an all-inclusive nothing-but-atoms-and-the-void picture. It can be done, but it comes at a high price (think many worlds and Bohmian nonlocality), and, most of all, we will likely not have learned one deep thing about nature in the process.

I'll be the first to admit that this characterization of the two camps is oversimplified. The interviews in this book display all the nuances and fill in all the blanks I couldn't capture here. For example, there are reconstructionists who emphasize the importance of giving an ontological account, or who lean toward an ontological concept of information. There are also people who embrace epistemic or informational attitudes toward the quantum formalism but stay clear of the business of reconstruction.

Be that as it may, I think an overall dichotomy is evident from the interview responses, and it makes for a recurring, overarching theme in this book. It's a dichotomy that's not altogether new, but one that has definitely become increasingly pronounced as ever more people approach foundational questions through reconstructions and information theory. The interview responses also make clear that we're witnessing not just a superficial methodological difference, but a separation that runs much deeper—all the way down to fundamentally distinct attitudes toward theory building, toward what a physical theory is, can be, and should strive to be. The puzzles of quantum mechanics, it seems, spur people into two diametrically opposed forms of action. That's not to imply a profound incompatibility or antagonism between the two camps. They each have their merits and shortcomings, and perhaps they are best regarded as complementary, as Bacciagaluppi suggests (page 202). Yet these approaches and their proponents still correlate to radically different temperaments.

If this book has shown one thing, then it is that the field of quantum foundations is humming with more activity than ever. The subject has clearly outgrown its popular image as some sort of idle philosophical exercise without cash return, as feel-good poetry not befitting a scientifically-minded person eager for clear answers. From technical results to experiments, quantum foundations has come of age.

Are we today any closer to answering the grand questions? Or are we simply caught in a web of more opinions, approaches, and infinitesimal increments of understanding? I think what can be said with reasonable confidence—something that I've already hinted at in the prologue and that the interviews in this book have hopefully amply demonstrated—is that we have acquired a more nuanced grasp of those grand questions than a contemporary of, say, Bohr or Einstein had. We have also assembled a larger, and continuously expanding, toolbox for tackling these questions. At the same time, at the deeper level, we may still feel stuck in a morass of the kind that had already stopped the founding fathers of quantum mechanics in their tracks. This may be frustrating, but it is also an enduring testament to the theory's depth and enigmatic beauty.

GLOSSARY

This glossary lists some of the key terms appearing in this book. A much more detailed discussion of many of these terms can be found in the *Compendium of Quantum Physics: Concepts, Experiments, History and Philosophy*, edited by Daniel Greenberger, Klaus Hentschel, and Friedel Weinert (Springer, 2009). The *Stanford Encyclopedia of Philosophy*, online at http://plato.stanford.edu, is also an authoritative source of information. It has comprehensive entries—some written by our interviewees—on staples such as EPR, the Bell and Kochen–Specker theorems, the measurement problem, entanglement, quantum information, decoherence, quantum logic, and the common interpretations (Copenhagen, Everett, collapse theories, Bohmian mechanics, and modal and relational interpretations).

beable A term coined by John Bell for the observer-independent ontological entity that, in Bell's view, a physical theory ought to make reference to. Bell intended the term and concept of beables as a counterbalance to the prevalent notion of a primacy of observables and observation in quantum theory:

> In particular, we will exclude the notion of "observable" in favour of that of "*be*able." The beables of the theory are those elements which might correspond to elements of reality, to things which exist. Their existence does not depend on "observation." Indeed observation and observers must be made out of beables.

Beables are a hobbyhorse of adherents of Bohmian mechanics and, to a lesser extent, collapse theories—theories that had enjoyed Bell's personal endorsement.

Bell–Kochen–Specker theorem See KOCHEN–SPECKER THEOREM.

Bell's inequalities First derived by John Bell in the 1960s, these mathematical expressions show that no local hidden-variables theory—as defined by Bell in terms of a set of locality assumptions—can fully reproduce the predictions of quantum theory (Bell's theorem). A Bell inequality involves combinations of expectation values for measurements on a bipartite system prepared in an entangled quantum state. If the probability functions used to calculate these expectation values are assumed to obey certain locality conditions, then the expression will be bounded from above. If, however, the expectation values are computed using the usual rules of quantum mechanics, the bound can be violated. Experiments have so far ruled in favor of quantum mechanics, though loopholes remain. See Question 8, *Bell's Inequalities*, for more.

Bell's theorem   See BELL'S INEQUALITIES.

Bohmian mechanics   A hidden-variables interpretation of quantum mechanics, developed by David Bohm in the 1950s as a modification of Louis de Broglie's original pilot-wave proposal. Bohmian mechanics describes the deterministic motion of particles along determinate trajectories. The distribution of the trajectories is given by the quantum equilibrium distribution $|\psi|^2$. This choice ensures that statistical predictions agree with those of standard quantum mechanics. While the wave function is transformed via the Schrödinger equation, the particle positions evolve according to the so-called guiding equation. The wave function acts as a "guiding field" that generates a velocity field followed by the particles. There are also versions using nonequilibrium initial distributions and de Broglie's original equation of motion. Therefore, the more general term "de Broglie–Bohm theory" is sometimes used. See also HIDDEN-VARIABLES INTERPRETATION and PILOT-WAVE THEORY.

Born rule   One of the axioms of standard quantum mechanics. In its most elementary form, it states that the probability of finding the value o_i in a measurement of an observable with eigenstates $\{|o_i\rangle\}$ and spectrum $\{o_i\}$ is given by $|\langle o_i|\psi\rangle|^2$, where $|\psi\rangle$ is the state vector of the measured system immediately prior to measurement.

coherence   See SUPERPOSITION.

collapse postulate   One of the axioms of standard quantum mechanics. It states that a measurement (introduced axiomatically in standard quantum mechanics) instantaneously changes the quantum state of the measured system into one of the eigenstates of the measured observable. See also BORN RULE.

collapse theory   An umbrella term for theories that add to quantum mechanics an explicit mechanism for wave-function collapse. As such, they make predictions different from standard quantum mechanics for certain situations. Collapse can be implemented by adding stochastic terms to the Schrödinger equation, or by postulating the occurrence of instantaneous, stochastic wave-function "hits" (or by combining these ideas). A well-known collapse theory is the GRW THEORY.

Copenhagen interpretation   An umbrella term for a variety of viewpoints associated with members and disciples of the "Copenhagen circle" of Niels Bohr, Werner Heisenberg, Nathan Rosenfeld, and others. Don Howard has argued that "[u]ntil Heisenberg coined the term in 1955, there was no unitary Copenhagen interpretation of quantum mechanics." According to Jan Faye, "today the Copenhagen interpretation is mostly regarded as synonymous with indeterminism, Bohr's correspondence principle, Born's statistical interpretation of the wave function, and Bohr's complementarity interpretation of certain atomic phenomena." It has also become popular to throw wave-function collapse, positivism, subjectivism, and the fundamental role of the human observer into the mix, even though such concepts are mostly alien to the spirit of Bohr's own philosophy, which focused on the complementarity principle and the irreducibility of classical concepts.

de Broglie–Bohm theory See BOHMIAN MECHANICS.

decoherence A quantum-mechanical process whereby interactions of a quantum system with its environment lead to uncontrollable and practically irreversible entanglement between the two partners. Decoherence explains why it is so difficult in practice to prepare certain quantum states and to observe interference effects—especially in the case of mesoscopic and macroscopic systems, for which decoherence is extremely fast and virtually inescapable. Decoherence is an application of the standard quantum formalism to open quantum systems; as such, it is neither an interpretation nor a new theory. Yet it is often invoked in foundational discussions, for example, when addressing aspects of the measurement problem. It's also a cornerstone of Everett-style interpretations. Decoherence is a lively subject of experimental investigation and a feared enemy of quantum computers.

density matrix See QUANTUM STATE.

dynamical-reduction theory See COLLAPSE THEORY.

Einstein–Podolsky–Rosen paradox See EPR PARADOX.

EPR paradox An argument presented in a seminal 1935 paper by Albert Einstein, Boris Podolsky, and Nathan Rosen, claiming to demonstrate the incompleteness of quantum mechanics. See page 162 for a brief introduction.

entanglement A genuine quantum phenomenon whereby two systems become "quantum-correlated." Formally, two systems are said to be entangled if they cannot be afforded with their own state vectors. Entanglement is sometimes described as a process by which systems lose their individuality and fuse into a quantum-mechanical whole ("quantum holism"), but there is disagreement about whether this metaphysical picture is actually appropriate. Suffice to say, entanglement implies that there exist physical properties that can be measured on the composite system but not be inferred from measurements on the subsystems. Entanglement underlies classic quantum paradoxes, such as EPR and Schrödinger's cat, and is a cornerstone of quantum information theory.

Everett interpretation Also known as the relative-state interpretation of quantum mechanics, it was proposed in the 1950s by Hugh Everett, then a Ph.D. student of John Wheeler's. Everett wanted to address the measurement problem and rid the theory of its system–observer dualism. He disposed of the collapse postulate and tried to show that nonetheless—even when no particular measurement outcome is singled out—our *subjective* experience of definite measurement outcomes (as well as their correct quantum statistics) could be recovered. Everett emphasized the principle of relativity of quantum states: each component in the uncollapsed superposition state at the conclusion of a von Neumann measurement describes a correlation between a definite state of the system and a definite state of the observer, with the latter state then interpreted as *relative* to the system's being in a particular state. Serious gaps in Everett's argument motivated later efforts to develop Everett's ideas into a coherent, satisfactory interpretation; see MANY-WORLDS INTERPRETATION.

gedankenexperiment A thought experiment (from the German word *Gedanke*, meaning "thought"). Famous examples relevant to the theme of this book are SCHRÖDINGER'S CAT and WIGNER'S FRIEND.

GRW theory A collapse theory postulating a spontaneous, stochastic spatial localization of the wave function. Named after its inventors GianCarlo Ghirardi, Alberto Rimini, and Tullio Weber. See also COLLAPSE THEORY.

hidden-variables interpretation An interpretation of quantum mechanics that adds to the wave function additional variables that specify the physical state of the system more accurately than the wave function alone could do. To avoid a clash with the predictions of quantum mechanics, the hidden variables must remain experimentally inaccessible. A well-known hidden-variables interpretation is BOHMIAN MECHANICS. See also Question 8, *Bell's Inequalities*.

interference In quantum mechanics, the phenomenon that observed distributions of events may have a distinctly nonclassical shape in (typically) space or time. The most famous example is the spatial interference pattern observed in the double-slit experiment with particles. Classically, the expected pattern would be two partially overlapping peaks (the sum of the contributions from each individual slit). The observed quantum-mechanical pattern, however, has an oscillatory shape ("interference fringes"). The formal account of interference rests on the fact that a quantum superposition represents a linear combination of probability *amplitudes* rather than actual probabilities. This means that the corresponding probability distribution contains additional crossterms ("interference terms"), which modulate the classically expected distribution.

Kochen–Specker theorem A no-go theorem that, together with Bell's theorem, imposes severe constraints on the structure of a viable hidden-variables theory. Derived by Simon Kochen and Ernst Specker in 1967, it may also be read as a powerful argument against naive realism, by implying that measurements cannot in general be construed as simply revealing objectively preexisting properties of the world. Specifically, the theorem proves that in quantum mechanics, it is not possible in general to assign values to a set of observables defined for a quantum system of Hilbert-space dimension greater than two such that (1) all these values are definite at all times, and (2) the value assignment is independent of how the value is eventually measured—say, independent of the choice of other co-measured observables ("noncontextuality"). Some authors prefer the term "Bell–Kochen–Specker theorem," arguing that the derivation of the Kochen–Specker theorem shares a key step with the (earlier) proof of Bell's theorem.

many-minds interpretation See MANY-WORLDS INTERPRETATION.

many-worlds interpretation An interpretation that develops the basic ideas of Everett's relative-state interpretation into the full-blown picture of a single quantum universe—represented by an all-encompassing wave function—containing a myriad of constantly branching, effectively classical worlds. "Our" observed world then corresponds to one such branch. Many-worlds interpretations were popularized by Bryce DeWitt in the 1970s and by David Deutsch in the 1980s. A variant is the class of "many-minds" interpretations proposed by David Albert and Barry Loewer, Dieter Zeh, Michael Lockwood, and others. See also EVERETT INTERPRETATION.

measurement problem The difficulty of reconciling the smooth, linear, reversible Schrödinger evolution of quantum states with the occurrence of definite events in the world of our experience. The measurement problem is one of the classic problems in the foundations of quantum mechanics. See Question 7, *The Measurement Problem*.

modal interpretation A class of interpretations of quantum mechanics. One characteristic feature is the definition of rules that permit the assignment of a definite value to a system even when the system is not in an eigenstate of the corresponding observable. The first modal interpretation was proposed in the 1970s by Bas van Fraassen. There, a system is described by the following two different states. (1) The *value state*, which specifies the values of physical quantities possessed by the system at a given time. (2) The *dynamical state*, which determines the evolution of the system—that is, the possible future value states. It coincides with the ordinary quantum state vector, but it never collapses.

no-cloning theorem A theorem of quantum mechanics showing that it is impossible (except by sheer luck) to duplicate an unknown quantum state.

no-go theorem An umbrella term for theorems that demonstrate an incompatibility between what quantum mechanics *allows* us to do and what we'd *like* to do—be it implementing particular actions, constructing particular hidden-variables models, or continuing to believe in particular worldviews. For examples, see BELL'S INEQUALITIES, KOCHEN–SPECKER THEOREM, NO-CLONING THEOREM, and NO-SIGNALING THEOREM.

nonlocality In the context of quantum mechanics, this term chiefly has two meanings. (1) The impossibility of describing correlations between outcomes of local measurements, performed at two different locations, in terms of a local hidden-variables model. (2) Actual physical action-at-a-distance, where the physical situation in one region instantaneously influences the physical situation in another, arbitrarily distant region.

no-signaling theorem A theorem showing that quantum mechanics does not enable us to use entangled quantum states for the instantaneous transmission of information between distant partners.

philosophy The art of skillfully questioning and analyzing subtle yet fundamental matters that the man on the street either takes for granted or does not regard as having practical bearing on his survival. The term is also sometimes employed by tough-minded scientists to dismiss issues that they do not regard as having any practical bearing on their survival.

physics The art of skillfully observing, analyzing, and quantifying patterns and relationships in the universe, and of formulating laws that capture and correctly predict these patterns.

pilot-wave theory A hidden-variables interpretation presented by Louis de Broglie at the 1927 Solvay meeting. De Broglie derived an equation for the motion of particles, each endowed with definite position and momentum values (which are the hidden variables), and demonstrated how interference effects could be understood on the basis of such particle trajectories. De Broglie's theory was later revived by David Bohm and developed into BOHMIAN MECHANICS. See also HIDDEN-VARIABLES INTERPRETATION.

PR box A model for studying the properties of (hypothetical) "superquantum" theories. Named after its inventors Sandu Popescu and Daniel Rohrlich. See Jeffrey Bub's introduction to PR boxes, page 68.

QBism Christopher Fuchs's term for his research program of elucidating the larger metaphysical implications of Quantum Bayesianism.

QBit A variant spelling of *qubit*, preferred and promoted by David Mermin. See QUBIT.

Quantum Bayesianism At the core, the view that quantum states encapsulate the subjective degrees of belief of an agent and are nothing but a tool the agent uses in navigating the world he is immersed in. Developed by Carl Caves, Christopher Fuchs, and Rüdiger Schack, the approach is grounded in personalist Bayesian probability theory and is nourished by insights from quantum information theory. See also QBISM.

quantum computer A device that exploits the laws of quantum mechanics to speed up a computation. There are several known quantum algorithms for solving certain problems faster than any classical (i.e., nonquantum) computer could do. The most famous examples are Shor's algorithm for factoring large numbers and Grover's algorithm for finding an element in a list. The heart of a quantum computer is an array of qubits, which can be physically realized in various ways (photons, trapped ions, two-level atoms, nuclear spins, coupled quantum dots, and so on). Gates are implemented via unitary operations acting on the qubits; one- and two-qubit gates are sufficient to perform any quantum computation. (There's an alternative equivalent approach—called measurement-based, or cluster-state, computation—which proceeds from a highly entangled initial state and implements the computation via a series of projective measurements.) Building a quantum computer is one of the holy grails of quantum science and engineering; to date, only proof-of-principle devices containing a handful of qubits have been realized.

quantum gravity An area of research devoted to finding a satisfactory physical theory that would unify quantum mechanics and general relativity. String theory is currently the most popular approach, followed by loop quantum gravity. All of the existing theories, however, have their problems, and so the field is best described as work in progress.

quantum information theory A recasting of quantum mechanics as a theory concerned with the flow, processing, and manipulation of information. It provides a new lens for looking at the structure and capabilities of quantum mechanics. It has also led to practical offshoots, such as protocols for completely secure communication. There is, moreover, the promise of a quantum computer. In absence of a clarifying hyphen, the term "quantum information theory" may be read as both "a theory of quantum information" and "a quantum-mechanical information theory." This is a healthy ambiguity. See also QUANTUM COMPUTER and Question 9, *Quantum Information*.

quantum state The mathematical object for describing the state of an individual quantum system. So-called pure states are represented by complex vectors or functions ("wave functions") in a Hilbert space; they provide, at least according to standard quantum mechanics, a complete description of the physical state of an individual system. Mixed states, formally represented by density matrices, are used in the following two situations. (1) To represent a classical, ignorance-interpretable probability distribution (ensemble) of pure states, one of which is actually realized by the system. (2) To encapsulate the statistics of all possible measurements that can be carried out on a system that is entangled with another system. In this case, the mixture is *not* ignorance-interpretable, because the presence of entanglement prohibits the assignment of a pure state to the system. See also WAVE FUNCTION and Question 4, *Quantum States*.

qubit Short for *quantum bit*, it refers to any quantum system with a two-dimensional Hilbert space. It is a prominent player in quantum information theory and the building block of quantum computers. While a classical bit has a value of either 0 ("off") or 1 ("on"), the state of a qubit will in general be a linear superposition of the form $\alpha |0\rangle + \beta |1\rangle$.

reconstruction A rederivation of the structure of quantum mechanics from a set of fundamental principles. Several such principles have been suggested to date; the challenge is to find principles that are sufficiently basic *and* uniquely specify quantum theory. Reconstructions are work in progress and may be considered either a complement or an alternative to the program of interpreting quantum mechanics. See Question 10, *Reconstructions*.

relative-state interpretation See EVERETT INTERPRETATION.

Schrödinger equation An equation specifying the evolution of the quantum state of an isolated, unmeasured system.

Schrödinger's cat A thought experiment devised by Schrödinger in 1935. It can be seen as a particularly vivid illustration of the measurement problem. The setup consists of a cat confined to a box together with an unstable atom that, at the moment of its decay, triggers a hammer breaking a vial of poison. According to quantum mechanics, the state of the atom is at all times described by a superposition of "not decayed" and "decayed." Unitary evolution leads to entanglement between all systems present, resulting in a seemingly grotesque superposition of two states that our experience deems mutually exclusive: one component of the superposition contains a live cat (together with an undecayed atom, untriggered hammer, and the vial intact), while the other component describes a dead cat (together with a decayed atom, triggered hammer, and the poison released). The second part of the paradox is established when an outside observer opens the box to look at the cat. According to the collapse postulate, such an act of observation will instantaneously reduce the superposition to one of its components. In Schrödinger's words, the "indeterminacy originally restricted to the atomic domain becomes transformed into macroscopic indeterminacy, which can then be *resolved* by direct observation." This raises the question of the state of the cat *before* the observer opens the box. See Arthur Fine's *The Shaky Game: Einstein, Realism, and the Quantum Theory* (Chicago, 1996) for an in-depth analysis of the history of Schrödinger's cat paradox and Einstein's influence.

SQUID An abbreviation for *superconducting quantum interference device*. A SQUID is a macroscopic quantum system consisting of a ring of superconducting material interrupted by thin, insulating barriers called Josephson junctions. At low temperatures, pairs of electrons of opposite spin condense into bosons ("Cooper pairs") and tunnel through the junctions. This leads to the flow of a persistent, resistance-free "supercurrent," which induces a magnetic flux threading the ring. In 1980 Anthony Leggett suggested that SQUIDs could be used to create quantum superpositions of macroscopically distinct flux states. In 2000 coherent superpositions of microampere supercurrents traveling in opposite directions around the loop were experimentally observed by Jonathan Friedman et al. and Caspar van der Wal et al.

state vector A normalized complex vector in a Hilbert space, representing a pure quantum state. See also QUANTUM STATE.

superposition A (pure) quantum state that is written as a linear combination of other (pure) quantum states. Such a quantum-mechanical superposition is often referred to as *coherent*, to emphasize the fact that it defines a new physical state of an individual system—rather than a statistical ("classical") distribution of the component states, with one of the states realized in the system. See also INTERFERENCE and SUPERPOSITION PRINCIPLE.

superposition principle A kinematical concept of quantum mechanics, grounded in the linearity of Hilbert space. It states that any linear combination $\sum_n \alpha_n |\psi_n\rangle$ of quantum states $|\psi_n\rangle$ is again a valid quantum state. See also SUPERPOSITION.

von Neumann measurement A formal scheme describing the entangling interaction between two quantum systems. It is often used to formalize a "measurement-like" unitary interaction between a system and an apparatus, both treated as quantum systems. Since no definite outcome is singled out at the conclusion of a von Neumann measurement, the scheme is sometimes referred to as *premeasurement* and serves as a classic—albeit not the most general—illustration of the measurement problem.

wave function   A complex vector or function in a Hilbert space, representing a pure quantum state. In traditional parlance, wave functions are mostly associated with a continuous function of real parameters that refer to the relevant degrees of freedom of the system (usually position, momentum, or spin). A wave function describes a probability amplitude; its mod-squared value $|\psi(x, t)|^2$ specifies the probability of finding the value x in an appropriate measurement at time t (Born rule). See also QUANTUM STATE.

wave packet   A wave function that is peaked in the relevant variable. An example is a coherent state, which is narrowly peaked in both position and momentum space.

Wigner's friend   A variant of the Schrödinger-cat gedankenexperiment, devised by Eugene Wigner in the early 1960s. The cat is replaced by a human observer ("Wigner's friend") inside a sealed laboratory. The decay of the atom triggers now merely a flash of light. The observer is instructed to assign a definite quantum state depending on whether she has seen a flash. On the other hand, from the perspective of a second, outside observer, the contents of the laboratory will evolve into a superposition of states associated, in particular, with different states of *consciousness* of Wigner's friend. For Wigner, this was a particularly absurd and unacceptable state of affairs. For more, see page 90, Časlav Brukner's answer to Question 11 (page 217), and Christopher Fuchs's answer to Question 5 (page 114).

INDEX

Note: A page reference to a person's name means that the corresponding text makes significant mention of that person—say, in the form of a quote, an anecdote, or a substantial description and acknowledgment of the person's work.